A NATION OF STEEL

STUDIES IN THE HISTORY OF TECHNOLOGY

Merritt Roe Smith, Series Editor

Experiment with "hot top" process for casting steel ingots at U.S. Steel's Gary Works, 1926
Courtesy Calumet Regional Archives, Indiana University Northwest.

A Nation of Steel

THE MAKING OF MODERN AMERICA
1865–1925

Thomas J. Misa

THE JOHNS HOPKINS UNIVERSITY PRESS   BALTIMORE AND LONDON

© 1995 The Johns Hopkins University Press
All rights reserved. Published 1995
Printed in the United States of America on acid-free paper

04 03 02 01 00 99 98 97 96 95   5 4 3 2 1

The Johns Hopkins University Press
2715 North Charles Street
Baltimore, Maryland 21218-4319
The Johns Hopkins Press Ltd., London

ISBN 0-8018-4967-5

Library of Congress Cataloging-in-Publication Data
will be found at the end of this book.

A catalog record for this book is available from the British Library.

For Ruth, who was present at the creation, and Christopher, who has "a thing about trains"

Contents

Figures and Tables

Figures

Tables

Preface

This book is about how people create technologies, and how technologies shape the world. Accurately explaining, wisely understanding, and carefully managing the dynamics of technology and social change are among the most urgent tasks that await us as citizens of a technological age. Our conceptual framework for these tasks is just emerging. Many of the necessary management and policy instruments are still on the drawing board, and others are yet to be conceived. Nevertheless, we now possess a series of powerful insights into the dynamics of technology and social change. Together, these insights offer the realistic promise of improving our ability to modulate the complex process of technical change, if we so choose.[1] This book may help in a modest way to refine these tools and concepts, and to indicate the urgency of the task. It analyzes how Americans created a modern industrial technology and how that technology in turn shaped their nation's social, economic, and institutional development. Much of the modern world takes durable form through technology, often creatively and sometimes cruelly outlasting those who brought it to life. Across time, our technologies are not only socially constructed but society-shaping. Where do they come from?

In our scientific age, pundits and publicists have often asserted that modern technology is simply the insights of modern science coupled to the craft tradition. This view of technology as applied science has served as a powerful myth for legitimating science policy (give scientists money, and technological innovations will automatically result), but this view is worse than useless for comprehending the dynamics of technical and social change. The myth must be abandoned if we are to see clearly the problems and possibilities of our technologies. From Renaissance automata to latter-day aeronautical theory, technologists have used distinctive modes of framing, revising, and solving problems. These modes cannot be deduced from scientific method as taught in textbooks or from scientific practice as carried out in university, industry, or government laboratories.[2] Furthermore, any reasonable description of technical change must consider how and why institutions have fostered and developed technologies. This entails consideration of the modern corporation and government, especially the military.[3] For example, battleships were just one of several means by which the military fostered technical change in the steel industry.

Such a conception of technology as a product of institutional and cognitive dynamics is a step forward, but a closer analysis of these elements is necessary for an understanding robust enough to inform the management of technical change. Science is part of the puzzle. Precisely how scientific knowledge interacts with technical processes and institutional structures remains an open question. This gap in understanding is not for lack of effort. Early attempts to quantify the impact of science on technology (or vice versa) yielded little insight into the many and often subtle interactions between them. Later studies have piled anecdotes upon examples, occasionally offering models for the range of science-technology interactions.

One theme of this book—an investigation of the dynamics of science, technology, and society—results in a "thick description" of how scientific knowledge was generated, deployed, and shaped by technical and industrial imperatives. My analysis may suggest a more realistic understanding of the forces at play. My findings concerning the utilities of scientific knowledge in the steel industry include the conventional wisdom (scientific knowledge can be used to rationalize production and inform innovation) as well as several surprising results. For example, scientific knowledge can be used to jigger standards of competition, to bolster military procurement policies, or to fix broken rails and rusty washing machines. At times, change in the structure of companies or industries can directly create opportunities for applying and generating scientific knowledge.[4] On this line of argument, then, science must be rejected as an independent variable: it does not reliably serve as any cause of or precondition for technical innovation. The science-policy myth must go. At least one individual held this view decades ago; in the 1960s this investor sold his steel stock the moment he heard of the corporation's plan to spend $10 million on a research center.[5]

Technology too must be rejected as an independent variable.

However often technologies appear to cause changes, technical change itself is frequently the *result* of underlying changes in the availability of raw materials, in the structure of firms, and in social forms and economic conditions. As Peter Gay writes, "For the historian to make sense of change is to unriddle its causes, whether he sees them emerging from the hand of God, the imperatives of technology, the conflict of classes, or the will of individuals." An example suggests the varying distances and natures of causal factors.[6] The abrupt emergence of steel skyscrapers in about 1890 (in one country only, one might add) probably required the presence of rapid and safe elevators, which had been available for a dozen years, as well as a certain intensity of urban land use and suitably permissive building codes. Yet elevators and land values and white-collar work were only distant and not immediate determinants.[7] Most immediately, steel-framed skyscrapers were the result of steelmakers' efforts to get out of the stagnating market for steel rails by developing a new steel suitable for urban structures. This mix of social, technical, and economic conditions did not exist outside the United States. The result was a distinctive sociotechnical form—the American skyscraper.

The results presented in this study also call into question an assumption of industrial policy making: that an entire industry (or worse, an entire economy) is a meaningful unit for analysis and action. A second broad theme guiding this study is that the dynamics of intra-industry sectors are the most powerful immediate determinants of technology and social change. I do not share the view of corporate culture as uniform throughout a company. In studying the dynamics of companies, I found that the culture—that is, the overall pattern of behavior, decision making, and expectations—varied significantly within the company. Certain regularities, however, could often be identified across companies in similar (user) sectors. For instance, the armor shops at the Carnegie and Bethlehem concerns (located at opposite ends of Pennsylvania) had far more in common—technically, organizationally, and culturally—than did, say, the armor and rail shops at Bethlehem (located side-by-side) or those at Carnegie (separated by the Monongahela River). Similarly, what I identify as the culture of reckless mass production could be found in the largest rail mills of Chicago as easily as in those of Pittsburgh—but not in the structural shops of those same mills. Companies recognize this phenomenon implicitly. When one steelmaker was beginning to manufacture automobile sheets, a new and demanding activity, the work had to be specially done on Saturdays when regular production was shut down. Around 1910, to effect a policy of stability, U.S. Steel purged its ex-Carnegie executives who were excessively production-minded. One could easily write a history of the distinctive *mentalités* within the steel industry.[7]

The striking similarities within user sectors across company boundaries suggest that the dynamics of consumers should be at center stage. Alfred Chandler, who has done so much to advance the understanding of the dynamics of producers, observed that the imperatives of an increased number of consumers might compel a business firm to alter its structure (for example, by creating special sales branches).[8] In this book I take the argument one step further. The final and most tantalizing theme guiding this study is how societal needs and priorities are and can be connected to technical change, and specifically how social and economic forces are channeled into technical change by the dynamics of intra-industry sectors. To situate the relationship between consumers and producers, we must somehow simplify the inherent complexity of a broad study. Otherwise, it would surely be intractable to analyze, as in this study, six technical processes and three distinct bodies of scientific knowledge in five distinct industrial contexts.[9]

The attempt to bring order to this complexity begins with the observation that not all consumers used all technical processes. During the first generation of transcontinental railroads, from the 1870s to the 1890s, nearly 90 percent of all Bessemer steel was manufactured into rails; after about 1880 virtually all rails were made of Bessemer steel. Bessemer steel was found in the United States to be unreliable and defective when used for structural shapes, so that after about 1890 skyscrapers, bridges, and subways were constructed of so-called basic open-hearth steel. Bessemer steel was similarly found wanting in the second generation of the railroad industry, beginning about 1900, when heavier locomotives and trains placed more severe strains on the rails; open-hearth steel was increasingly used there too. Fifty-ton ingots of acid open-hearth steel were forged into the foot-thick armor plates for battleships. An entirely different steel, made in 100-pound batches in clay or graphite crucibles, was forged into tool steels for metal-cutting factories. Finally, automobile manufacturers relied on yet another steelmaking process, the electric furnace, for producing a range of high-tech alloy steels. Iron with various refinements remained the material of choice for certain purposes, including automobile cylinders and home appliances. As one steelmaker put it, each process has its particular field.[10] To understand the central dynamics of producing and using steel, one need not investigate crucible steel rails, or Bessemer steel automobile alloys. But one must investigate

Bessemer steel rails, structures made of basic open-hearth steel, armor made of acid open-hearth steel, crucible tool steels, and alloys, used in automobiles, that were produced in the electric furnace.

The strong and identifiable dynamics of these user sectors are crucial to an understanding of the dynamics of technology and social change. The dynamics of a particular sector depend on the nature of its technical process (for example, skill-dependence, economies of scale, and custom-, batch-, or mass-production orientation); the expectations of quantity and quality of product (including producers' responses to consumers' changing requirements); the organization of consumers (whether in concentrated markets like those for railroads or battleships, direct-consumer markets like those for automobiles, or intermediate cases like the structural trade and metal working); and finally the organization of producers (including who has the power to make and implement decisions). Public policies sometimes made a big difference.

The key point is the interaction of consumers and producers. Over the course of this study I argue that the relationships between producers and consumers are the single most important determinant of the dynamics of technology and social change. Borrowed from studies in the economics of technical change, my conception of user-producer interactions encompasses how users and producers are internally organized, how and why they develop modes for communicating with each other, and how and whether strategic decisions about innovation are made. The concept also includes straightforward economic considerations, such as price signals and demand structures.[11] Such a concept is preferable to the notion of 'market,' which narrowly focuses on economic transactions and wrongly assumes atomistic actors. In this regard, neoclassical economics is of no help in analyzing technical change, which violates the static world presumed by economic theory. A broader concept is necessary to analyze the function and behavior of cartels, which persisted in the markets for rails, structures, armor plate, and iron ore at various times during the period covered by this study.[12] The concept of user-producer interactions also covers relationships between business and government.

The organization of this book is by sectors of steel use. This approach places the history of steelmakers and users at the center of U.S. social, economic, and institutional development. On the most detailed level the book is about technical processes and the social and economic conditions necessary for their development and deployment. On the broadest level it is about the building of transcontinental railroads and the settlement of the West, the building of steel-framed urban centers, the development of the navy's ocean-going battleships, imperialism, relationships between government and business, the turn-of-the-century merger movement and the ascent of big business, the factory reforms that placed white-collar experts in control of the economy, and finally the emergence of consumer culture along with the automobile. The book suggests that modern America took institutional and physical form as a nation of steel.

This study proceeds over roughly a half century and across most of the United States, with sideways glances across the Atlantic. It describes activities employing a million or more workers. It portrays inventors at work, engineers wrestling to get those inventions to work better, and workmen laying rails, puddling iron, and rolling steel sheets.[13] It shows scientists in their laboratories and on the shop floor, as well as executives in conferences, naval officers at artillery test ranges, politicians politicking, financiers financing. This study does not present top-down economic analysis or bottom-up narrative of every major steel company. It does outline the industry's economic history in prices, production volumes, and shifting demands. It also presents narratives of certain companies (some whose histories are well known, like Carnegie, and many whose histories are not) at key moments of challenge and change.[14] Geographically, the study examines the building of railroads across the continent and then the virtual rebuilding of railroads along the corridors where traffic was heaviest. Construction of the first generation of steel-framed skyscrapers takes us to Pittsburgh, Chicago, and New York. The building of armored battleships takes us predictably to Washington and unexpectedly to London, Paris, Essen, and Rome. The reform of factories returns us to Pennsylvania, and following the imperatives of automobiles we make stops in Detroit and Denver.

Chapter 1 on the dominance of rails (1865–85) begins and ends with railroads, for they were the single most important consumer of steel in the era before automobiles. The year 1865 saw General Robert E. Lee's surrender at Appomattox, as well as the first commercially successful Bessemer plant in the United States and Andrew Carnegie's retirement from the Pennsylvania Railroad. For steelmakers as well as railroads, rails were the key item. Only the preparation of the roadbed cost the railroads more than the purchase of rails. Rail consumption was highest in three massive building campaigns of transcontinental railroads, which peaked in 1872, 1882, and 1887. Responding to wide shifts in demand, the industry transferred and adapted the Bessemer process to yield the reckless mass production of steel rails. This emphasis on high-volume, low-quality production bedeviled later efforts to do anything else with Bessemer steel, setting in motion a series of events that distinguished the U.S. steel industry in size and character from its European rivals. Railroads also shaped the industry's dominant mode of scientific knowledge, in part by emphasis on chemical process control and more blatantly by their stance in an apparently esoteric but commercially crucial classification dispute.

Chapter 2, on the structure of cities (1880–1900), analyzes the emergence of steel-framed urban structures, especially bridges, transit systems, and skyscrapers. It offers a fresh look at the first generation of skyscrapers in Chicago and New York. Railroads tended to promote the growth of urban centers, and steel structures were a necessary condition for the rising density of urban life and its increasingly multidimensional patterns of movement—the key factors in comprehending the form and growth of cities.[15] Neither iron nor Bessemer steel (nor wood nor masonry) was wholly suitable for urban structures. The best efforts to expand wrought-iron production failed to meet rising demand, as a case study of Phoenix Iron demonstrates. Even worse, beams made of Bessemer steel (sometimes from rail mills) broke unexpectedly. A solution resulted from the collaboration of Pittsburgh steelmakers with Chicago architects to develop a new steel suited for structural purposes. Construction and steel production are inseparably linked, asserted one prominent skyscraper builder. The demands of the one furnished the incentive for the colossal scale upon which the other was developed.[16] Further, city-specific public policies, specifically building codes, directly influenced the building process and the steelmakers.

Chapter 3, on the politics of armor (1885–1915), portrays the technical shaping of domestic and international politics. Before about 1890 the diversity of effective armor types permitted the U.S. Navy to maintain effective control over its rearmament effort. The navy conducted scientific tests that proved its battleship armor was superior to those of the rival great powers, while they conducted similar tests that proved just the opposite. In effect, each navy fabricated objective evidence that it possessed superior armor and hence a superior navy. After 1895, however, an international technological convergence on one truly superior type of armor destroyed this assumption and transformed business-government relations. The United States and the other great powers all effectively lost control over their rearmament efforts. Political forces in each of the countries proved incapable of disciplining an international armor-patent pool and producer cartel whose dynamics effectively determined domestic prices for armor. This remarkable situation resulted from the undeniable superiority of case-hardened armor, which was originally invented in New Jersey, further developed in Germany, and eventually controlled in London. Possession of the new type of armor became the sine qua non of military power. As the naval arms race intensified before World War I, the battleships of the belligerent nations became protected by technically identical armor.

The chapter opens with the Battle of Jutland (1916), the first and last great engagement between battleships. The inconclusive result of the battle, in the face of the British navy's numerical superiority over the German fleet (thirty-seven warships to twenty-one), convinced many Britons that defective British armor was at fault. Obviously, given the technical convergence, this was not the case.[17] This mistaken view was partly a result of the secrecy surrounding armor manufacture. In the United States, armor prices fifteen times as high as rail prices helped the Carnegie and Bethlehem companies to consolidate their leading positions in the industry.

Chapter 4, on the merger of steel (1890–1910), deals with the contingent nature of structural change, specifically change in the steel industry's largest consumer and largest producer.[18] First, it shows that mergers—specifically the merger of mergers in 1901 that resulted in the nation's first billion-dollar corporation, U.S. Steel—were neither the inevitable product of economic logic nor the free creation of business entrepreneurs. J. P. Morgan's leading role in reorganizing bankrupt railroads in the 1890s led directly to his effort to stabilize the steel industry and thus to conflict with Andrew Carnegie. Meanwhile, the purchases of iron ore lands by Carnegie's company led to efforts to destabilize the railroad industry and thus to conflict with Morgan. With "competitors . . . tired of dancing to the music of his bagpipes," the purpose of the U.S. Steel combine was, as one contemporary understood, "to eliminate MR. CARNEGIE from the trade."[19]

Second, the chapter explains how structural change in the railroad sector concurrently affected the industrial application of science. The consolidation of the railroad industry in the 1890s and its revival in the next decade brought unprecedented heavy loads—and new heavy rails. Railroads and rail mills alike were baffled when these 90- and 100-pound rails wore out prematurely and broke anomalously. The rail crisis eased when structure-oriented metallurgists visually and graphically demonstrated that rail-rolling temperatures should be much lower. This was the first serious

U.S. application of the twenty-year-old study of physical metallurgy (metallography). The account provides a new vantage point on the rise of big business.

Chapter 5, on the reform of factories (1895–1915), provides a detailed analysis of Frederick W. Taylor's invention of high-speed tool steel, focusing on the period from his first publication on shop management to his death. The context of Taylor's tool steel experiments includes the craft orientation of the crucible steel and metalworking industries, particularly military contracting. The consequences of high-speed steel involved the larger process of factory reform that culminated with scientific management and the efficiency craze. The case also demonstrated a method of engineering research—parameter variation—that became a model for professional engineers.[20] The chapter then narrates the battle to control the high-speed steel patents and concludes by assessing Taylor's vision of science and professional engineering.

Chapter 6, on the imperative of automobiles (1905–1925), analyzes the automobile industry as a leading user of steel. It begins in 1905, the year when the United States surpassed France to take world leadership in car output[21] and when the Society of Automobile Engineers was founded. The chapter begins with the automobile engineers' singular success in devising and implementing standards for steel that have persisted to today. Because of their challenging technical problems and novel consumer orientation, automobiles were the final sector to alter basic patterns in the steel industry. New alloy steels and heat-treatment technologies were a response to the technical problems. The continuous mass production of steel sheets was the culmination of nearly twenty years of engineering experimentation at the American Rolling Mill Company, a medium-sized specialty producer with a proven track record of developing latent market niches. Finally, the automobile industry ensured the dramatic rise of electric steelmaking. Between 1915 and 1925 makers of alloy steel adopted electric furnaces to provide large volumes of high-quality steel for the automotive industry. Electric furnaces yielded an exceptionally clean steel. They also conserved the expensive alloying elements that other steelmaking processes wasted. The requirements of the automobile industry brought the steel industry to its recognizably modern form.

Together, these six chapters aim to provide an understanding of the most significant changes in the making and using of steel. The balance of narrative and analysis takes in technological innovation in the broadest sense, including the invention, transfer, development, and diffusion of each of the key technical processes; how the industry used and deployed systematic knowledge; and how the structure of the industry evolved over time. By explicitly considering consumers and their use of steel, the study also examines the relationship between technology and social change.

Chapter 7, on the dynamics of change, departs from the narrative of sectors to pick up four strands of analysis. It first presents evidence that the patterns traced out in this study have been durable ones, and that what is notable, even frightening, about the industry since the 1930s is the absence of fundamental change driven by vigorous user-producer interactions. The analysis next spotlights two practical areas—economics of technical change and labor-management relations—implicated in the steel industry's postwar decline. Then, the chapter makes explicit and develops the conceptual framework underlying this study. Finally, it gathers all these strands together to locate this study in the larger project of explaining, understanding, and managing the vital processes of technical change.

E patel 1

Acknowledgments

M any people have contributed to this study. At MIT Leo Marx, David Noble, Merritt Roe Smith, Langdon Winner, and Ken Keniston kindled my interest in technology as a social process. My interest in steel began at the University of Pennsylvania with a graduate seminar project on the industrial context of knowledge, under the guidance of Robert Kohler. Judith McGaw forced me to think through the structure of the book in a new way. And Thomas Hughes has been encouraging and supportive in ways far beyond the compass of this study. For financial support, I am indebted to the University of Pennsylvania (for Mellon, Dean's, and Annenberg graduate fellowships during 1981–87), to the American Historical Association (for a Beveridge grant for research in American history during 1985–86), to the University of Illinois Foundation (for a Rovensky fellowship in business and economic history during 1985–86), and to the Hagley Museum and Library (for support of research in 1985, 1988, and 1991–92).

I am further indebted for opportunities to present my work before seminars at Twente University, Indiana University, University of Trondheim, University of Oslo, Technische Universität—Berlin, the TNO Center for Technology and Policy Studies, the National Museum of American History, the University of Pennsylvania, and MIT, as well as in conference papers for the Society for the History of Technology (presented in Pittsburgh and Madison) and for the Society for Social Studies of Science (in Troy, N.Y., and Amsterdam). Paul Barrett, Bernard Carlson, Thomas Hughes, Maury Klein, Arie Rip, Johan Schot, Francis Sejersted, Steve Usselman, and Ulrich Wengenroth have each clarified my thinking, in various ways, or saved me from one blunder or another. Dean C. Allard,

John Arrison, Lance Metz, Tom Peters, and Sara Wermiel graciously shared their ongoing research and accumulated wisdom. Peter Kreitler helped complete my quest for an image of the Flatiron Building. I am especially indebted to Alex Roland, who made insightful comments on an early draft of the manuscript, and to Ed Todd, who gave me detailed comments on later drafts.

At the Johns Hopkins University Press, I have had the good fortune of sound counsel (and patience) from Henry Tom, meticulous editing from Lys Ann Shore, and efficient flexibility from Lee Sioles. I appreciate, too, Larry Cohen's permission to reprint portions of chapters 1 and 4, which appeared in edited volumes published by MIT Press.

My research has taken me to libraries in several countries. Such a study could not have been conducted without the active help of many archivists; those at four libraries were especially helpful. I have often been asked why I have done so much research in Delaware rather than in, say, Pennsylvania. The reason, simply put, is the superb collection of archival materials at the Hagley Museum and Library. Michael Nash, Marge McNinch, and Chris Baer of the Hagley Museum and Library have been of great help. Not far away, in Hoboken, N.J., Jane Hartye was of particular assistance with the Frederick W. Taylor Collection at Stevens Institute of Technology. Across the Atlantic, the staff at the Eisen-Bibliothek in Schaffhausen, Switzerland, have addressed many queries. Heinrich Lüling and Clemens Moser made my stay there particularly productive and enjoyable. And finally, at the Strathclyde Regional Archive in Glasgow, an overworked yet friendly archivist whose name I never knew asked me (when I called up a certain box of bridge-builder William Arrol's papers) if I wished "a dooster." Puzzled, but not wishing to offend, I said yes. I quickly found that a dusting cloth was indeed essential to approach the task, for this box contained business papers intermingled with finely divided coal: the prototypical experience with dusty documents.

A NATION OF STEEL

The transcontinental railroad across the Nevada desert, 1868: construction on the Central Pacific Railroad at Humboldt Plains.

Courtesy Department of Special Collections, Stanford University Libraries; A. A. Hart Collection, PC 2, #317.

CHAPTER ONE

The Dominance of Rails

1865–1885

E arly in the morning of 28 April 1869, eight Irish track layers
—Michael Shay, Patrick Joyce, Michael Kennedy, Thomas
Dailey, George Wyatt, Michael Sullivan, Edward Kieleen, and
Fred McNamara—faced a full day's work, and then some. Before
them stretched a smoothly graded path in the Utah desert, the
final product of a sizable army of Chinese graders who had shov-
eled, chiseled, blasted, and bored their way nearly 700 miles from
Sacramento over the Sierra Nevada, then across the desert toward
a mountainous peninsula on the north shore of the Great Salt Lake.
The graded path stretched away beyond their range of vision. Fifteen miles in the distance lay Promontory Point. If the Irishmen
could deliver on the boast of Charlie Crocker, the stoutest and most
stubborn of the Central Pacific's promoters, to lay ten miles of rail
in a single day, they would claim an unbeatable record. With about
twelve miles to go, the Union Pacific's track crew had caught up
with the grading crew and could make at best a half-mile each day.
Just two years earlier, laying an entire mile of track had represented
an extraordinary day's work, but building the transcontinental railroads had transformed the art of laying rails. In time the crews of
both roads were laying two, three, and even four miles of track day
after day. For its part, the Union Pacific was now resting on its lau-
rels, having some months earlier staged a marathon from daybreak
into the night that laid an unheard-of 7¾ miles of track.[1]

No grading bottleneck faced the Central Pacific, and the company planned no marathon. "We must organize," Crocker had directed James Strobridge, who had pushed the track-laying crews
at breakneck pace for months. "You don't suppose we are going to
put two or three thousand men on that track and let them do just as

they please?"[2] In sending the entire railroad car by car across the Sierra Nevada, the Central Pacific's supply staff had developed an exquisite system that loaded sixteen flatcars with every item needed for two miles of track. Now, five such trains had been moved to the front, positioned on the main line or convenient sidings. Ties had already been distributed along the graded path. A crack team of 850 track layers, enticed by the challenge and by the prospect of quadruple wages, stood ready. A locomotive derailment the day before had delayed the attempt, but now all was ready again.

At 7 A.M. Crocker raised his hand, the locomotive of the lead supply train blasted its whistle, and the Central Pacific's machine made of men went into action. Chinese laborers began unloading the first train: kegs of spikes and bolts, bundles of fishplate fasteners, rails. "In eight minutes, the sixteen cars were cleared, with a noise like the bombardment of an army," wrote one correspondent.[3] As soon as the supply train was out of the way, six-man gangs lifted onto the track the "iron cars," small flatcars fitted with rollers to speed the on-and-off movement of rails, and began loading each with sixteen rails plus necessary hardware. Two horses pulled each iron car to end of track, where they were unloaded by Chinese laborers. Meanwhile, a three-man team with shovels had gone ahead to align the ties with the surveyor's stakes at track center.

At end of track stood the eight strapping Irishmen with tongs to grab the rails. A portable track gauge—a wood-framed device to ensure that 4 feet, 8½ inches separated each new rail pair— was moved ahead by two additional men. The Irishmen worked two men at the head and tail of each rail, on both sides of the iron car. As the forward pair seized each 30-foot rail, the rearward pair helped guide it over the rollers and onto the cross tie. As soon as the rail was in place, one gang started the spikes, eight to each rail, and bolted fishplates at each rail joint; another gang finished the spiking and tightened the bolts. As end of track moved forward, ten yards at a clip, track levelers followed in its wake, lifting ties and shoveling dirt under them as needed. At the rear of the column, which eventually stretched out some two miles, were 400 tampers with shovels and iron bars, to give the roadbed a firm set. Whenever a workman grew tired or faltered, a fresh replacement took his place.

Progress that day was phenomenal. The Union Pacific's observers timed the column proceeding down the tracks at 144 feet per minute, a pair of rails every 12 seconds. Another observer witnessed 240 feet of rail laid in 1 minute 20 seconds. At a pace that one might average on a day's walk, the Iron Horse was moving across the desert. "I never saw such organization," marveled an army officer. "It was just like an army marching over the ground and leaving a track built behind them."[4]

At 1:30 P.M. Crocker gave the signal for an hour's lunch break. James Campbell, superintendent of the division, ran the camp train in and served five thousand dinners to workers and guests. "That morning," recalled Strobridge, the construction superintendent, "we laid six miles in six hours and fifteen minutes, and although we changed horses every two hours, we were laying up a sixty-foot grade, our horses tired and could not run."[5] After an hour's rest the crews returned to action. For almost an hour the most important task, conducted with hammer, wood blocks, and eyesight, was bending rails for the upcoming curve. As the shadows lengthened, the last supply car was unloaded, then the final few iron cars. By 7 P.M. the 10 miles of new track, plus 56 extra feet, was complete. The effort had consumed 25,800 ties, 28,160 spikes, 14,080 bolts, and 3,524 rails. But the most prodigious feat of the day was that of the eight Irishmen, who waved off the replacement crew offered them at lunch and stuck out the entire grueling day. Each four-man team hefted nearly 1,800 rails, every one of which, at 56 pounds per yard and 30 feet in length, weighed 560 pounds; together, the eight men handled that day just over 2 million pounds of rails.

The eight heroes' mark was never broken by the Union Pacific, and the restless tempo of railroad building soon yielded to the irresistible pressure of running the railroad. In just 40 minutes superintendent Campbell ran an engine over the new 10-mile track to prove its soundness. Within two days, at a more leisurely pace, the two rival crews laid the last rails that took their ends of track to Promontory Point—690 miles from Sacramento and 1,086 miles from the Missouri River. There they waited for the bigwigs to arrive. On 10 May, as Maury Klein relates, "the wrong people came to the wrong place for the wrong reason."[6] Absent from the ceremony were the most important promoters of each railroad, who were in New York or California on business. Nor was any government representative present for this most impressive of the nation's internal improvements. Neither of the railroads had wanted to meet there; both had hoped to build more mileage and thereby claim more construction bonds. Only a congressional resolution in early April designating the point of junction prevented an open feud. After Leland Stanford of the Central Pacific and Thomas Durant of the Union Pacific had both taken awkward swipes at the Golden Spike, missing it entirely and whacking the rail, the telegraph mes-

sage went out to the world: "THE LAST RAIL IS LAID! THE LAST
SPIKE IS DRIVEN! THE PACIFIC RAILROAD IS COMPLETED!"

If less easily cast in a heroic mold, the machine made of iron
that lay behind the transcontinental railroad was no less impressive. In the one year of 1869 the nation had laid more miles of rails
than ever before, nearly five thousand. Most of those rails were
produced in the United States. Imports, previously a dominant
source of rails, were reduced if not eliminated by the domestic-
only provision of the Pacific railroad legislation. Indeed, domestic
rail production had grown handsomely with the expansion of Pennsylvania's iron industry and with rail rerolling operations scattered
from Massachusetts to Illinois. Output of new rails began climbing
after 1857 when John Fritz's three-high rail mill cut the amount
of arm-breaking labor needed to form square bars of iron into finished rails with their distinctive inverted T shape. As more and
more mills adopted Fritz's invention, the tally of rails climbed ever
higher. Between 1855 and 1865, scarcely dented by the Civil War,
rail production increased 250 percent.[7]

Yet, however much labor had been saved in the rolling mill, still
more was necessary before the iron was suitable for rolling in the
first place. Each rail began in the blast furnace as a mixture of
limestone, coal, and iron ore that was melted down, or smelted,
into crude iron and waste slag. Iron directly from the blast furnace, known as pig iron, was a brittle product suitable for casting
into pots, stove plates, and the like, but totally unsuitable for rolling into rails or anything else. Refining of the crude pig iron was
accomplished by a skilled workman at a puddling, or boiling, furnace. There a half-day of careful stirring and heavy manipulation
yielded a large ball of pasty and fibrous iron, which was hammered
or rolled into "wrought iron" bars that could be sent to the rolling mills. The rub was that expanding the output of wrought iron
required additional skilled workers; capital for production equipment alone was insufficient. The most important workers were the
skilled puddlers, an independent lot, the "aristocrats" of the industry. Inevitably, puddlers organized the early iron unions, with a
propensity for effective strikes. A strike at the Cambria mill in 1867
briefly stalled the Union Pacific in its march across the continent.[8]

By the early 1870s, with railroad construction at fever pitch,
there emerged tell-tale signs of an industrial system pressed beyond
its limits. The price for iron rails jumped by $15, to $85 a ton, while
the price of steel rails climbed all the way to $120 a ton. Although
the land-grant railroads were limited to domestic rails, their competitors were not. From 1867 to 1872, increased demand for rails

was met mostly by increased imports because domestic iron production had stagnated. In 1872 more steel rails were imported than were produced in America; together, iron and steel imports constituted one-third of the six thousand miles of track laid that year.[9] These figures raised the nagging question of who, or what, could fill the gap between domestic production and domestic consumption? Clearly, the United States needed a process to mass-produce steel for rails in these years of extensive railroad building. Like most of the steel rails themselves, the process was imported from England and transformed by a new geographical and economic context.

Inventing the Process

*An invention must be nursed as a mother nurses her baby,
or it inevitably perishes.*—HENRY BESSEMER

The Bessemer process was the most important technique for making steel in the nineteenth century. From the start it was bound up with powerful institutions; in this respect, its transfer and development by American railroads continued a pattern set out by its English inventor, Henry Bessemer. Although at least five English or American inventors devised processes for making steel by blowing steam or air through molten iron, Bessemer's process became the most widely used. His technical ingenuity was only part of the reason for his success. A close examination of Bessemer's inventive activities shows that his process was inspired by association with powerful institutions, especially the military; that his efforts to develop and commercialize his process rescued it from failure and obscurity; and that his considerable flair for publicity and patronage was as important to his overall success as his original technical conception. "The Messrs. Bessemer, the great steel makers, [have] long been known to keep their own leading article writer . . . Mr. Bessemer's investing in what some might deem a luxury was a powerful natural step on his part," wrote the *Engineer*. "Few of the many investments Mr. Bessemer has made during his career have been more promising." "Bessemerizing" was understood by his peers as a social as well as technical achievement.[10]

As a young man, Bessemer proved himself adept at lining up prominent patrons behind his mechanical inventions. The son of a French émigré who had talked his way into the English mint, then retired to the countryside, Bessemer came to London in 1830 at age seventeen and made a living by executing artistic castings, printing specialty items, and making metal dies. By the time he was twenty-three his experiments in electrometallurgy had attracted

the attention of Andrew Ure, the prominent philosopher of indus-
trialism. About the same time, Bessemer made his first notable
invention, for the Inland Revenue Office. He devised a dated stamp
that prevented the fraudulent reuse of revenue stamps on official
documents. The invention reportedly brought the treasury an additional £100,000 per year and eventually a knighthood for Bessemer
in 1879.[11] In the meantime Bessemer's inventive career flourished.
Windsor Castle tendered the first order for his novel machine-made
embossed velvet, and the pioneering Coalbrookdale Iron Company
placed the first order for his machine-made bronze powder. In fact,
a handsome income from manufacturing powdered bronze, which
he conducted with great secrecy for some forty years, supported his
inventive activities. At Baxter House, in the St. Pancras district of
London, Bessemer set up an experimental factory and laboratory.
His work there, when added to his earlier efforts, netted Bessemer
thirty-four patents on new methods for casting and setting type;
for manufacturing paints, oils, varnishes, sugar, and plate glass; for
constructing railway carriages, centrifugal pumps, and projectiles;
and for ventilating coal mines.[12]

Bessemer had achieved modest success as a mechanical inventor
when a military problem led him to a new activity, the manufacture
of iron. "At the time when the Crimean War broke out," he wrote,
"the attention of many persons was directed to the state of our
armaments." Bessemer identified heavy rifled ordnance as a key
problem, and he devised a way of imparting spin to projectiles shot
from a smooth-bore gun. When the British War Office rejected his
idea, Bessemer took it to France. In 1854, after demonstrating the
revolving projectile at Vincennes, Bessemer heard from the officer
in charge, Commander Claude Etienne Minié, the rifle inventor,
that he mistrusted firing the heavy 30-pound shot from the available 12-pounder cast-iron guns. Minié asked whether a new type
of gun could be made to withstand such heavy projectiles. This observation, by Bessemer's account, "instantly forced on my attention
the real difficulty": withstanding the "heavy strains" caused by the
projectiles. He returned immediately to Baxter House determined
"to study the whole question of metals suitable for the construction
of guns." Within three weeks he filed the first in a famous series of
patents on iron and steel.[13]

Assisted at Baxter House by his brother-in-law, Bessemer began
his gun-metal experiments by melting various mixtures of pig iron,
blister steel (a refined and expensive metal), and scrap in a standard iron-melting reverberatory furnace, producing enough metal
to cast a model gun.[14] Then Bessemer's experiments took an unexpected twist:

FIG. 1.1. Bessemer's early steelmaking experiments, 1854–55. *Top left:* Section of crucible with blow-pipe. *Top right:* Section of vertical converter. *Below:* Sections of vertical converter with upper chamber. *From Henry Bessemer,* Autobiography *(London, 1905), pl. XIII.*

Some pieces of pig iron on one side of the bath attracted my attention by remaining unmelted in the great heat of the furnace, and I turned on a little more air through the fire-bridge with the intention of increasing the combustion. On again opening the furnace door, after an interval of half an hour, these two pieces of pig still remained unfused. I then took an iron bar, with the intention of pushing them into the bath, when I discovered that they were merely thin shells of decarburised iron . . . showing that atmospheric air alone was capable of wholly decarburising grey pig iron, and converting it into malleable iron without puddling or any other manipulation. Thus a new direction was given to my thoughts, and after due deliberation I became convinced that if air could be brought into contact with a sufficiently extensive surface of molten crude iron, it would rapidly convert it into malleable iron.[15]

Guided by this radical thought that air *alone* could refine iron, Bessemer built a crucible with a blowpipe in its center (fig. 1.1). Into the crucible he poured about 10 pounds of unrefined pig iron. He then placed the apparatus into a hot furnace; after 30 minutes of blowing air into the metal, he found that the crude iron had become malleable iron. This experiment proved that air could decarburize pig iron, turning it into a useful product, yet the furnace surrounding the crucible still consumed a large amount of fuel. Bessemer's real insight was to get rid of the furnace entirely. For this he built a 4-foot-tall, open-mouthed cylinder with openings, or tuyères, to blow air into the metal from the bottom. As Bessemer related it, at first the blast bubbled quietly through the seven hundredweight of molten pig iron in the vessel, with the opening emitting some sparks and hot gases. Suddenly, after 10 minutes, white flames shot forth. "Then followed a succession of mild explosions, throwing molten slags and splashes of metal high up into the air, the apparatus becoming a veritable volcano in a state of active eruption." Unable to approach the out-of-control converter, Bessemer and his assistants could only watch until the eruption quieted after 10 minutes more. On tapping the converter and casting an ingot, Bessemer identified the metal as wholly decarburized malleable iron—the product he desired. And except for melting the crude iron in the first place, no fuel was used.[16]

In the following weeks Bessemer sought to tame the violent blow. To deflect the flame he placed a cast-iron grate above the opening, but the white-hot blast melted it. He then tried lessening the blast by decreasing the number of tuyères, their diameter, or the pressure of the air. These changes indeed reduced the heat, in one case leaving an entire converter full of solid iron. Resigning himself to a violent blow, Bessemer sought to contain it in a converter featuring a cone-shaped upper chamber, which strikingly resembles an earlier furnace he devised to melt optical glass (fig. 1.2). This familiar arrangement harmlessly vented the hot gases. With this vertical, fixed, two-chamber converter, Bessemer achieved the successful conversion of crude iron to steel.[17]

Bessemer quickly began to publicize his achievement. He showed his converter to George Rennie, president of the mechanical sciences section of the British Association for the Advancement of Science (BAAS), where a decade earlier James Neilson had showcased his hot blast furnace, the invention that gave iron smelting its characteristic form and name. Rennie was duly impressed and invited Bessemer to address the upcoming BAAS meeting in Cheltenham. There, on 13 August 1856, Bessemer presented "The Manufacture of Malleable Iron and Steel without Fuel." If this seemingly absurd concept elicited snickers before the talk, few remained afterward. It was "a true British nugget," stated the

FIG. 1.2. Bessemer's optical glass furnace, c. 1850.
From Bessemer, Autobiography, *pl. IX.*

FIG. 1.3. Publicity for Bessemer's first converter, from the *Illustrated London News*, 30 August 1856.

From W. M. Lord, "The Development of the Bessemer Process in Lancashire, 1856–1900," Newcomen Society Transactions 25 (1945–47): 163–80.

engineer James Nasmyth. The next day's report in the *Times*, reproducing Bessemer's 2,000-word essay, attracted the attention of Edward Riley, chemist of the Dowlais Iron Company, the leading Welsh firm. Two weeks after the BAAS talk Dowlais purchased the first license for £10,000, just as the *Illustrated London News* featured Bessemer's converter in action (fig. 1.3). Early in September Bessemer staged an exhibition at Baxter House for "some seventy or eighty of the most eminent persons connected with the manufacture of iron."[18]

Trial converters were erected in the following months by Dowlais Iron, Messrs. Galloway (Manchester), Govan Ironworks (Glasgow), and Butterley Iron (Derbyshire), but their trials were disastrous. Many found the new metal too brittle to roll or forge into useful shapes; at the rolling mill at Dowlais, for instance, "the ingots . . . were crushed into rough gravel like powder, showing a total want of malleability."[19] To diagnose the problem, Bessemer consulted with the prominent metallurgists T. H. Henry and John Percy, while the Dowlais chemist continued his investigations. Bessemer had by happenstance used a gray Blaenavon pig iron that was exceedingly low in phosphorus content. Nearly all other pig irons available in Britain contained significant amounts of this element, and they yielded an obviously inferior product. Bessemer first tried to improve common grades of British pig iron by using different fluxes and by passing various gases through the liquid metal. When this line of experiments failed he decided to switch to Swedish pig iron, a premium grade often used in making Sheffield's fiendishly expensive crucible steel. Finding and matching suitable pig irons was a hit-or-miss proposition for several years, until the key role of phosphorus—and especially its relation to brittleness in steel—was recognized in the early 1860s. It was from bitter experience that Bessemer later remarked, "An invention must be nursed as a mother nurses her baby, or it inevitably perishes."[20]

Another problem was that Bessemer had announced his process as if it were trouble-free when it manifestly was not. Backers of the several commercial ventures were especially annoyed that their license payments had not brought them a workable process. Their persistent difficulties eventually forced Bessemer to refund some £32,500 in licensing fees. A complete commercial failure was at hand. What saved Bessemer was his experience in developing and commercializing the manufacturing ventures in embossed velvet and bronze powder, and the money he had realized from them. Furthermore, he had retained full rights to his patent. These resources together gave Bessemer the insight and means to build a steel works of his own in Sheffield (fig. 1.4). In 1858 he formed a partnership with his brothers-in-law Robert Longsdon and William Allen and his machinery suppliers William and John Galloway of Manchester. For its first year the steel works produced an odd product. A stream of the molten steel was run into a vat of water and solidified as pellets; then a mixture of these pellets from several batches was remelted in one of Sheffield's famous clay crucibles. Only after a visit in 1859 from the Swedish producer Göran Göransson, the first licensee to make commercial quantities of the new steel, did Bessemer begin producing steel directly from the converter without the crucible remelting. Although the details of his contribution remain unclear to this date, Göransson may have helped Bessemer turn the corner. After losing £1,800 in its first two years of operation (1858–59), the Sheffield works made increasingly handsome profits.[21]

Yet there is no gainsaying Bessemer's own substantial efforts. He helped commercialize in Cumberland and Lancashire newly discovered beds of nonphosphorus hematite ore, which could easily

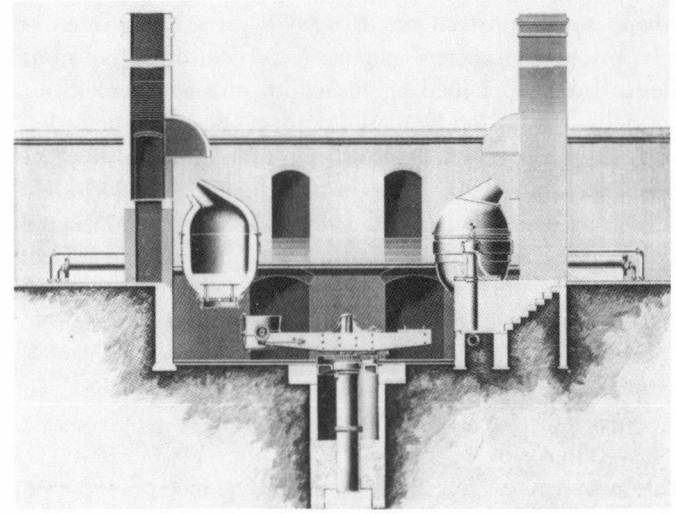

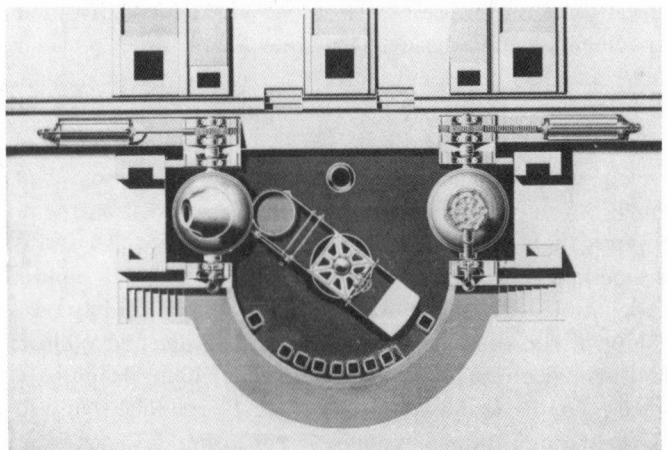

FIG. 1.4. Plan of Bessemer plant at Sheffield, 1858.
From Bessemer, Autobiography, *pls. XVII and XVIII.*

and economically be smelted and converted into a malleable steel
that was easily rolled or forged. He also worked to devise a firebrick
lining to contain the white-hot metal in the converting vessel. For
the Sheffield works in 1858 Bessemer adopted a tiltable converter,
a major innovation again prefigured in the optical glass furnace.
The physical arrangement of the stationary vertical converter had
required that the air blast begin before the crude iron entered the
converting vessel and last until the converted steel left; otherwise
the metal would simply run out the bottom. The problem was that

any delay in casting the converted steel into ingots could result in the air blast cooling the metal, and there was always the specter of the cold converter filled with solid metal. Bessemer's solution was to tilt the converter so that the air blast was needed only for the duration of the converting blow (fig. 1.5). With this arrangement, minor delays could no longer ruin the batch of steel. With new sources of low-phosphorus pig iron, a more durable vessel lining, and the tilting converter, Bessemer had achieved technical if not yet commercial success.[22]

Bessemer's commercial success was a product of his ability to arrange financing, retain patent control, and fend off rivals. Again the publicity machine was fired up. Bessemer's exhibit at the London International Exhibition in 1862 attracted the attention of John

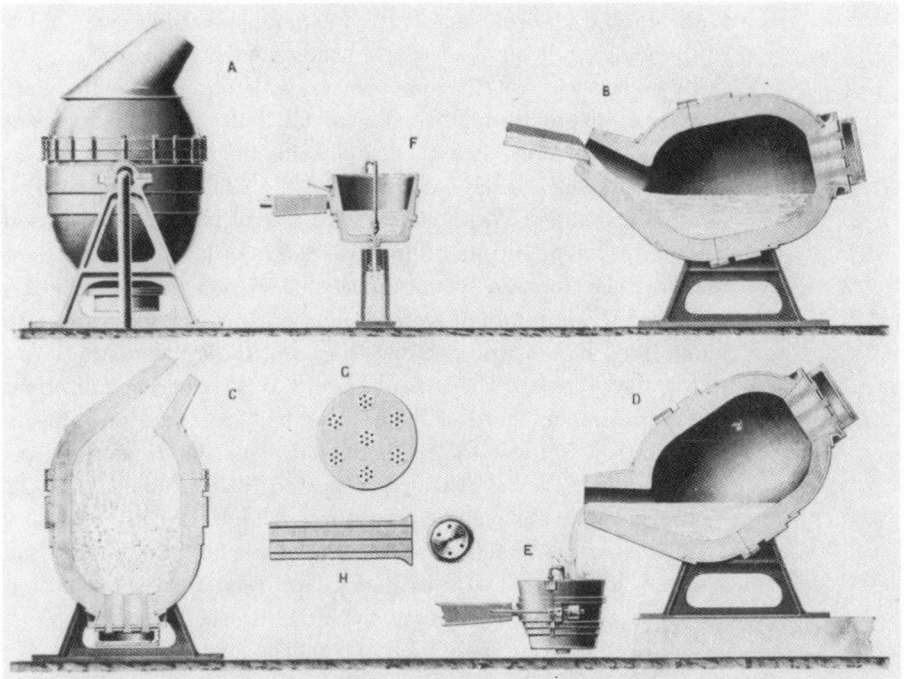

FIG. 1.5. Bessemer's tilting converters, 1858. (a) vertical converter; (b) tilted converter being filled with crude iron; (c) converter during air blast; (d) converter tilting to pour out converted steel into (e), the casting ladle; (f) cross-section of ladle pouring steel into ingot mold; (g) cross-section of converter bottom, showing the air openings, or tuyères; (h) close-up cross-section of tuyère.

From Bessemer, Autobiography, *pl. XV.*

Platt, an engineer and industrialist who was also the member of Parliament from Oldham. Impressed by the process (and by the profits finally coming to the Sheffield works), Platt quickly assembled an investment syndicate that offered to purchase a one-fourth share of the patent. Wary of losing control, Bessemer pressed for a more advantageous deal. The two parties soon came to agreement that the syndicate would make an investment equivalent to a one-fourth share for £50,000, but that Bessemer and Longsdon would retain complete control of the patents, thus retaining the right to grant licenses and to raise or lower royalties. Three weeks after the exhibition's opening, the deal was concluded after dinner at the Queen's Hotel in Manchester, when each of the ten investors personally handed £5,000 to Bessemer and Longsdon. This timely infusion of capital added momentum to commercializing the new process. Over fourteen years Bessemer's original five partners in the Sheffield works racked up profits totaling 8,100 percent, while the members of the Platt syndicate made a still handsome 520 percent.[23]

Bessemer was not the only person to experiment with blowing air or steam through molten metal, but his ability to coopt potential rivals was unmatched. Following the BAAS paper Bessemer had declined a buyout offer from the Ebbw Vale Ironworks, another prominent Welsh concern. Instead, Ebbw Vale purchased a vaguely similar and therefore rival patent of Joseph G. Martien, and later the company's own furnace master gained several patents for blowing air into and over various mixtures of scrap iron. In about 1864 Ebbw Vale decided to exploit these patents by forming a new joint stock company with a very large capitalization. When Bessemer heard of these plans he threatened not only to file a lawsuit but also to set his publicity machine to work covering London's financial district with posters, placards, and handbills to prevent the stock subscription from going forth. In the event a compromise was arranged between Bessemer and the principals of Ebbw Vale, Abraham Darby and Joseph Robinson. It turned out that Ebbw Vale did not actually own the furnace master's patents, which Bessemer purchased for £5,000. The new company took out a regular license and paid standard royalties, except that Bessemer deducted £25,000 from the first royalties in exchange for the Martien patents. Bessemer now possessed his principal rival's patents, he had another licensee, and despite the stiff price, he netted some £50,000 or £60,000 from the deal. Thus, as Bessemer put it, "happily was removed the last barrier to the quiet commercial progress of my invention throughout Europe and America."[24]

Developing the Technology

*If we compare the two men we see strikingly exemplified the
difference between the inventor and the industrious engineer.
Where Bessemer left the process which bears his name,
Holley's work began.*—R. W. RAYMOND

Like most new technologies, Bessemer's process contained many
latent possibilities. In England and Europe, Bessemer converters
produced a moderate-quality, general-purpose steel used widely
for structures, merchant steel sold for general purposes, and rails.
As modified by the basic, or Thomas, process, European converters
could take high-phosphorus pig iron and turn it into an accept-
able steel. In the United States, however, the Bessemer process
took a different path determined by domestic natural resources and
by the decisive influence of a leading consumer. To begin, there
was never an "American" Thomas process, since American ores did
not contain enough phosphorus to make the chemistry go.[25] More
important, the nation's fever for westward expansion produced a
boom in transcontinental railroad building and demands for iron
and steel that surpassed the dreams of European steelmakers.[26] It
was during the second of three great spurts of railroad construction
(fig. 1.6) that Bessemer production in the United States once and for
all exceeded that in Great Britain. This peculiar demand structure,
spiky and cyclical, encouraged the development of a rail-making
process that could produce vast volumes quickly on command.[27] In
response, American steelmakers developed the latent features of
the Bessemer process that facilitated maximum production while
ignoring other features that were needed for maximum quality. The
result was the reckless mass production of Bessemer steel rails.
This technology was unique to the United States and peculiarly
adapted to the extensive phase of railroad building that lasted from
about 1870 to 1890. As we shall see in chapters 2 and 4, these
acquired characteristics bedeviled later efforts to do anything else
with Bessemer steel.

To a remarkable extent, the shape and definition of Bessemer
technology in the United States was achieved through the efforts of
Alexander L. Holley. An early immersion in railroading prepared
him for a brilliant and prolific steel-engineering career: he designed
eleven of the first thirteen Bessemer plants in the United States.
Born in 1832, Holley was the son of a successful cutlery manufacturer and lieutenant-governor of Connecticut. As a young man
he saw in his father the established connection between politics,
industry, and the wider world. Holley's own easy familiarity with

people of all classes gained him entrée into steel mills and engineering circles on both sides of the Atlantic, and he served as a model for mechanical engineers seeking to combine professional values, technical knowledge, and social graces. Inevitably, he was one of the organizing spirits behind the American Society of Mechanical Engineers (ASME).

Holley arrived at such a position through detailed experience with two of the principal mechanical technologies of his era, railroads and steel manufacture. In the seven years after graduating from Brown University in 1853, he designed locomotives for the steam-engine builder George Corliss and for the New York Locomotive Works, published two books based on firsthand knowledge of European railways, and served as technical editor for the *American Railway Review*. Between 1858 and 1863 he also wrote two hundred articles for the *New York Times*, mostly on engineering subjects. Holley learned the craft of technical publicity not by hiring a "leading article writer" as Bessemer had done but by being one. In 1862, while working for Edwin Stevens, the founding manager and longtime treasurer of one of the nation's oldest railroads, the Camden & Amboy, Holley began gathering information about shipbuilding, armor plate, and armament in Europe. He visited

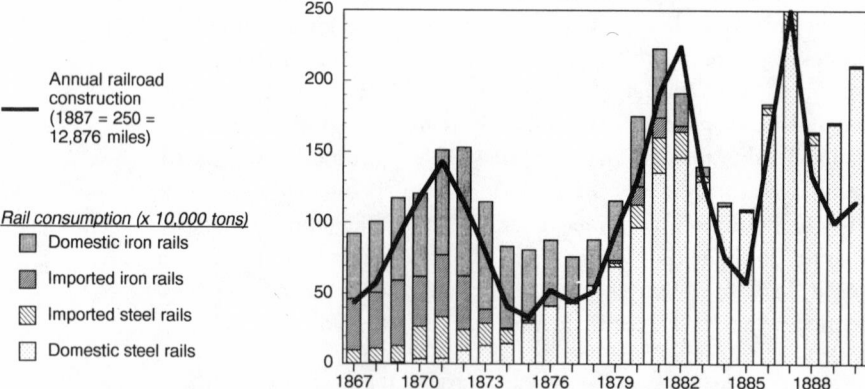

Annual railroad construction (1887 = 250 = 12,876 miles)

Rail consumption (x 10,000 tons)
Domestic iron rails
Imported iron rails
Imported steel rails
Domestic steel rails

FIG. 1.6. U.S. rail production, imports, and construction, 1867–90. Actual rail total for 1887 included 2,373,335 tons domestic steel, 154,370 tons imported steel, and 23,062 tons domestic iron.

Here and in other graphs in this book, in order to show the relationship between two sets of data (here, rail construction and rail consumption), an index is used for one of the data series and a convenient scale for the other.

Data from AISA, Annual Statistical Report *(1887 and 1931)*.

Bessemer's works in Sheffield, where he observed the new process as a prospective licensee. "The Bessemer process of making steel," Holley observed in a widely acclaimed book based on this trip, "promises to ameliorate the whole subject of ordnance and engineering construction in general, both as to quality and cost."[28]

When he returned to the United States in 1863, Holley found several parties interested in Bessemer technology. The assistant secretary of the navy had just urged the prominent engineer John Ericsson to establish a steel plant with the new technology. When Holley consulted with Ericsson about realizing the steel process, Ericsson put the young man in touch with the ironmaster John F. Winslow and banker John A. Griswold, who had jointly sponsored Ericsson's ironclad ship, the *Monitor*. These two businessmen from Troy, New York, provided Holley with steadfast support for the steelmaking venture. To secure an American license for Bessemer's patents, Holley crossed the Atlantic again in the summer of 1863 and arranged to pay Bessemer £10,000. Winslow, Griswold, and Holley set up a partnership and began planning a Bessemer plant at Troy. The following year Holley was once again in England, as he put it, "to finish my education in the Bessemer process."[29]

In the mid-1860s Troy sustained the gathering storm of railroad building in two distinct ways. The immediate region had nurtured, educated, or given railroad experience to a surprising number of the Union Pacific's and Central Pacific's leading figures, including Leland Stanford and Charlie Crocker. More prosaically, it had also become a center for rerolling rails.[30] The Erie Canal, the Hudson River, and a growing railroad network linked the region to consumers and raw materials. Before Holley's effort, however, Troy lacked steelmaking facilities. After obtaining promising results from initial experiments in an abandoned gristmill, Winslow and Griswold approved a full-scale plant located with access to both the Hudson River and the Hudson River Railroad. When it opened with a public demonstration in May 1867, the mill would scarcely have been recognizable to Henry Bessemer (fig. 1.7).[31]

"Holley seems to have at once broken loose from the restraints of his foreign experience," noted an observer. "Where Bessemer left the process which bears his name, Holley's work began," stated another. One of Holley's assistants, who had copied tracings of machinery and fixtures sent by Bessemer's office, noted that "very few of the features . . . of Mr. Bessemer's practice as thus given were embodied in the American work." Another assistant, Robert W. Hunt, whose career spanned the entire first generation of the industry, recorded the details. At Troy Holley did away with the deep

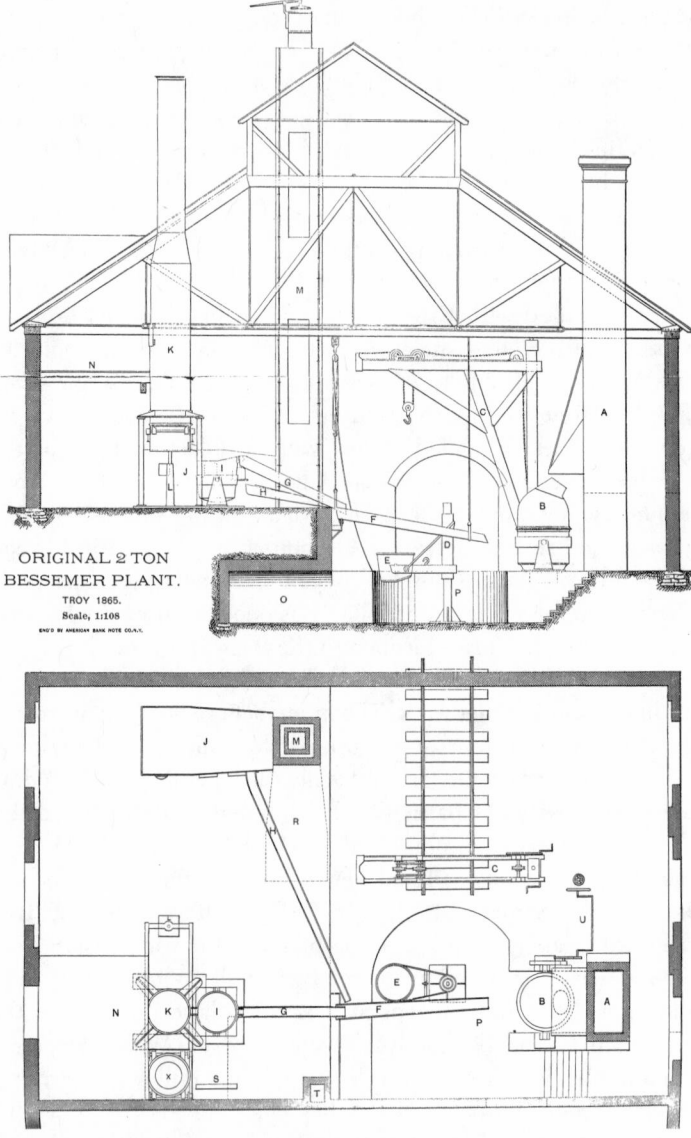

ORIGINAL 2 TON
BESSEMER PLANT.
TROY 1865.
Scale, 1:108
ENG'D BY AMERICAN BANK NOTE CO./N.Y.

FIG. 1.7. Troy Bessemer plant, plan of 1865.
*From Robert W. Hunt, "The Original Bessemer Steel Plant at Troy," ASME Trans.
6 (1885): 62–63.*

casting pits of the British mills and raised the converting vessels to get working space under them on the ground floor. He replaced the British counterweighted cranes with less expensive, top-supported hydraulic ones, adding a third ingot crane to expedite the flow of material around the pit. He modified the ladle crane and worked all the cranes and the vessels from a single control station. He located the converting vessels more conveniently to the casting pit and melting-house. He substituted cupolas for reverberatory furnaces, and introduced the intermediate or accumulating ladle, which was used to weigh each charge of melted iron before it went into the converter.[32] Holley eventually designed a dozen other Bessemer steel rail mills—each one more capable of maximum quantity, to fit the insatiable demands of transcontinental railroad building, but less capable of maximum quality.

Even before railroads made the huge, spiky demands that determined the characteristics of successful steel production technology, they helped structure the emerging industry. In the early 1860s the leading rail mills organized a patent pool around John Fritz's three-high rail mill, which helped boost rail output prodigiously; this activity was soon extended to Bessemer technology.[33] Through the Bessemer Association, which formally was again a patent pool, railroads met periodically with rail mills to agree on prices and divide contracts among the rail mills.

The Bessemer Association was the product of a complicated legal battle that threatened to stall the industry's growth. Holley and the Troy entrepreneurs faced a challenge from a group headed by Eber Brock Ward, who had investments in transportation, minerals, banking, and iron ventures across the Midwest. In May 1863 Ward organized a company to exploit the legal strength of William Kelly's "air boiling" patent. Kelly was a sometime ironmaster who had used suggestive experiments in Kentucky in the 1850s to successfully fight an American patent-interference case against Bessemer. He had thus guaranteed that his patent covered the concept of blowing air through molten metal.[34] In addition to Kelly's patent, Ward controlled the American rights to Robert Mushet's key patents for treating converted and decarburized Bessemer metal with manganese, which improved its mechanical properties. These patents informed the design of the Eureka Iron Company south of Detroit, the site of a new air-blowing steel works (table 1.1). Another Ward holding, the North Chicago Rolling Mill Company, stood ready to roll Eureka's steel into rails. On the other hand, the Troy group possessed the American rights to Bessemer's patents, which included the tilting converter. Given this split, neither group could construct a state-of-the-art steel works without infringing on the other. As Holley put it, "Litigation of a formidable character was imminent."[35]

The legal block was resolved through a series of out-of-court settlements beginning in 1866 and lasting three years. The precise course of these negotiations was kept out of the public eye and remains unknown, but the outcome was clear enough. The Kelly and Bessemer patents were pooled, and the proceeds from licensing them were split. The Troy group received 70 percent of the proceeds, and the Ward group, 30 percent. To administer the patent pool the two groups set up the "Trustees of the Pneumatic or Bessemer Process of Making Iron and Steel," or the Bessemer Association. This organization—reorganized as the Pneumatic Steel Association, the Bessemer Steel Association, and the Bessemer Steel Company—did not operate plants directly but rather provided a means for licensing the pooled patents, collecting royalties, and dividing the proceeds. For the immediate future the Bessemer Association assured that no long legal battle would check growth. From 1866 to 1877 it licensed eleven plants, to all the major rail mills including Cambria (fig. 1.8).[36]

After 1877 the Bessemer Association prevented others from legally acquiring the technology it controlled, which then included

FIG. 1.8. Cambria Bessemer plant, 1876.
Courtesy of Hagley Museum and Library; Acc. 80.300.

several key mechanical patents of Holley. It did this by sharply restricting the number of licensees. Additional licensees meant additional competitors, and the established concerns wanted neither. Further, in the early 1880s Holley's detailed reports on state-of-the-art steel technology, from both sides of the Atlantic, circulated only to association members. By controlling access to technology in these ways, the Bessemer Association amputated the invisible hand of a competitive market. Moreover, its visible hand of administrative coordination lasted through the Bessemer era and beyond. From 1877 until 1915, with the exception of the depression decade of the 1890s, the price of steel rails was largely determined by the Bessemer Association and its successors.[37] These restrictive practices can be understood only by recognizing that railroads and steel mills were not atomistic actors meeting in a classical free market; rather, even beyond the market-killing coordination of the Bessemer Association, the users and producers of rails could be owned or controlled by the same (railroad) corporation.

The Pennsylvania Railroad was the outstanding example of the virtual fusion of user and producer that could result in this industrial sector. A large number of employees with "second" careers, the prevalence of insider contracting, and strategic personal investing by its top officers all contributed to this result. Andrew Carnegie was only the most famous of its middle managers who had successful second careers in closely related fields. Before becoming steel producer par excellence, Carnegie climbed the Pennsylvania's corporate ladder and proved himself an able manager of its western division. After leaving the railroad in March 1865 with twelve years of service, he enjoyed close relations with Thomas A. Scott and J. Edgar Thomson, the Pennsylvania's vice president and president, respectively. Thomson, Carnegie said, was the "great pillar in this country of steel for everything." Scott and Thomson lent financial backing to Carnegie's many ventures, including the Union Iron Mills, the Keystone Bridge Company, and the Edgar Thomson Steel Company, discussed below. J. N. Linville, formerly a Pennsylvania Railroad bridge engineer, was an early member of Keystone's board and later its president. Another Pennsylvania Railroad manager, Samuel M. Felton, after serving as president of the Philadelphia, Wilmington & Baltimore Railroad, capped his career as president of the Pennsylvania Steel Company. The Pennsylvania Steel Company and the Pennsylvania Railroad were more than namesakes. The railroad formed the steel company with $600,000 in capital and promptly awarded it a $200,000 contract for steel rails.[38] The railroad's desire for a captive steel plant de-

TABLE 1.1. Bessemer works in United States, 1864–76

First Blow	Company	Location	Context
September 1864	Eureka Iron Company	Wyandotte, Mich.	Rolling mill
February 1865	Winslow & Griswold	Troy, N.Y.	Rolling mill
June 1867	Pennsylvania Steel	Harrisburg, Pa.	Railroad
May 1868	Freedom Iron & Steel[a]	Lewistown, Pa.	Rolling mill
October 1868	Cleveland Rolling Mill	Newburgh, Ohio	Rolling mill
July 1871	Cambria Iron	Johnstown, Pa.	Rolling mill
July 1871	Union Iron[b]	S. Chicago, Ill.	Rolling mill
April 1872	N. Chicago Rolling Mill[c]	Chicago, Ill.	Rolling mill
March 1873	Joliet Iron & Steel[d]	Joliet, Ill.	Rolling mill
October 1873	Bethlehem Iron	Bethlehem, Pa.	Iron mill
August 1875	Edgar Thomson Steel	Pittsburgh, Pa.	New mill
October 1875	Lackawanna Iron & Coal	Scranton, Pa.	Iron mill
1876	Vulcan Iron	St. Louis, Mo.	Iron mill

Source: Robert W. Hunt, "A History of the Bessemer Manufacture in America," AIME Trans. 5 (1876): 201–16.

[a] Of the twelve Bessemer plants built in 1865–76, Holley did not design this one; lacking suitable pig iron it failed after one year.

[b] Owned by the owners of Cleveland Rolling Mill.

[c] Owned in part by E. B. Ward.

[d] Used blowing engine, converters, and hydraulic cranes purchased from Freedom Iron and Steel's abandoned works.

termined its otherwise disadvantageous location on the road's main line near Harrisburg.[39] There, Holley built his second Bessemer steel rail mill (table 1.1).

The railroad's influence extended beyond such directed financing; it also provided a model for managerial structure. In his formative years Andrew Carnegie gained experience with the Pennsylvania's pioneering of modern managerial systems. In 1853 he joined the Pennsylvania as a telegraph operator, and three years later he followed his mentor Tom Scott to the road's central shops and office at Altoona, just as the transportation superintendent, Herman Haupt, was implementing a military-inspired managerial system. Detailed reporting and accounting from employees at all levels, as well as careful division of managerial responsibilities into so-called line and staff functions, was the essence of the Haupt plan. Under this system Carnegie worked as superintendent of the western division for six years beginning in 1859, with a brief tour of duty managing railroads in Washington, D.C., during the Civil War. Then, as well as later in his industrial enterprises, Carnegie understood the importance of getting the right people and organizing them in the right way. "I am neither mechanic nor engineer, nor am I scientific. The fact is I don't amount to anything in any industrial department," wrote Carnegie. "I seem to have had a knack of utilizing those that do know better than myself."[40] For his entry into the manufacture of Bessemer steel rails, aptly named the Edgar Thomson works in a none-too-subtle bid for the good will of his anticipated leading customer, Carnegie capitalized on all his railroad connections.

Railroad executives brought to Carnegie's new steel venture effective managerial models, the ability to mount large-scale financing, and not least, detailed knowledge of the Bessemer industry's leading consumer. Besides Carnegie at least five railroad executives—David Stewart, John Scott, William Shinn, Edgar Thomson, and Thomas Scott—helped plan and finance the new mill. Carnegie himself, flush with commissions from selling railroad securities to European investors, subscribed the largest share ($250,000) of the total capital ($700,000) in the partnership formed on 5 November 1872. William Shinn, who concurrently served as secretary-treasurer of the steel mill and as vice president of the Allegheny Valley Railroad, imported the railroad's system of cost accounting. Before cost accounting, Carnegie wrote, "we were moles burrowing in the dark." Cost accounting revealed "what each man was doing, who saved material, who wasted it, and who produced the best results."[41]

Construction crews broke ground at the 106-acre site eleven miles up the Monongahela River from Pittsburgh in April 1873, but that year's panic upended the mill's financing and halted the work. The panic also undid the mesh of railroad financing, pushing Tom Scott to the verge of bankruptcy and probably contributing to the death of J. Edgar Thomson in May 1874. To get the project back on track during that summer, Holley and Carnegie met in London with Junius S. Morgan, the international banker, to secure a $400,000 bond issue. The mill finally began full operation in September 1875, producing that month just over 1,200 tons of steel rails. Two months later, facing especially stiff competition from the established Cambria, Pennsylvania, Joliet, and North Chicago mills, Carnegie knew that the new mill was a trump card. "Having faith in our ability to mfr. cheaper than others I do not fear the result of a sharp fight," he wrote to William Shinn.[42] The mill would indeed be a fitting tribute to the railroad president and to the essential linkage between railroads and steel mills. It achieved stunning financial success: the startup month's profits of $11,000 represented 19 percent annual rate of return on capital, and annual profits of

$1,625,000 in 1880 topped the mill's entire first cost. This success flowed directly from the mill's innovative design.[43]

Holley compressed a full decade of experience into his design for the plant, his ninth since Troy. Its twin 6-ton converters (inside diameter 6 feet, height 15 feet) featured his patented detachable bottom, which sped the replacing of refractory brick exactly where the Bessemer blast was most corrosive. Every eight blows, nearly twice each working day, the spent bottom was unbolted and a fresh one attached; the converter itself stayed hot (fig. 1.9). Pig iron melted in three cupolas reached the converters via two 12-ton tilting ladles. Two independent blowing engines, each with twin 20-ton, 20-foot-diameter flywheels, translated the expansive power of steam into a mighty blast of air. Steam from twenty cylindrical boilers, each 5 feet across and 15 feet tall ("a particularly excellent and imposing feature of the plant," boasted Holley), powered the whole. Notably, the rail mill and the converting mill each cost roughly $200,000. "From the very start," wrote one observer, "Holley was convinced that the Bessemer process always meant blast-furnaces, blooming-mills, and rolling-mills."[44]

Previously Holley had been forced to fit Bessemer mills into

FIG. 1.9. Edgar Thomson Bessemer converters with Holley bottom.
From A. L. Holley and L. Smith, "The Works of the Edgar Thomson Steel Company (Limited)," Engineering (London) 25 (19 April 1878): 296.

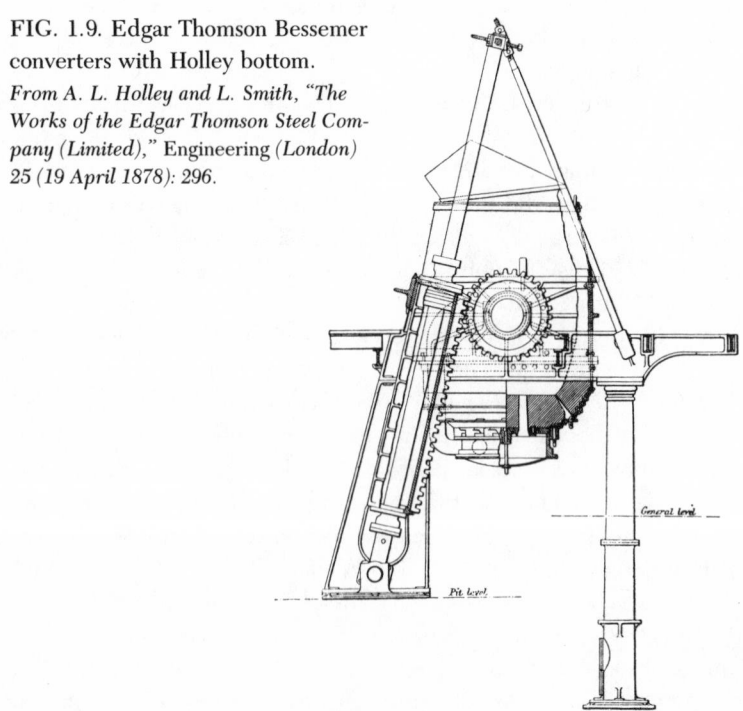

preexisting plant layouts, but for the Edgar Thomson works "the buildings were made to fit the transportation" (fig. 1.10). Holley, said one colleague, "began at the beginning with them, taking a clean sheet of paper, drawing on it first the railroad-tracks, and then placing the buildings and the contents of each building with prime regard to the facile handling of material; so that the whole became a body, shaped by its bones and muscles, rather than a box, into which bones and muscles had to be packed."[45] The transportation arteries included the Monongahela River, which linked the plant to the Ohio and Mississippi rivers, and the Baltimore & Ohio and Pennsylvania railroads running adjacent to the plant. Once raw materials arrived, Holley's design expedited their flow through the works. "As the cheap transportation of supplies of products in process of manufacture, and of products to market, is a feature of the first importance," Holley noted, "these works were laid out, not with a view of making the buildings artistically parallel with the existing roads or with each other, but of laying down convenient railroads with easy curves." "Coal is dumped from the mine-cars, standing upon the elevated track . . . directly upon the floors of the producer and boiler-houses. Coke and pig iron are delivered to the

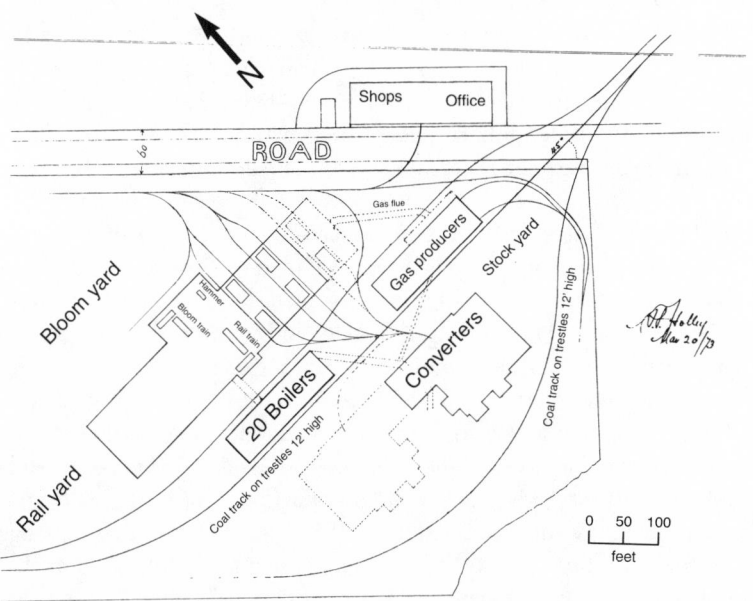

FIG. 1.10. Edgar Thomson steel works, plan of 1873.
Courtesy of National Museum of American History, Smithsonian Institution; ALH-SI.

stock-yard with equal facility. The finishing end of the rail-mill is accommodated on both sides by low-level wide-gauge railways . . . There is also a complete system of 30-inch railways for internal transportation." Ingots passed by rail from the converting mill to the rolling mill not directly but in a wide arc, which cleared space for inspecting and sorting merchant bars and for locating the boilers. "The ingots need some cooling before they can be charged," Holley explained. "The less cooling the steam gets the better."[46]

The efficient flow of heat itself preoccupied Holley's engineering mind. Why burn extra coal, he reasoned, when a white-hot blast went up the chimney from the Bessemer converter? At Troy and Harrisburg, Holley had experimented with schemes to capture this wasted heat but had not hit upon a satisfactory solution.[47] In 1879 he devised for the Edgar Thomson works an ingenious heat exchanger that used the converters' white-hot blast to preheat incoming air for the cupola furnaces that in turn melted pig iron for the converters. Holley revamped the hot-blast stove design of Brown, Bailey, and Dixon, tripling the areas of the heating surface and air passage in the exchanger.[48] Including the internal piping, floor plates, and 27,000 firebricks, the exchanger weighed over 200 tons and required four 15-inch I-beam trusses for support. An array of butterfly valves directed cold air into the twin heat exchangers (one for each converter) and returned hot air to any of the three cupola furnaces. Such an arrangement—in effect, a continuous feedback loop of heat circulating between the converters and the cupola furnaces—began transforming the Edgar Thomson's Bessemer mill from a batch to continuous process.

A fully continuous Bessemer process was soon realized by William R. Jones. "Captain" Jones had been seasoned by twenty-five years of making iron and steel at Crane & Thomas, Port Richmond, and Cambria, and in all those places he had shown an uncommon ability both to inspire loyalty and to achieve high rates of production. When he was hired by Carnegie, on Holley's advice, to superintend the new Edgar Thomson works, a select crew of two hundred men followed him from Cambria. His appreciation of the shop-floor realities of steelmaking made him an early if unsuccessful advocate of the eight-hour work day. Furthermore, until his tragic death in a blast furnace accident in 1889, Jones contributed a string of key inventions, including an improvement on Holley's detachable converter bottom and a novel "straightener" that smoothly and accurately curved hot rails. His most notable and notorious invention—the subject of patent litigation for years—was his patented metal mixer, which made the Bessemer process into a fully continuous process. From the pioneering days at Troy onward, bars of pig iron had been remelted in cupola furnaces and then fed to the converters (fig. 1.11, top). After 1882, to quicken throughput as well as to cut fuel and labor costs, the Edgar Thomson's cupola furnaces were bypassed altogether and molten pig iron flowed directly from blast furnace to converter. With this "direct" method, however, the converters operated unpredictably owing to small variations in the pig iron (fig. 1.11, middle). If there was a high level of silicon or carbon, say, the converters overheated; if the level was low the converter could chill solid. To achieve chemical control over

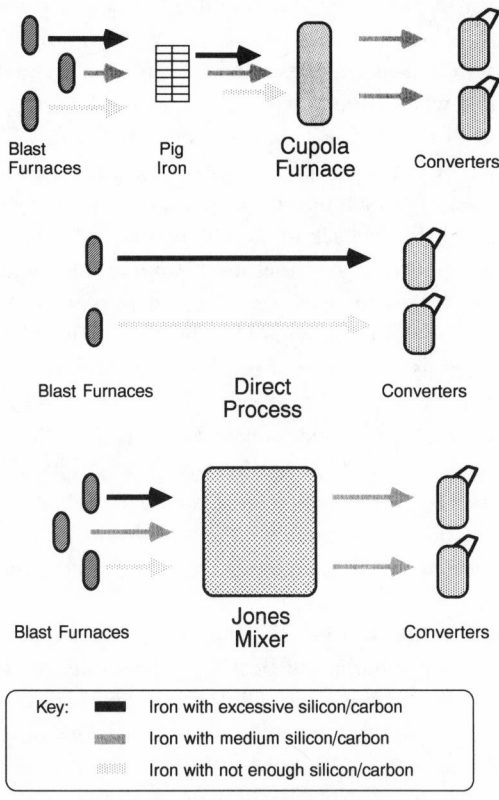

FIG. 1.11. Edgar Thomson steel works, Jones mixer of 1887. In the original Bessemer practice, unavoidable irregularities in crude pig iron were dealt with by mixing several complementary pig irons and remelting them in a "cupola furnace" before charging the converter (*top*). Such irregularities were a severe problem for the "direct process" (*middle*) because the pig iron remained fluid and no mixing took place. The Jones mixer (*bottom*) minimized these irregularities without remelting.

the production process, the mill was forced to double its staff of chemists as well as employ "special men . . . in the Bessemer department at high salaries" to judge and correct these troublesome variations.[49]

Difficulties in controlling the converters persisted until the summer of 1887 when Captain Jones installed his first metal mixer. Situated between the blast furnaces and the converters, the huge firebrick vessel temporarily held and mixed together up to 100 tons of molten pig iron; as needed, 10-ton charges were tapped to fill the converters (fig 1.11, bottom). With the direct process, iron charged to the converters had varied as much as 2 percent in silicon content; with the Jones mixer in place, the variation dropped to only a few tenths of a percent. The chemical stabilization worked wonders. "The Bessemer operations at once assumed the uniform character prevailing at the time when cupolas were used," stated James Gayley, superintendent of the Edgar Thomson's blast furnaces. "The extra men that had to be employed previously were dispensed with, the quality of the steel was much improved, [and] the waste of steel in chill heats . . . was brought back to normal amount."[50]

In the American context, then, mature Bessemer technology reflected systematic design to maximize the production of steel rails. Holley's integrated design provided for the rapid flow of large amounts of raw materials and finished products, and his heat exchanger introduced the concept of continuous flow. With the Jones mixer Bessemer steelmaking became a continuous process by design. "The Jones process consists in a continuous operation," Gayley observed. "This receiving vessel [the mixer] after being once filled is kept filled, or practically so, and . . . the withdrawals are balanced by new charges, which in itself has the element of continuity provided for."[51]

It is a commonplace that the Edgar Thomson works became a prototype for the mass production of Bessemer steel rails. Its design inspired a new plant at the North Chicago Rolling Mill Company, among others. Holley just inked in new dimensions on the Edgar Thomson's blueprints for Vulcan Iron's Bessemer mill at St. Louis. As Carnegie put it, "A perfect mill is the road to wealth."[52] Railroads, as we have seen, were a dominant force in forming the structure and shaping the technology of the Bessemer steel industry. The technological knowledge that grew up with that industry, a new chemically oriented metallurgy, also bore the direct stamp of the railroads.

Shaping Technological Knowledge

"A bar of steel" is, in the present state of the art, a vastly less definite expression than "a piece of chalk."—ALEXANDER HOLLEY

The engineers, entrepreneurs, and railroad officials who introduced the Bessemer process did more than found a new industry. They also upset the traditional craft-based control of iron- and steelmaking, initiating events that shaped the emerging science of steel. At Bessemer plants especially, craft-oriented ironworkers were pushed aside when managers hired workers without experience in the iron trade (who had no "bad" habits to unlearn). Conflict sometimes flared up between the craft-oriented ironworkers and the university-trained chemists who took their place.[53] Conflict also erupted within the community of science-oriented metallurgists and professional engineers. Most vexing, as the quotation from Holley suggests, was the problematic nature of "steel."

Since the 1850s, mechanical, physical, and chemical methods had been available to define iron and steel. Data from mechanical tests were the easiest to obtain. The army's huge manufacturing arsenal at Watertown, Massachusetts, became the leading center for mechanical testing, and the mechanical engineer Robert H. Thurston became the leading spokesman for this approach. For its advocates, mechanical strength was revealed when sample bars were gripped between giant jaws and subjected to torsion, tensile, and compressive forces. "Iron" had certain mechanical properties of bending, elongating, and stretching before breaking, while "steel" had other properties. Physical tests, such as measurements of density, were a second possibility. Advocates of this approach argued that desirable mechanical properties could be correlated with the density of reference pieces of metal, and unknown samples could be classified by density alone. Most believed that metal became stronger as density increased (being free from holes or inclusions), although some argued for an optimum density signaling maximum strength.

Mechanical or physical standards were inadequate, however, for those concerned with the iron and steel industry on a regional or national level. While density measurements, for example, might compare batches of pig iron made in the same blast furnace from the same iron ore, density did not distinguish iron from different regions or different ores. Similarly, while mechanical testing could identify that a particular piece of metal was desirable, it could not reveal how it was made or how it could be reproduced. The ever vigilant Alexander Holley soon recognized that mechanical and physical tests were insufficient to construct general rules about the useful properties of iron and steel.

Holley observed that while engineers and machinists often complained that they could not obtain a certain quality of steel, many thousands of tons of existing steel were entirely suitable. The problem was to identify *which* thousands of tons. "In order that engineers may know what to specify, and that manufacturers may know not only what to make, but how to compound and temper it," he stated, "the leading ingredients of each grade of steel must be known." Chemistry was the central issue for Holley. "The manufacture of pig iron for the Bessemer," he had advised the Troy backers, "require[s] the immediate professional attention of a thoroughly educated metallurgical chemist." The difficulties of manufacturing Bessemer steel, he observed, "were not chiefly mechanical" but stemmed from "a chemical stumbling block." "What a conflict of the elements is going on in that vast laboratory!" he wrote of a converter blast.[54] Analytical chemists who met this demand for chemical knowledge included J. Blodget Britton, whose Ironmasters' Laboratory was established in Philadelphia in 1866 "exclusively" for analyzing iron ores, pig iron, steel, and other metallurgical materials, and James Curtis Booth, whose laboratory was also in Philadelphia and who trained many prominent figures in science and industry, including Joseph Wharton.[55]

Beyond its significance in controlling production, chemistry also promised to standardize the steel industry. As the principal consumers of Bessemer steel, railroads selected and defined its essential properties. Charles B. Dudley, the Yale-trained chief chemist of the Pennsylvania Railroad, persistently advocated chemical specifications for steel rails, though his effort was controversial. Although larger geographical domains of distribution probably required some standard to guarantee quality at a distance, the specific chemical form of the standard owed much to the railroads.[56]

The contentious issues behind scientific standards came into sharp focus during an acrimonious debate on a deceptively simple question: what is steel? The debate in the mid-1870s, though conducted in scientific language, had immense commercial implications. The question was really: who would be deemed to make the valuable commodity "steel," and who would be left making "iron"? Participants openly articulated their respective commercial and professional interests; more precisely, they identified their opponents' interests. Makers of high-temperature steel supported what was known as the fusion classification, while makers of low-temperature steel, along with university metallurgists, affirmed the carbon classification.[57]

The origins of the controversy lay in the economic and technical instability and uncertainty of the iron and steel industry. "Great excitement this evening in Philadelphia," recorded one diarist upon the failure of the financial house of Jay Cooke, whose speculation in railroad bonds collapsed in September 1873.[58] The resulting panic and ensuing depression pushed down prices and subdued overall economic activity for several years. The failure of capital markets meant the collapse of railroad construction. Since railroads consumed more than half the total iron produced and imported, this resulted in a severe slump in the iron and steel trade. In early 1874 rail mills were running at less than one-third capacity, with some 21,000 rail-mill workers thrown out of full-time employment. Not until late in the decade did the iron and steel industry recover.

An ongoing changeover from iron to steel compounded the economic crisis. Railroads were increasingly adopting steel over iron rails, and as the price gap between steel and iron narrowed, the iron industry was evidently in trouble. The production of iron rails peaked in 1872 at 809,000 tons; it fell steadily across the next seventeen years to a mere 9,000 tons (fig 1.6). Total iron output fell after the 1873 panic and did not recover until the end of the decade. In contrast steel production grew continuously and vigorously. In 1870 total steel output stood at 69,000 tons; by 1880 it topped 1.2 million tons. At this time Bessemer steel made up 86 percent of all steel produced, and rail mills consumed 83 percent of all Bessemer steel.[59]

Even steelmakers were experiencing unsettling economic shifts. In 1873, when there were six Bessemer producers, steel rails sold for $120 per ton. Five years later, with the addition of three new Bessemer mills and the rebuilding of several older mills, steel rails sold at $42 per ton (fig. 1.12). Iron rails held a significant if shrinking proportion of the trade. (Prices for new iron rails soon disappeared from trade statistics, signaling their end as a viable commodity; at the same time trade statistics began tracking the prices of used iron rails for scrap and rerolling.) Facing such economic and technical uncertainty, the leading manufacturers of Bessemer steel moved to alter how the trade determined what was iron and what was steel. Because "steel" rails cost more per ton than "iron" rails, their move sparked a scientific dispute that had direct commercial consequences.

The traditional method to distinguish iron from steel relied on carbon content. Carbon was a critical ingredient because in small amounts it imparted resilience, strength, and most important the capacity for being hardened upon sudden cooling, or quenching, from high temperatures. "Wrought iron" contained essentially no carbon, "steel" contained from 0.2 to 1.0 percent carbon, and "cast iron" contained 2 percent or more carbon. Steel could be hardened by quenching, wrought iron could not. Before 1880 metallurgical textbooks invariably gave this carbon-based definition of steel.

The spread of the Bessemer process made possible an alternative classification. While wrought iron came from puddling or boiling furnaces as a pasty semisolid mass, the extreme heat of the Bessemer converter completely melted, or "fused," its products. Metal that had been completely melted was free from the slag, cinders, and carbon flecks that characterized wrought iron. In one account, the resulting "homogeneous" product had qualities that were "universally recognized" if "not readily described." In kicking off the debate Holley articulated the fusion classification: "Steel is an alloy of iron which is *cast while in a fluid state* into a malleable ingot." By the fusion classification, if the metal had been completely melted—regardless of its carbon content—it became "steel"; if not it remained "iron."[60]

Advocates of the carbon classification rallied behind a young metallurgist, Henry M. Howe. Howe, like Holley, came from a socially prominent family, but one that was reformist rather than establishment. Howe's childhood home on Beacon Hill in Boston was filled with the cultured friends of his parents. His father was

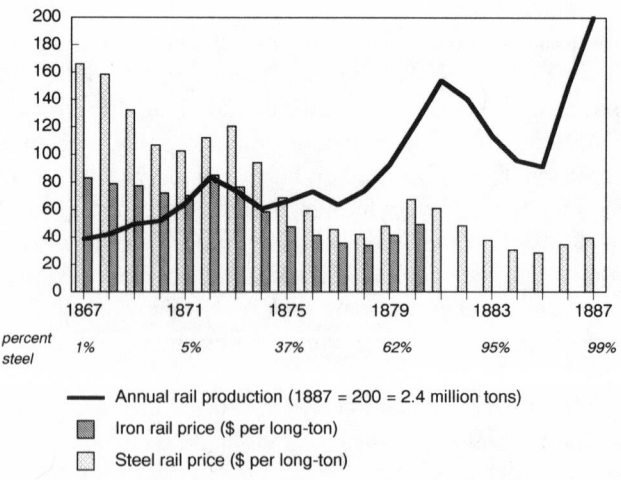

FIG. 1.12. Iron and steel price and rail production, 1867–87.
Data from AISA, Annual Statistical Report *(1887 and 1931).*

head of the Perkins Institute for the Blind, and his mother was a prominent suffragette and the composer of the "Battle Hymn of the Republic."[61] Born in 1848, Howe attended Boston Latin School, Harvard College, and the newly founded Massachusetts Institute of Technology. After completing his education in 1871, he gained steelmaking experience at the Bessemer works in Troy, at Joliet Iron and Steel's new Bessemer works in Chicago, which he superintended, and at the Blair Iron and Steel Works in Pittsburgh. Howe's arguments in the debate implicitly supported the manufacturers of low-temperature iron and steel, whose products the fusion classification would reclassify as "iron." Howe also upheld the metallurgical tradition that defined steel by hardness and resiliency. In this respect, as well as in his attempt to keep metallurgy from being shackled to raw economic interests, Howe might be seen to represent the scientist's viewpoint as opposed to that of the engineer or manufacturer.[62] This neat typology, however, fails to account for the variety of his later activities. Howe went on to design and build two metallurgical works for the Orford Nickel and Copper Company—an activity for which Holley was admired as an engineer—and he also served as vice president of a specialty steel-manufacturing firm.

Howe advanced his case for the carbon classification in the *Engineering and Mining Journal*. He advocated correlating the mechanical properties universally associated with steel—resilience and ability to be hardened—with the carbon content of a sample. Carbon content would then become an index to these desirable properties: "steel" would have the carbon content (0.2–1.0 percent) of reference samples of steel. He then attacked the fusion classification. It had become "fashionable" to label as "steel" all products of the Bessemer converter and open hearth, without regard for carbon content or mechanical properties. In addition, in a veiled reference to Holley, he stated that "cultivated and intelligent engineers" claimed that "the distinction between wrought iron and steel should be based on homogeneousness and freedom from slag, and that hardness, tensile strength, resilience, and the power of hardening have nothing to do with it." Howe suggested how this "confusion" arose. When low-carbon iron from a Bessemer converter or open-hearth furnace was cast, the resulting ingots had not been worked or "wrought," and could not be "wrought iron." Since these ingots looked and felt like steel, some believed the easiest way was "to call the whole product steel, and not bother about mechanical tests, or split hairs about physical properties." The same reasoning, he observed, "would justify a jeweler in selling brass as gold or strass [flint glass] as gems." Howe explained, "It is possible that some manufacturers, being human, were influenced by the consideration that steel was vaguely associated in the minds of the public with superiority and was in general higher priced than wrought iron, to sell that part of their product as steel which a strict adherence to the then recognized distinction between steel and wrought iron would have compelled them to call wrought iron." [63]

Holley sharply rebutted Howe's view. In so doing he launched the metallurgical community into a full-scale controversy that was contained by the American Institute of Mining Engineers, founded in 1871. In an article titled "What Is Steel?" he stated curtly that "the general usage of engineers, manufacturers, and merchants, is gradually, but surely, fixing the answer to this question." He disputed Howe's arguments point by point, contending that while the fusion classification was already—or nearly—in place, the carbon classification was "arbitrarily devised" and "must bear the demerit . . . of upsetting existing order and development." [64]

Holley gained support from industrialists using high-temperature steelmaking processes (Bessemer, crucible, open hearth). James Park, Jr., an early Pittsburgh investor in the original Ward-Kelly company, and a cofounder of the Black Diamond crucible-steel firm, attacked Howe and supported the fusion classification. Another fusion advocate was William Metcalf, also a maker of crucible steel. After graduating in 1858 from Rensselaer Polytechnic Institute, Metcalf had returned to his native Pittsburgh as assistant manager and draftsman at the Fort Pitt Foundry, where within a year he rose to general superintendent. In the late 1860s he had helped organize the firm of Miller, Metcalf and Parkin, which owned the Crescent Steel Works. As managing director, he specialized in fine crucible steels until 1895 when the Crucible Steel Company bought out his firm. In various ways Park, Metcalf, and Holley were all committed to high-temperature steelmaking.

No high-temperature steelmakers stood on the other side. Those rallying behind Howe and the carbon classification included Thomas Egleston, head of the School of Mines at Columbia College; Frederick Prime, professor of metallurgy at Lafayette College; Benjamin W. Frazier, professor of metallurgy at Lehigh University; Frank Firmstone, superintendent of the Glendon Iron Works; and Eckley B. Coxe, a prominent anthracite mining engineer. John B. Pearse became an unexpected ally. As a chemist for Pennsylvania Steel since 1868 and its manager since 1870, Pearse "ought" to have supported the fusion classification. But in June 1874 he resigned his position at the steel works to become commissioner and secretary of the Pennsylvania state geological survey; in October 1875 he attacked Holley and supported the carbon classification. A year later he became general manager of the South Boston Iron Company. Advocates of the carbon classification shared at least one formal characteristic: "steel" did not appear in the title of their affiliations.

The carbon advocates soon identified the commercial interest behind the fusion classification. Frederick Prime pointed to Holley, then president of the American Institute of Mining Engineers, who "belongs to a group composed of himself . . . and many manufacturers of Bessemer and open-hearth steel, who propose to overthrow the definition I have given as the current one. With energy worthy of a better cause . . . he gives his definition, pronounces *it* to be the current one, and claims that 'several high metallurgical authorities and clever writers have of late proposed to disturb this natural and somewhat settled nomenclature.'(!)" Howe expanded Prime's argument that commercial interests motivated this gambit: "Certain mechanical engineers and manufacturers, many or most of whom were pecuniarily interested in attaching the name steel to the new products, because that name was associated in the mind of the public with superiority, have called these new products steel, in full face of the fact that they had none of the essential qualities of steel and all of the essential qualities of wrought iron." Howe tabulated the conflicting results of using the two classifications for standard iron and steel products, making the point that the fusion classification redefined the three low-temperature steels (blister, puddled, shear) as "iron" while all products of high-temperature processes (Bessemer, open hearth, crucible) became "steel."[65]

The fusion advocates also did their part to identify the professional interests behind the carbon classification. Holley and Metcalf portrayed the carbon advocates as elitists and autocrats. Metcalf complained that "the few, the men of science" were arbitrarily enforcing an "ancient" meaning of steel, and he chafed at their assumption of authority and superiority. "The names of new materials and processes," added Holley, "are not fixed by the arbitrary edicts of philosophers." Yet this was no simple division between science and engineering. Holley and Metcalf both claimed the mantle of science, and both argued that their classification was more scientific than that of the "high metallurgical authorities and clever writers."[66]

Logic alone does not unravel or explain these debates. Both sides claimed priority for their classification. Both maintained that the opposing classification was arbitrary or confusing. And both classifications had technical merit. The ability of "steel" to be hardened, the focus of the carbon advocates, was a property of real significance. Similarly, the fused "steels" had important properties, such as freedom from slag and other inclusions, that unfused "steels" did not possess. Nevertheless, although they contained similarly low percentages of carbon, wrought iron (unfused) and mild steel (fused) were two entirely different products.[67]

Instead, as the disputants readily identified, behind the debates stood conflicting commercial and professional interests. Holley and other high-temperature steelmakers supported the fusion classification, which defined their products as the higher priced "steel." Howe and the low-temperature steelmakers attempted to retain the carbon classification, which preserved their professional integrity as well as premium prices for their products. The outcome of this controversy determined the contours of steel metallurgy for the rest of the century.

Undesirable consequences haunted the metallurgical community so long as this controversy persisted. While in the abstract two rival classifications could coexist, several practical problems emerged that challenged the metallurgical community's legitimacy as experts who dealt in reliable knowledge and objective facts. Import duties were one such problem. In May 1878, after an eighteen-month lobbying blitz headed by William Sellers, the machine-tool magnate who had recently reorganized Midvale Steel, the secretary of the treasury reclassified as "steel" the imported Siemens-Martin metal, a fused product that had entered the country under the (lower) iron tariff to the detriment of American steel manufacturers. Thereafter, perhaps to the chagrin of the scientific metallurgists, "collectors of customs" were "to make the proper classification." The controversy spilled over the Atlantic in another way. American and European metallurgists initiated a joint effort to develop a unified nomenclature of iron and steel, but differences in industrial practices as well as linguistic problems paralyzed the effort. Finally, there was "a heavy suit pending in the United States courts, turning upon the question whether steel is steel or iron."[68] For all these reasons the need to resolve the controversy became urgent. In the end the American iron and steel community adopted the fusion classification and retained chemical methods.

As the panic of 1873 had sparked the debate, the reviving economy of the early 1880s helped extinguish it. By 1880 orders for iron and steel had surpassed even predepression levels, and the price gap between iron and steel rails had virtually closed as Carnegie's Edgar Thomson works, along with a half-dozen other high-volume mills, turned out steel rails in unprecedented numbers. The resulting drop in steel prices had driven out iron rail producers. After 1877 the leading Bessemer producers' licensing policy restricted additional competitive pressures. One way or another, the instability that had plagued the trade had been resolved.

The triumph of the fusion classification owed much to the technological needs of the railroads, still the largest consumers of steel. Railroads had found that rails rolled from the completely melted or fused metal, even with carbon content similar to wrought iron, were less likely to crack open than rails made from unfused metal. The superintendent of Pennsylvania Steel noted of steel rails that "their homogeneity is their distinguishing characteristic." By supporting the fusion classification, railroads ensured that the metal fitting their specific technological needs would be available and uniform. Railroad financiers no less than railroad managers and steel-mill owners appreciated the advantages that the wondrous title of "steel" conferred. Adopting steel rails, noted one financial analyst, was an effective strategy to inflate the value of rail stock for speculative purposes.[69]

By taking up managerial positions, metallurgical chemists themselves contributed to the momentum of the Bessemer process and chemical metallurgy. Experts in process control, they rose quickly as managers, as several biographies illustrate. Robert W. Hunt took a course in analytical chemistry at the Philadelphia laboratory of James Curtis Booth, his only formal education, then supervised the pioneer Bessemer operations at Wyandotte and Troy, where he served as Holley's lead assistant. At Cambria's Bessemer works he established the first chemical laboratory associated with an iron and steel company in America. In 1867 he rolled the first commercial order for steel rails, delivered to the Pennsylvania Railroad. Thereafter he held a succession of managerial posts, eventually heading an important Chicago rail-inspecting firm (see chapter 4). Booth's teaching laboratory also trained John B. Pearse, who entered the laboratory with a bachelor's degree from Yale in 1861, stayed two years, then studied in Germany at the Freiburg School of Mines for another year. As noted earlier, he started out as the chemist for the Pennsylvania Steel works and within three years was promoted to general manager. Another Pennsylvania Steel chemist, Edgar C. Felton, who joined the company in 1880 after graduating from Harvard, rose through the ranks to become company president in 1896.

The rise of chemists into management could be high indeed. James Gayley, a graduate of Lafayette College, served for three years as chemist to the Crane Iron Company in the Lehigh Valley of Pennsylvania, then achieved fame as superintendent of the Edgar Thomson mill's blast furnaces. He capped his career in 1901 when he became first vice president of the U.S. Steel Corporation. Gayley's successor as vice president, David Garrett Kerr, started out as laboratory boy at the Homestead works, then became a chemist for the Edgar Thomson blast furnaces. Gayley's boss, the second president of U.S. Steel, William E. Corey, began his career as a student in the chemical laboratory of the Edgar Thomson steel works.[70]

Finally, what of the issues that divided the advocates of the carbon and fusion classifications? Although vigorously contested, the differences between Holley and Howe were mediated by practice. Holley consistently advocated the use of chemical composition to standardize the *varieties* of steels; the fusion classification served only to distinguish steel from wrought iron. Although Howe also advocated chemical methods to classify steels, he failed to establish carbon content as the single method to distinguish steel from wrought iron. It was in this regard that the fusion classification triumphed.

By 1880 the debate over the method to classify steel had ended, and the once problematic categories of "iron" and "steel" were stable. "Steel" was a pourable fluid from the Bessemer, open-hearth, or crucible process; wrought "iron" was a spongy product of the puddling, boiling, or so-called direct process. Both were subject to chemical standards.[71] Economic, technological, and sociological factors together shaped a metallurgy based on the fusion classification (for defining "steel" and "iron") with a strong chemical component (for classifying the varieties of "steel").

In the years after 1880 the consensus on "steel" was maintained less by written authority than by the daily practice of thousands of steelmen. Only after 1900 did the authors of metallurgical textbooks grant canonical status to the fusion classification.[72] To the end, Howe maintained that the fusion advocates would have failed if "the little band [of his fellow carbon advocates], which stoutly opposed the introduction of the present anomaly and confusion into our nomenclature, [could] have resisted the momentum of an incipient custom as successfully as they silenced the arguments of their opponents."[73] The "incipient custom" was, of course, that of the railroads, Bessemer steelmakers, and their allies, such as Alexander Holley. It is no accident that Holley was a staunch advocate of modern railroading, the Bessemer process, and the fusion classification—in that order.

Building Transcontinental Railroads

Lo, soul, seest thou not God's purpose from the first?
The earth to be spann'd, connected by network,
The races, neighbors, to marry and be given in marriage,
The oceans to be cross'd, the distant brought near,
The lands to be welded together.
—WALT WHITMAN

Through the 1880s, if not beyond, the mass production of steel in the United States was predicated on the mass consumption of steel rails. In turn mass consumption required a sustained building campaign by the railroads. By the 1860s the northern and eastern states already had a primary rail network in place, a factor often cited in the North's success in prosecuting an industrialized war against the South. The defining demand for rails arose only with the impressive mileage that was laid down in the trans-Mississippi West after 1880 (fig. 1.13). In this respect the peculiar character of Bessemer steelmaking in America was ultimately a product of the nation's western expansion, "the lands to be welded together."

Given their substantial investment—and considering that the outlay for rails was exceeded only by the staggering cost of cutting, filling, and grading the route—the major railroads had to ensure sufficient traffic and revenues. Demand could not be left to chance. For the major eastern trunk lines, such as the Pennsylvania, the New York Central, and the Baltimore & Ohio, the problem was controlling competition (see chapter 4). For the transcontinental railroads pushing beyond the Mississippi River, or even more boldly beyond the Missouri, the problem was creating traffic. A railroad such as the Chicago, Burlington & Quincy, pushing across Iowa in the 1860s, Nebraska in the 1870s, and Wyoming and Montana in the 1880s, had a comparatively easy time, for along its route the Burlington planted communities that provided eastbound freight in Chicago-bound commodities as well as westbound freight in farm machines and consumer goods. Roads such as the Santa Fe faced a more difficult task. The Santa Fe passed through arid lands in Arizona and New Mexico that were impossible to farm without irrigation. As promoters of settlement, and thus providers of traffic, the land departments of the transcontinental railroads were the final link in the technological system that began with the Bessemer steel rail mills.

Indeed, railroad "colonization" schemes imprinted some durable characteristics on the West. The towns west of Lincoln, Nebraska, line up in alphabetical order—Crete, Dorchester, Exeter, Fairmont, Grafton, Harvard, Inland, Juniata, Kenesaw, Lowell—each one planned and plotted and sold by the Burlington railroad to ensure traffic. One railroad official became so proficient at organizing towns he claimed the task took him only twenty minutes. Nebraska filled up with certain European immigrants, and not with others,

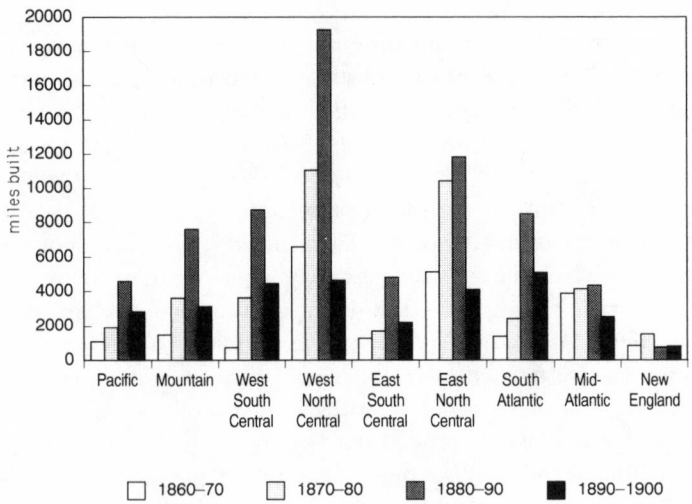

□ 1860–70   □ 1870–80   ▨ 1880–90   ■ 1890–1900

FIG. 1.13. Railroad building by region, 1860–1900. "Pacific" is California, Oregon, and Washington; "Mountain" is Arizona, Colorado, Idaho, Montana, Nevada, New Mexico, Utah, and Wyoming; "West South Central" is Arkansas, Louisiana, Oklahoma (including Indian Territories), and Texas; "West North Central" is Iowa, Kansas, Minnesota, Missouri, Nebraska, North Dakota, and South Dakota; "East South Central" is Alabama, Kentucky, Mississippi, and Tennessee; "East North Central" is Illinois, Indiana, Michigan, Ohio, and Wisconsin; "South Atlantic" is Delaware, District of Columbia, Florida, Georgia, Maryland, North Carolina, South Carolina, Virginia, West Virginia; "Mid-Atlantic" is New Jersey, New York, and Pennsylvania; "New England" is Connecticut, Maine, Massachusetts, New Hampshire, Rhode Island, and Vermont.

Data from Dorothy R. Adler, British Investment in American Railways, 1834– 98 *(Charlottesville: University Press of Virginia, 1970), 193 n.15; Department of Commerce and Labor, Statistical Abstract of the United States 1911 (Washington, D.C.: Government Printing Office, 1912), 280.*

as the Burlington saw fit. "I have so poor an opinion of the French & Italian immigrants for agriculturists that I shall not issue any circulars in their languages," wrote the Burlington's land officer. "My efforts will be most confined to Germans, Scandinavians, English, Welsh, and Scotch, as they make good farmers." Russians, Canadians, and Bohemians were also among the favored peoples. The classic crop of Kansas, hard red winter wheat, arrived with a large colony of Russian Mennonites. The colonists settled lands of the Atchison, Topeka & Santa Fe, whose land officer thereupon became known as "the Moses of the Mennonites." Finally, in most contests with Native Americans the railroads proved irresistible. "The construction of the road," stated the Central Pacific's general solicitor, "virtually solved the Indian problem."[74]

The force of railroads and steel did not transform the Reconstruction South, however. The region's leading ironworks, Tredegar Iron of Richmond, attempted to establish a Bessemer plant in 1866. As one of its officers wrote at the time, "Demand continues brisk— particularly from our railroads, which though torn up again & again during the war, are still, for the most part, financially sound and can look forward to a promising future." The Bessemer effort failed, nevertheless, when Tredegar was unable to attract northern capital. Worse, Tredegar became shackled by worthless (northern) railroad bonds after the 1873 depression. Lacking the capital to convert to steel, it passed into receivership and decline within the decade. Regional leadership in metal manufacturing passed to Birmingham, Alabama, which flourished as a center for iron- and steelmaking beginning in the 1880s after the Alabama & Chattanooga and the Louisville & Nashville railroads linked it to the region's rich mineral lands.[75]

Difficulties in raising the large amounts of capital needed for comprehensive regional systems hampered railroad construction and system building across the South, in large measure because the region's own investors had insufficient funds. Inexorably, northerners stepped into the breach—despite pockets of regionally oriented capital in Richmond, as well as entrepreneurs, promoters, and visionaries active across the region. In the early 1870s the Pennsylvania Railroad made a concerted if short-lived effort to build a regional system from Richmond and Charleston in the east to Memphis in the west. Later in the decade the Illinois Central began to establish a lasting presence in the South by purchasing a defaulted southern railroad that ran from Cairo, Illinois, to New Orleans, thus connecting the Great Lakes with the Gulf of Mexico. Nevertheless, according to historian Mark Summers, "it was parochialism, not sectionalism or nationalism, that left the most serious marks on railroad policy" in the South. Before the war, parochialism took the form of a patchwork of different railroad gauges that preserved local control and prevented the efficient through-shipment of goods. In 1861 eight changes of cars due to gauge were necessary to complete a trip from Philadelphia to Charleston. After the war, parochialism was scarcely affected by the northerners' imposition of Reconstruction. While communities in both the North and South initially competed enthusiastically for railroads, in time southerners became increasingly alarmed at railroads whose corporate form seemed to embody foreign dominance and whose ownership in fact passed increasingly into northern hands. Taking into account the major southern railroads (100 miles or more in length), northerners controlled 21 percent of the region's mileage in 1870, 48 percent in 1880, 88 percent in 1890, and 96 percent in 1900.[76]

Not only in the South did railroads pass into corporate hands. The second most famous railroad event of 1869, after the Golden Spike at Promontory Point, hinted at the corporate structure that the transcontinental rail boom would assume. Previously most railroad companies north and south had been local corporations, owning perhaps a few hundred miles of track that connected at its terminals with other lines. In 1869 Cornelius Vanderbilt gained control of the Lake Shore & Michigan Southern, united it with his New York Central, and secured a through line from Manhattan to Chicago. The advantages of having a long trunk line under one ownership were soon noted by Vanderbilt's rivals, and those with sufficient resources followed his precedent. The chief rival, the Pennsylvania Railroad, grew into a national concern by purchasing smaller roads. In 1880 its system comprised forty-four smaller companies and operated 3,773 miles; by 1890 it had added twenty-nine more companies and operated a total of 5,000 miles. In 1890 only 609 of the total 1,705 railway organizations remained independent; most were subsidiaries or feeders to one big line or another. In effect nearly two-thirds of the country's railroad companies had been absorbed by the other one-third. The process of consolidation under northern corporate control culminated in the reorganization of the industry in the 1890s (see chapter 4).[77]

To a striking extent, steelmaking in the United States was created for a single product: Bessemer steel rails. Because railroad officials promoted, funded, and even founded early Bessemer steel works, in addition to consuming almost all the new steel, their influence was immense. Railroads helped train the first generation of steel executives in modern managerial models and accounting practices.

Their influence can further be traced to the scientific knowledge developed and deployed by the industry, for railroads decisively influenced the fusion-chemical controversy that established the science of steelmaking. Later developments reflected this legacy, most notably as producers of Bessemer steel rails struggled fitfully to stabilize their prices.

Railroads of Bessemer steel rails furthermore shaped the pattern and pace of national development. With the primary rail network in place, manufacturers east of the Mississippi River were connected in all seasons and in all weather to expanding markets in the West; the frozen canals, spring freshets, and late summer droughts that had imposed seasonal schedules on manufacturing and shipping were reduced to local annoyances. Machines running year-round increasingly constituted the dominant mode of industry. In turn, cheap foodstuffs flowed from the West into manufacturing and exporting cities, fueling growth. The flow of grain, hogs, lumber, and other commodities was so vast that an immense commercial superstructure was erected for shipping, trading, storing, and financing it. These new activities required an unprecedented number of white-collar professionals, thus creating demand for office space in the commercial core of cities. In no American city was the dynamic of growth stronger than in Chicago—the terminus for east- and westbound routes that by 1893 totaled one twenty-fifth of the *world's* rail mileage.[78] There, in a complex partnership with the Pittsburgh mills, the next fundamental development in steel took place.

Construction of the steel-skeleton Flatiron Building, New York, 1901, the first New York commission of Chicago architect Daniel H. Burnham.
Courtesy the Library of Congress; LC-D401-14278.

CHAPTER TWO

The Structure of Cities

1880–1900

On the morning of 2 February 1925, Paul Mueller was a nervous man. He would soon step into the witness box, for the most important trial a structural engineer could imagine, and he knew his professional judgment and personal integrity were at stake. Ominous cracks had appeared in certain walls of the Auditorium Building, an extraordinary block-long complex whose opening in 1889 and later success had secured the permanent fame of Louis Sullivan and Dankmar Adler and had also done much to bolster Chicago's claims to high culture. The architects had labored for three years on a million details and had gotten most of them right, even elegant. At the gala opening the building's promoters had arrayed sensational publicity: the presence of President Benjamin Harrison, a performance by an Italian diva, the ritual lineup of local donors and dignitaries. Now, thirty-six years later, the building was showing its age, and Mueller knew he would be called on to pinpoint structural problems and, more important, fix the responsibility for repairs. In fact, the building itself was at stake, for there were grand plans to raze the Auditorium and erect in its place a twenty-two-story skyscraper with better financial prospects.[1]

Of the dozen testifying experts Mueller was neither the most famous nor the most prolific. For fame no one could top Ernest R. Graham, partner of the legendary Daniel H. Burnham and more recently founder of the architectural firm of Graham, Anderson, Probst & White. For sheer output Mueller also trailed Louis E. Ritter, who had been chief structural engineer for William Jenney and William Mundie, the pioneers of steel-skeleton buildings, during the busiest years of their partnership (1893–98). Ritter had since designed forty-five buildings in Chicago and elsewhere.[2] Nevertheless, no one had Mueller's special experience, for he had joined Adler & Sullivan just as the temperamental partners started work on the Auditorium project in 1886, and its structure had been his principal charge. If he marveled at the organic form of Sullivan's ornamentation ("flowers and frost, delicate as lace and strong as steel"),[3] he must have admired even more the daring of Adler's structure. It enclosed a four-thousand-seat theater, a seventeen-story office tower, and a grand hotel whose ornate bar had become a signature of metropolitan life in America.[4] Mueller had at his command the details of the moving platforms in the theater, the vaulted arches of the hotel ballroom, the ingenious prestressed foundation of the heavy office tower, and a thousand structural details (fig. 2.1). He had carefully reviewed his and Adler's calculations of loading, floor by floor, nearly column by column. It unnerved him, however, when the attorney inquired whether the Auditorium was built of iron or steel. If it was built of steel, the structural calculations he had done and Adler had approved looked proper; if iron, these calculations were bold to the point of recklessness.

Although Mueller did not press the point, the question of iron or steel in 1889 was not the simple one it appeared to be in 1925. In 1889 the structural trade—architects, fabricators, builders, suppliers—was in the midst of a changeover from iron to steel; that year steel accounted for just 56 percent of structural shapes.[5] If the Auditorium had been built five years earlier the question of iron or steel would never have arisen. Columns would have been cast iron, a brittle, hard material well suited for compressive strains; beams and girders would have been wrought iron, a tough and fibrous material well suited for tension strains. But the Auditorium took shape during 1886–89, just when steel began challenging iron as a viable structural material—the twin result of oversupply in the steel rail industry and unprecedented demand for urban space.

Although Mueller recalled when a Carnegie man began talking up the advantages of steel, first he had to deal with an oblique but devastating criticism. Structural engineer Ritter had testified that up to 1890 or 1892, riveted steel columns had been unknown, and everyone had used cast iron columns, "such as you have in the Auditorium." Cast iron was an uncertain material to use in structures, Ritter had observed, since it gave little warning of failure and had little tensile strength. The failure of a column was not a matter of direct compression, he explained, but of bending. Steel adapted well to such loads and strains, and in addition steel columns could be riveted to one another and to steel girders and beams, forming a rigid structure. By contrast, Ritter had concluded, with the "old cast iron construction" columns simply rested one on top of another.[6] Mueller was alarmed. It sounded as if the whole Auditorium, all 100,000 tons of it, rested on building blocks, strong enough only until the first wind came along. He decided to tell how the Auditorium's metal structure had come to be, and why it was better if it was steel.

Mueller presented to the court his credentials as a structural expert. He had graduated from the German school of mining and civil engineering in the Saar, and had come to Chicago in 1881, where he worked for three different iron fabricators and engineers. Upon joining Adler & Sullivan in 1886, he testified, "the first thing I did for them was to start on the diagrams for the Auditorium." Adler soon made him foreman of the whole office, and the next decade was busy and satisfying. Besides the Auditorium, he had worked on the Standard Club; the McVickers, Hooley, and Pabst theaters; plans for a hotel in Salt Lake City; and the Schiller Building, whose combined theater and office space took an idea or two from the Auditorium. Then there was Adler's acoustical consulting for Carnegie Hall in New York. Later, Mueller related, he had served as consulting engineer for Probst Construction during the 1893 Columbian Exposition.[7]

The Auditorium, as Mueller recounted, was a mammoth undertaking—an event. The project forced Adler & Sullivan to engage between thirty and forty draftsmen. For his personal assistants, Sullivan hired Frank Lloyd Wright in 1887 and George Elmslie a year later. Overseeing the seventy subcontractors became a task in itself. Captain Neiman, an experienced builder in the employ of Martin A. Ryerson, supervised construction. The theater, whose acoustics remain a marvel a century later, was Adler's own brilliant design. At least three consulting engineers—William Sooy-Smith, E. F. Osborn, and William G. McHard—were employed for the foundation, steam heating, and sanitation, respectively. Furthermore, for the structure itself, Adler & Sullivan had the services of two consulting engineers associated with the Carnegie concern: Edgar Marburg, who made stress drawings for the theater's complicated truss supports, and Charles L. Strobel, an expert in the new field of steel structures.[8]

The two Carnegie engineers arrived in Chicago because of the web of contracts for structural work. The principal contract for fabrication was let to Snead Iron, a Louisville firm that had worked with Chicago architects for years. Snead Iron made all the castings, including the columns, then shipped them to Chicago. It did not make steel or roll heavy structural shapes. The contract for these

was let to Carnegie Brothers & Company, while the contract for
the theater's bridgelike trusses and for part of the theater itself was
let to another Carnegie concern, the Keystone Bridge Company.
Strobel and Marburg's presence in Chicago signaled that Carnegie
understood that Chicago was the central location for steel structures. As "special agent" in Chicago, Strobel stood at the center of
this new field.[9] Strobel and Marburg both checked the drawings for
the cast-iron columns, Mueller told the court, and why was Ritter
trying to criticize the design in this way? Actually, what Mueller
wanted to highlight was the Carnegie mill and its steel beams.[10]

For years, structural work from Carnegie had been synonymous
with the Union Iron rolling mills. An impressive business organization had emerged in 1881 when Andrew Carnegie merged Union
Iron with the Edgar Thomson rail mill, blast furnaces, ore mines,
coal mines, and coke ovens. In that year Union Iron was capable
of rolling thirteen different sections of I-beams; the largest measured 15 inches deep and weighed 80 pounds per foot, and could

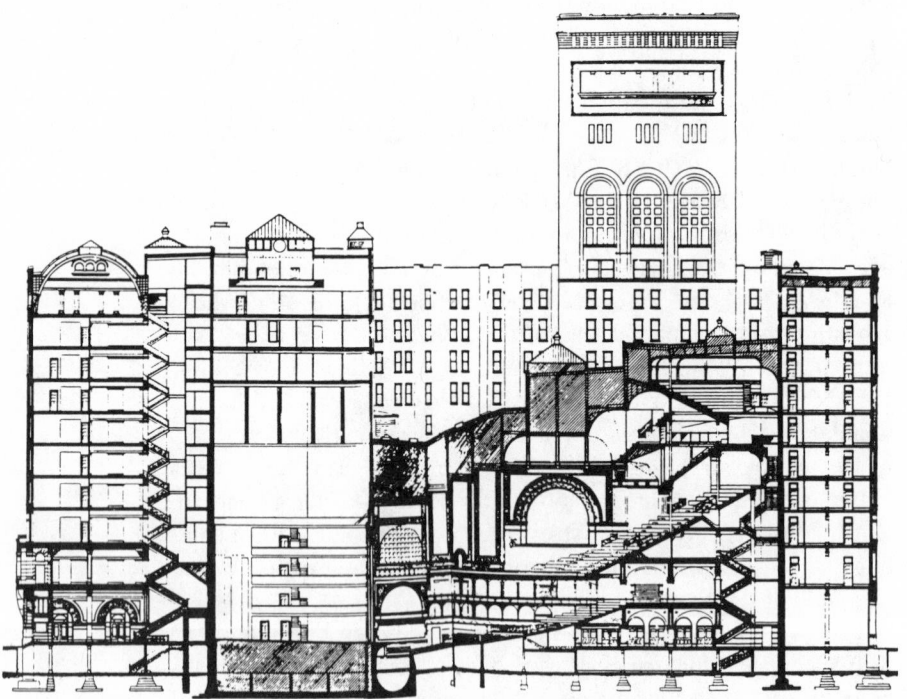

FIG. 2.1. Auditorium Building, Chicago, 1886–89.
From Dankmar Adler, "Theater-Building for American Cities," Engineering
Magazine 7 (1894): 723.

safely span 20 feet with a load of 20 tons. Two years later Carnegie purchased a new mill at Homestead and set about converting its Bessemer steel output from rails to structures. This was a key event in Mueller's mind. He allowed that the interior columns were cast iron (the Carnegie experts had signed off on that), but he defended his claim that all the beams, girders, and trusses were steel. Because the Auditorium project began *after* Carnegie's structural-shape rolling mill had changed over from wrought iron to Bessemer steel (in June 1885) Mueller could accurately confirm that the Auditorium's structures, other than its columns, were made of steel.[11]

Thus fortified, Mueller's recitation of structural details was well rehearsed and precise; applied to steel, the calculations were accurate and conservative. Then, in the midst of piling satisfying detail upon careful calculation, Mueller was interrupted. The attorney was evidently puzzled about just how Chicago architects had come to team up with Pittsburgh steelmakers. To clear up the matter he asked Mueller to evaluate a passage from Louis Sullivan's new book, *Autobiography of an Idea*. There Sullivan had explained the genesis of the steel-framed tall office building:

> It became evident that the very tall masonry office building was in its nature economically unfit as ground values steadily rose. Not only did its thick walls entail loss of space and therefore revenue, but its unavoidably small window openings could not furnish the proper and desirable ratio of glass area to rentable floor area.
>
> Thus arose a crisis, a seeming *impasse*. What was to do? Architects made attempts at solutions by carrying the outer spans of floor loads on cast columns next to the masonry piers, but this method was of small avail, and of limited application as to height. The attempts, moreover, did not rest on any basic principle, therefore the squabblings as to priority are so much piffle. The problem of the tall office building had not been solved, because the solution had not been sought within the problem itself—within its inherent nature.

Every problem contains and suggests its own solution, Sullivan had written, and the solution is invariably simple, basic, common-sensical:[12]

> As a rule, inventions—which are truly solutions—are not arrived at quickly. They may seem to appear suddenly, but the groundwork has usually been long in preparing . . .
> . . . So in this instance, the Chicago activity in erecting high

buildings finally attracted the attention of the local sales managers of Eastern rolling mills; and their engineers were set at work. The mills for some time past had been rolling those structural shapes that had long been in use in bridge work. Their own ground work thus was prepared. It was a matter of vision in salesmanship based upon engineering imagination and technique. Thus the idea of a steel frame which should carry *all* the load was tentatively presented to Chicago architects.

"The need was there, the capacity to satisfy was there, but contact was not there," Sullivan concluded. "Then came the flash of imagination which saw the single thing." The attorney paused in his recitation; was this account accurate? "That is pretty much the same flowing language" as an earlier extract, Mueller observed, "but correct."[13]

Following the success of transitional buildings like the Auditorium, the advent of a true steel-framed architecture in the early 1890s transformed the nation's steel firms as well as its cities. The effort to create a new steel for urban structures—in effect, to break the tyranny of technique imposed by the Bessemer steel rail—was a mammoth technical and scientific effort involving new linkages between producers and consumers of steel. This effort drew on nearly four decades of groundwork in building bridges, towers, and elevated trains, here illustrated through a case study of the Phoenix Iron company. In the mid-1880s, the Carnegie company successfully used engineering imagination to steal a march on the Chicago steel companies that had focused so thoroughly on steel rails that they were unable to make anything else. Still, at the turn of the century, the Carnegie concern itself stumbled when it failed to adopt a new steelmaking technique, the basic open hearth, that yielded a steel especially suited for structures. These cases—involving, in Sullivan's words, groundwork, engineering imagination, and technique—help explain how and why steel came to form the structure of cities.

Groundwork

*In every direction our customers are calling for steel,
and we must satisfy them or see them go elsewhere.*
—WILLIAM H. REEVES

Could the nation's urban structures have been constructed of *iron* rather than steel? The question, though counterfactual, is not frivolous. Cast iron was cheap, available in quantity, and well suited for compression applications like columns; in ultimate compressive strength cast iron was perhaps two-thirds stronger than steel. Cast iron formed the columns of the sixteen-story Unity Building in Chicago, built in 1892, and the seventeen-story Manhattan Life Insurance Building in New York, built in 1894.[14] Wrought iron might have furnished beams, girders, and other tension members for the nation's urban structures.[15] But a close consideration of the Phoenix Iron Company's unsuccessful efforts to expand production of structural wrought iron suggests that the answer remains no—because, as hinted in the case of railroads, wrought-iron production simply could not be expanded quickly and economically. The Phoenix Iron Company of Chester County, Pennsylvania, not only made metal and fabricated it into structures but also shifted its principal focus from railroads to urban structures. The company serves here as a microcosm of the structural trade during the changeover from iron to steel that haunted the Auditorium's builders.

Over the years Phoenix had won a modest reputation for innovation. In the 1850s it was the second firm, behind Peter Cooper and Abram Hewitt's Trenton Iron, to roll iron into I-shaped beams. David Reeves, Sr., incorporated the firm in 1855, serving simultaneously as the president of Cambria Iron and hiring the renowned John Fritz as its superintendent. Before the Civil War the general superintendent of Phoenix, John Griffin, patented a rolling technique that wove together slabs of fibrous wrought iron to yield especially strong beams. Griffin, like Bessemer, was drawn to military problems, and he used his special rolled wrought iron for an artillery field piece. Nearly half the Union Army's light artillery pieces, some fourteen hundred guns, were made to the Griffin design. Iron rails filled out the company's product line. In 1862 Reeves's son Samuel patented a prefabricated wrought-iron column (fig. 2.2). He later organized a bridge-building subsidiary whose fortunes neatly paralleled the boom in railroad construction.[16] In 1871 he succeeded his father as company president, and two years later Phoenix gained a showcase in Philadelphia with the contract for the Girard Avenue bridge. Experience in winning contracts from corruptible city officials was a valuable by-product of this bridge venture.[17] For the 1876 Philadelphia Centennial Exposition, Reeves offered the fantastic suggestion of building an observation tower a thousand feet tall. It was to be a skinny tube, 30 feet in diameter, braced by cables attached to its circular masonry base, and its daring height if not its banal form provided some inspiration for Gustave Eiffel's 1889 tower of the same elevation. In the event Reeves's firm built two columns of a modest 215 feet for the Philadelphia exposition.[18]

Such technical innovation did little to help Phoenix weather the 1873 panic and the depression that followed, which hit ironmakers severely. With the collapse of railroad construction Phoenix suffered the collapse of railroad bridge building as well as of the sale of rails. In response to disastrously low market prices for finished iron, the company closed its iron ore mines, damped up its three blast furnaces, slashed wages and salaries, and put on hold the huge new rolling mill under construction. The $1.15 million bonded debt taken on to finance the mill weighed heavily on the company directors. The idle mill, said one observer in 1875, was a "cemetery where millions of dollars are buried."[19]

Realistic proposals to elevate the trolley lines in the nation's most congested cities were a splendid opportunity for a company like Phoenix with expertise in railroad bridge building. Congestion was especially acute in lower Manhattan. A prototype elevated railway built by Charles T. Harvey in 1867–68 along a half-mile of Greenwich Street rested on a single row of specially designed, 14-foot-tall, 8-inch-diameter Phoenix columns. The success of Harvey's

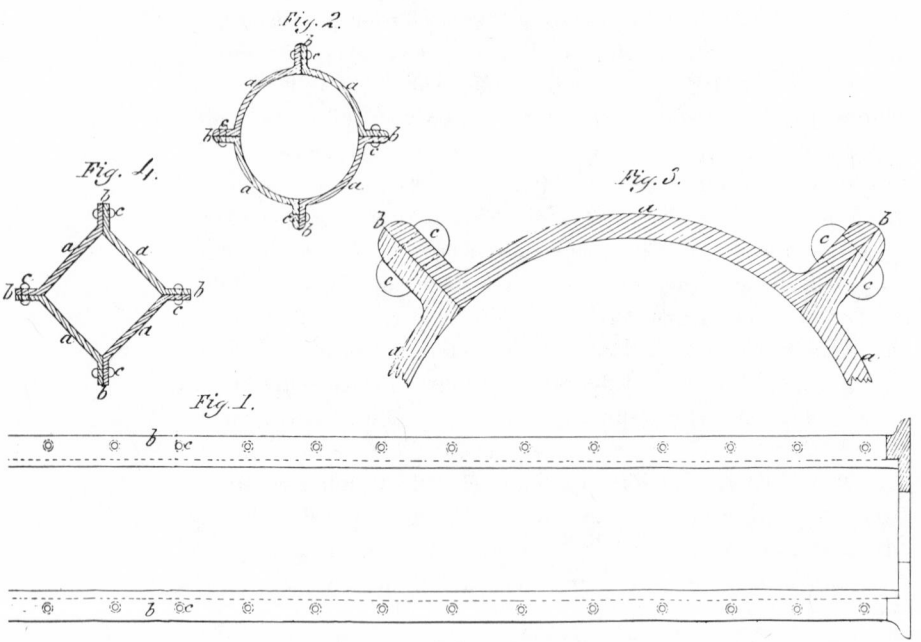

FIG. 2.2. Patented Phoenix column, 1862. From Samuel Reeves, Patent 35,582, "Improvement in Construction of Wrought-Iron Shafts or Columns" (17 June 1862).
Courtesy Hagley Museum and Library; Acc. 916, Box 71.

venture gave authority to the West Side Patented Railway Company to build twenty-five miles of elevated track. During 1868–70 the elevated line was extended as a one-legged structure to Thirtieth Street and Ninth Avenue, not quite four miles, but financial difficulties prevented further construction.[20]

In 1875 Reeves's company built at its works a 250-foot-long experimental elevated railway, which resembled nothing so much as a long railroad bridge complete with trestles. Two years later, when Phoenix slashed prices in an effort to keep its works full, the company gained its first significant contracts in New York: providing 150 spans for the Gilbert Elevated Railway's West Broadway line and 60 spans for the Metropolitan Elevated Railway's Fifty-third Street division. Phoenix took the entire 1878 contract, a total of some 460 spans, for the Metropolitan's extension farther north into Harlem along Eighth and Ninth avenues. In 1879 it gained the contract for 493 spans of the Manhattan Elevated Railway's East Side elevated line along Second Avenue from 49th Street to the 127th Street terminal at the Harlem River.[21]

Fabricating the heavy compound girders and special Phoenix columns for these projects presented a challenge. The company's million-dollar rolling mill was first used as a riveting shop for the compound girders. But a more severe problem was making sufficient iron in the first place. In February 1878, Superintendent Griffin estimated that the company needed to increase its iron-making capacity by 75 percent to meet the delivery schedule of the Gilbert contract alone; a month later, he wrote Reeves in exasperation, "I do not see where your 625 tons per week comes from." By 1879 more than half the mill's product was in large-section angle iron for elevated-railway building.[22]

Several obstacles frustrated Phoenix's attempt to expand its iron-making capacity. Its blast furnaces had been idle ever since a three-month strike in the anthracite coal fields during 1875; when coal shipments resumed Phoenix found it could purchase crude pig iron more cheaply than it could make the product.[23] By 1880, to meet the revived demand of its order book, Phoenix restarted two of its three blast furnaces, leased a puddle mill in neighboring Norristown, and purchased so-called muck bars to feed into its rolling mills. It also erected an 18-inch "column mill" to meet the elevated-railway orders. In 1881 Phoenix began feeding a small amount of old steel rails into its rolling mill.

Even with these alterations, the firm continued to depend on the workmen who operated the puddling and boiling furnaces that produced wrought iron, and they were a highly skilled and independent lot. During the summer, if the temperature rose much above ninety they simply walked out of the sweltering mill. Inevitably, iron puddlers—the "aristocrats" of the industry—were at the center of labor union activity at Phoenix. On 1 May 1873 the puddlers had marched off their jobs when the company refused their demand to increase tonnage rates by about 17 percent to match those of the rival Bethlehem. Although workmen began trickling back in mid-July, the strike did not end until 13 August when the union was dissolved and selected workers were rehired.[24] Nine years later, union activity returned. On 28 January 1882, the company shut down the puddle mills "to forestall the encroachment of the Labor Union" that the puddlers had again taken the lead in organizing. A company broadsheet circulated a month later estimated that three hundred out of two thousand workers had joined the union, and the firm asserted its absolute control over the works. "The Company," it warned, "reserves to itself the right to employ its own men, and to make its own rules and regulations, and no man will be employed who does not accord to it a recognition of these rights."[25] The lockout at Phoenix continued until 21 August.

While the lockout continued, Phoenix took two radical steps to reduce its dependence on skilled ironworkers. In February, at its remote Safe Harbor puddle mill, the company began experiments with the Du Puy system, a "direct" process that promised to transform iron ore directly into so-called muck bars suitable for rolling. The experiment ran for a year, until the puddle mill itself was closed.[26] A more determined effort was a four-year experiment with mechanical puddling that began in early March. Superintendent Griffin had been wanting to build mechanical puddling furnaces, and the lockout provided a convenient time to try out this idea. At $55,000 the mechanical puddler was a substantial investment, comparable to the combined charges for rebuilding the iron foundry and building a huge mechanical testing machine. After the mechanical puddler's first year of operation, however, Griffin had to concede that the results were unsatisfactory; a key problem was that "everyone connected with working them was without experience." Still, he was convinced of "the correctness of the principle" (of mechanical puddling) and looked forward to rising quantity and quality as well as falling costs.[27]

Despite Griffin's enthusiasm, the mechanical puddler's insignificant output scarcely diminished the company's dependence on its skilled iron puddlers. When Griffin died in 1883 the troubled process lost its most forceful advocate within the firm; later reports by William H. Reeves, who took over as general superintendent, detail a host of frustrations and unfulfilled technical promises. At the end of 1883, when the mechanical puddlers had made a mere 190 tons of 9½-inch-square muck bar, Reeves reported that "we have failed to make a successful run with these machines . . . thus far the experiment has ended in disappointment." He noted trouble with the machines' rotating mechanical drums, and difficulty in making good joints between the rotating drums and the flues and fire chambers. Fuel economy was appalling. In 1884 mechanical puddling of a ton of iron consumed approximately five times as much coal as hand puddling. A year later, when its fuel economy improved, the puddling machine's output of 469 tons simply paled in comparison with the hand puddlers and boilers' combined output of 15,700 tons. In 1886 the mechanical puddlers were shut down. Only after these two alternatives had failed did Phoenix commit itself to building a steel plant, which finally ended the dependence on skilled puddlers and boilers. The mechanical puddler was eventually scrapped and melted down in the new steel furnaces.[28]

Compared to the new rolling mill, which by 1889 had soaked up an investment of $1.78 million, building the open-hearth steel plant required a modest $300,000 funded by retained earnings. While it was building the steel mill Phoenix issued additional capital stock, but its purpose was simply to retire the massive bonded debt of the rolling mill. In 1890, when the company issued $1.5 million in consolidated mortgage bonds, the primary purpose was again to cut interest charges. Expansion of the steel plant in that year required an additional $120,000.[29]

Recruiting new skilled workers for the steel plant proved the most difficult challenge. Phoenix was expert in ironmaking, but steelmaking required competencies that the company did not possess. The most critical skills were those of the "melter." This was a highly skilled position paying ten times the day wage of a manual laborer. The melter had complete, day-by-day charge of the two 10-ton open-hearth furnaces. The two melters, who earned a handsome day wage "but only for actual days worked," were at the juncture between labor and management; they reported to the steel plant's chemist, who also oversaw the gas producers and the testing department and who held a junior salaried position at $125 per month (a sum just barely above the melters' monthly pay).

Phoenix sought melters as far away as Alabama and Ohio, and finally the steel plant's foreman, Fred Heron, hired Erastus Darcy and Nathaniel Shed after interviews in Pittsburgh. Heron also conducted special searches for the chemist, first and second helpers, ladle man, and the "gas man" who worked the Siemens gas producers. Apparently, in a decision that came to haunt him, Heron did not hire outside workmen for the other half of the steel plant: the blooming mill that did the first rolling of cast steel ingots into blooms or beam blanks. At the same time Phoenix was starting up the steel plant it was straining to expand iron production, so it faced a shortage of workers—the perennial problem of the wrought-iron industry.

The melter Nathaniel Shed gave a wonderfully detailed account of the steel plant's startup during March and April of 1889. The main difficulty in structural steel, he observed, was to produce a metal that would roll easily, without cracking on the flanges of beams. Such difficulties in rolling were traceable to excess sulfur, which could be neutralized by some (undisclosed) "proper treatment." The best open-hearth heats were made up of one-quarter pig iron with three-quarters Bessemer steel rail butts or beam scrap. Once the 20-ton charge of metal was melted, Shed added a half-ton of hematite ore as quickly as possible, within an hour. While adding the ore, he observed, "there should be a lively action throughout the metal; and before all the ore has been put in an energetic 'boil' should be visible all over the bath. The heat on the metal is now gradually increased, and the lively action is kept up for nearly an hour. With the decrease in carbon the intensity of the action diminishes and the slag becomes thicker and more viscous. At the end of one hour after the last addition of ore the bath becomes very quiet, the bubbles of gas formed are few, and break slowly through the slag. The metal, if exposed to view, presents a clear, silvery appearance." If the carbon level was right (about 0.16 percent), a test sample broke with a bright, coarsely crystalline fracture, showing considerable toughness. If the carbon level was slightly high, the gentle action in the hearth itself needed to be continued. If the carbon level was very high, and iron ore was added to lower it, Shed emphasized that sufficient time should pass until the batch again become "tranquil" before casting. Shed favored adding ferro-manganese, for improving the metal's mechanical properties, directly to the casting ladle, just before the metal was poured into the ingot molds. The steel, after cooling somewhat in the casting ladle, was "top poured" into ten ingots, 15 by 18 inches in size, weighing just over 2 tons each. Finally, Shed noted that the temperature of the ingots when rolled into blooms (9 inches square) was "a good red heat, noticeably cooler than is the usual practice."[30]

Unfortunately Shed's steel had several unpredictable and hence undesirable physical properties. Normally, higher carbon produced greater tensile strength, lower reduction in area, and less elongation. For Shed's steel, increasing the carbon from 0.13 percent (a medium-low value) to 0.22 percent (a rather high value) increased its tensile strength but yielded wildly erratic figures for reduction and elongation. Only eight in ten batches would satisfy *minimum* specifications for structural steel; only three in ten would meet *desired* tensile-strength specifications. It was not clear, moreover, whether Shed's steel met the overriding criteria of being "uniform in quality."[31]

Nor did Shed and his fellow melter make enough steel. In fact, after production fell disastrously in May (fig. 2.3a), Foreman Heron fired them. The open hearths were shut down for some minor repairs and alterations before Heron secured the services of two new melters, E. E. Robinson (from Birmingham) and Sam Jones (from the Pittsburgh district), who by mid-July had restored production to its earlier level.

Labor problems still plagued the new steel plant that summer. Beginning in late July and continuing through mid-August, Heron discharged no fewer than twenty middle-skill workers (fig. 2.3b). The foreman's letter book does not make clear why he discharged such a large fraction of his work force at the same time he was expanding production. The available evidence permits some tentative explanations. Two workmen who left with the foreman's permission are not included in the discharge numbers, and Phoenix did not lay off workers during temporary lulls (fig. 2.3a, weeks 14–16). None of the men Heron discharged appears on the list of tonnage-rate open-hearth workers, so they likely were from the blooming mill. If so, they must have been previous Phoenix employees whose experience was solely with making iron, since new workmen were not hired for that half of the steel plant. One might speculate that these ironworkers were endeavoring to obstruct the new steel technology that clearly threatened their craft skills. Inevitably, just as steel production ebbed, the firm once again tried to start up the mechanical puddlers.[32]

The die was cast, however, and it was made of steel. "In every direction our customers are calling for steel, and we must satisfy them or see them go elsewhere," General Superintendent William Reeves observed in June 1890. Reeves argued that "a sure proof of the necessity for the new plant" was that productive capacity could not satisfy the rising tide of orders. Accordingly, he directed the building of two more open-hearth furnaces. Finally, with the deployment of capital rather than the employment of labor, the company could expand production substantially. The open-hearth process had additional advantages. For years, the finishing mills

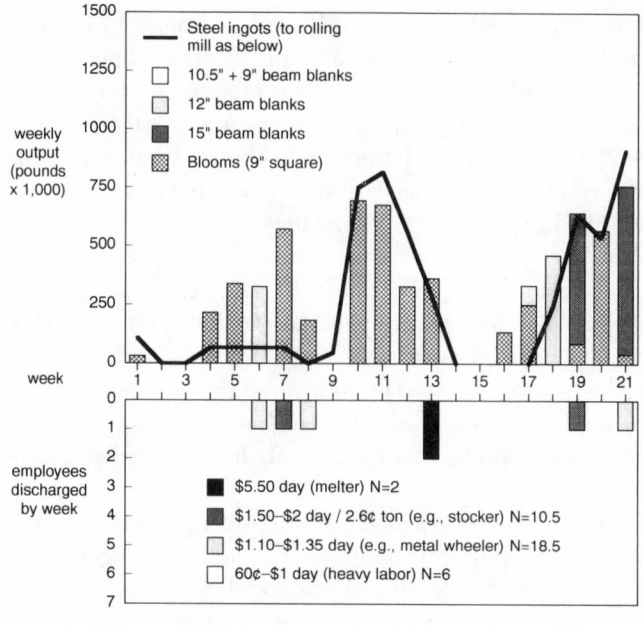

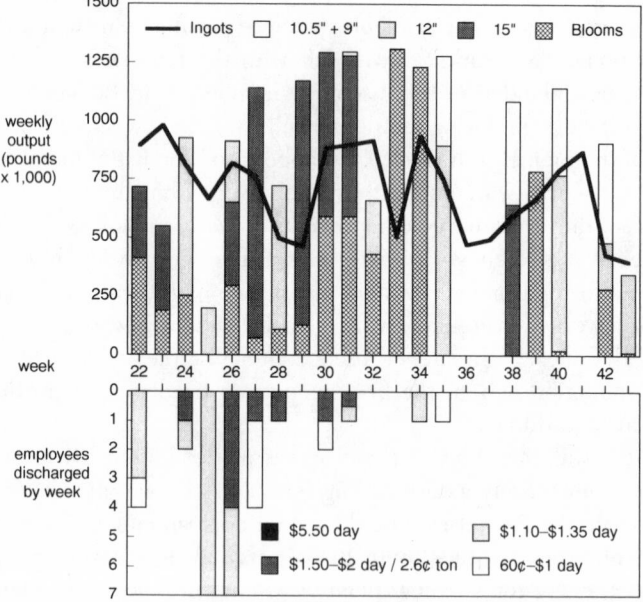

FIG. 2.3. Phoenix Steel startup, by week: (*top*) March–July 1889; (*bottom*) August–December 1889.

Data from Fred Heron, Correspondence–Manager–Steel Dept. (May 1887–January 1890), 267: Reports–weekly, PS.

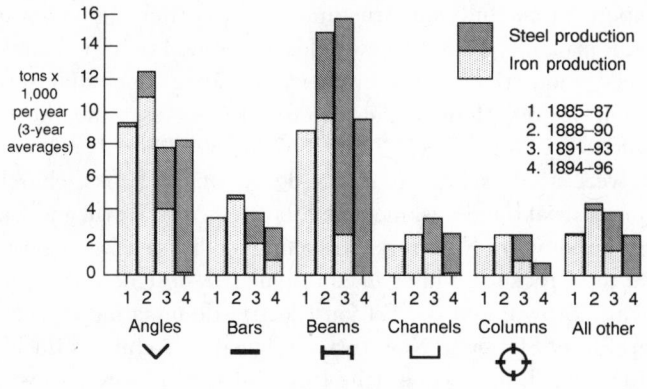

FIG. 2.4. Phoenix Steel production, by shape, 1885–96. Note the shift to steel in beams and columns by period 3 (1891–93).
Data from General Superintendent's reports (1872–1922), PI.

had been underused, and the extra steel output would finally fill them. Indeed, since the same mill could roll nearly twice as much steel as iron, Reeves believed that even with the expanded output the finishing mills would still have more capacity than the steel plant. Putting heavy machinery in the eye-bar shop, adopting heavier rolling machinery, and installing a heavy testing machine ("as the testing of steel bars for bridges is required by all Engineers") completed the changeover to steel. By 1892, 67 percent of the company's output was steel (fig. 2.4). In the panic year of 1893 Phoenix closed its puddling mill and without intending it left the iron business forever.

In April 1894, "owing to the great difficulty of obtaining proper material to make high grade metal in the acid open hearth," Phoenix converted one of its open hearths to chemically "basic" linings. The success of this experiment prompted the company to expand its production of basic open-hearth steel, and by July 1895 three of the firm's four open hearths were basic. During 1896–97, stated William Reeves, "the Puddle Mill was torn down and all the old material used up about the works. This closes our history as an iron producing plant."[33] Pulled by customer demand, pushed by falling iron prices, and thwarted by its inability to mechanize iron making, Phoenix had hit upon basic open-hearth structural steel. The emergence of this characteristically American steel is discussed in the final section of this chapter.

From the mid-1860s to the mid-1880s, Phoenix used wrought-iron beams and girders, and above all its patented columns, to compete successfully for structural contracts. Relying on wrought iron, it expanded into the new fields of elevated railways and commercial, industrial, and governmental buildings (including several federal Custom Houses, the Library of Congress, and the Government Printing Office). From 1887 to 1890 Phoenix columns of iron were used exclusively in two dozen offices, banks, churches, and industrial buildings, including the New York World and Union Trust buildings in New York, both designed by architect George B. Post; the Crocker Building in San Francisco, designed by architect A. Page Brown; and the Edison Electric Illuminating Company's Pearl Street Station in New York, designed by architects Buchman and Deisler. But only one—the Hoyt Building in New York—used Phoenix columns of steel.[34] Although Phoenix and Cooper-Hewitt had been leaders in structural iron, they became followers in structural steel. Other companies were able to roll larger or lighter sections of steel, while Phoenix and Cooper-Hewitt remained expert in rolling the largest sections of iron.[35] The potential of structural steel was captured more fully by Carnegie Steel and a remarkable group of Chicago architects.

Engineering Imagination

The standard specifications of the Carnegie Steel Company are usually used in the West.—W.L.B. JENNEY

In 1884, the year after Andrew Carnegie began converting his newly purchased Homestead mill to structural shapes, his company won a structural contract in Chicago for the Home Insurance Building. A local concern, the Dearborn Foundry, had received the fabrication contract; but just as with Snead Iron and the Auditorium, Dearborn did not manufacture heavy iron and steel structures, and this building would have quite a lot of both. It was to have nine stories and be supported by a system of iron beams and columns; at the urging of Carnegie Steel the top three floors would feature the pioneering use of Bessemer steel beams. With this effort Carnegie Steel placed itself at the center of Chicago building at a providential moment. The Home Insurance Building was architect William Jenney's first effort to master the new technology of iron and steel construction. "It was the major progenitor of the true skyscraper," writes architectural historian Carl Condit, "the first adequate solution to the problem of large-scale urban construction."[36] Whatever its priority as the first steel-framed skyscraper (vigorously debated ever since the building was razed in 1931),[37] the Home Insurance

Building was of surpassing importance as the first salvo by Pittsburgh steelmakers in taking over the Chicago structural trade.

The Home Insurance Building was a demanding commission. In 1883 Jenney was appointed architect for the New York company's new building on the northwest corner of Adams and LaSalle streets. The first story was to host a bank, and each story above was to be divided into the maximum number of small offices. Realizing the difficulties, the insurance company president, J. J. Martin, told Jenney that this would require a large number of moderate-sized windows and reduction of the piers between the windows, probably to such an extent that they could not be of masonry, especially in the lower stories. "How are you going to manage it?" Martin asked. Jenney replied that he was going to study the matter.[38] Speculation on the sources of Jenney's inspiration has ranged from a metal birdcage upon which his wife had piled heavy books, through the bamboo huts he had seen in the Philippines, to the towers constructed in New York to manufacture lead shot, to the strike that year by Chicago bricklayers.[39] Jenney's own account was disarmingly frank:

> It was evident that if the brick pier of the maximum size admissible was insufficient to carry the requisite load that something must be used that was stronger than masonry. Every architect and engineer under such circumstances had been in the habit of placing within the pier a cast iron column. I had done this myself, particularly in a large bank building in Indianapolis a few years before, but as the [Home Insurance] building was about 150 feet high a continuous column of that height would have a considerable expansion and contraction ["from the possible minus 20 in winter to the 100 and upwards in summer"] . . . The disintegration of the enclosing masonry might result.
>
> While studying over this problem it occurred to me that to overcome this difficulty it was only necessary to extend the lintels under the entire window heads entirely across from column to column and carry the outer walls story by story on the columns as well as the floors.[40]

In this way, the expansion and contraction of the 150-foot-tall metal columns was effectively divided into ten parts and hence rendered harmless.[41] With this design, the Home Insurance Company would have its array of small offices, their occupants would have maximum natural lighting, and Jenney would have a structural system that he and the five leading Chicago architects who trained in his office would elaborate into a distinctive "school" of architecture.

In the mid-1880s, Chicago was an exceptionally fertile environment for innovation in building. The 1871 fire had cleared away much of the central business district, and the hastily erected structures put up in the spurt of building before the 1873 panic left much to be desired. The depression of the 1870s had kept construction at very low levels, but by 1882 new construction activity began a sustained climb (fig. 2.5). Land values, population, manufacturing employment, bank clearings—virtually all the indices of urban density and intensity were up sharply. These developments culminated in a building frenzy during 1890–92, when, as in 1872–73, Chicago's total new building volume actually surpassed Manhattan's.

Yet Chicago had one immense liability with regard to the building of steel-framed skyscrapers: there was no local steel company specializing in structural shapes. The region had several iron-fabricating outfits, such as Snead and Dearborn, that made cast-iron columns and molded wrought-iron grilles, staircases, and other decorative elements. But these companies had no facilities to roll heavy I-beams or to produce steel. Three concerns in the Chicago area—south Chicago's Union Iron, the North Chicago Rolling Mill, and

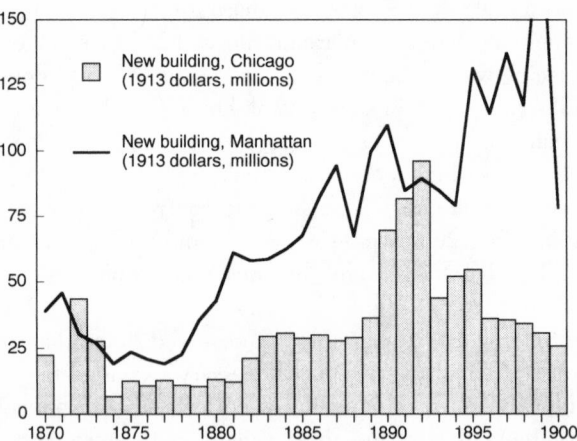

FIG. 2.5. Chicago and Manhattan construction volume, 1870–1900. The figures are adjusted to constant 1913 dollars. Chicago 1872 is for 9 October 1871 to 9 October 1872. Manhattan 1899 is $191 million.

Data from Homer Hoyt, One Hundred Years of Land Values in Chicago (Chicago, 1933), 474–75; Frank A. Randall, History of the Development of Building Construction in Chicago (Urbana: University of Illinois Press, 1949), 294–95; R. E. Lipsey and D. Preston, Source Book of Statistics Relating to Construction (1966), 23–25, 248.

Joliet Iron and Steel—had the metal-making capacity to produce heavy structural components. Each built new plants to produce Bessemer steel rails during 1871–73 (table 1.1); by 1885 their combined rail capacity of 500,000 tons a year, as Carnegie wrote, was "sufficient to put a light steel rail 'girdle round the earth.'" None of them, however, successfully diversified into structural shapes in the 1880s. When North Chicago announced its entry into structural shapes, while the Home Insurance Building was being erected with Carnegie steel, it had to admit that the Pittsburgh firm had stolen a march. In 1889 these three concerns merged to form Illinois Steel; the following year only 0.35 percent of the combine's output was in structural shapes.[42] The relative weakness of Chicago's makers of structural steel is all the more surprising given the dynamism of the city's structural trade.

The architects, builders, fabricators, and suppliers in Chicago formed a dynamic community with a track record of innovation. Some Chicago building techniques, such as balloon-frame housing, reflected the ready supply of Great Lakes timber; others, such as foundations using grillage and later caissons, responded to the city's soft and compressible soil. In 1884 Jenney's colleague Frederick Baumann published a pamphlet, "Concealed Iron Construction in Tall Buildings," which Jenney might have consulted in seeking a solution to the Home Insurance problem. Jenney himself treated his innovation as an evolution in the methods of iron and brick construction, at least until 1892 when he adopted a more possessive attitude owing to the patent claims being pressed by inventor Leroy Buffington.

Yet the pressing issue is not Jenney's individual priority but the question that puzzled the attorney in the Auditorium trial: how and why was the "contact," as Sullivan put it, established between Chicago architects and Pittsburgh steel firms? Why in the crucial transition decade of the 1880s was New York, the nation's largest city, a less innovative building environment than Chicago? And what role did "salesmanship based on engineering imagination" play in the Carnegie company's dominance of the structural-steel field? To shed light on these questions we must consider four factors: the fruitful analogy between railroad bridges and tall buildings, and the railroad-structural engineers who realized this conception; the professionalization of architects, and their willingness to transform architecture to embrace structural engineering; linkages between steelmen and architects that were actively nurtured and sometimes aggressively sought; and, finally, a comparison of public policies, specifically the building codes of New York and Chicago.

The striking resemblance of skeleton buildings to railroad bridges had both practical and promotional consequences. The two tiers of columns in a skeleton building, Jenney said, were much like one side of a railroad bridge stood on end.[43] The point is significant. The chief structural problem of the Auditorium Building, Condit writes, was the support of a great elliptical vault over the orchestra floor of the theater. "Adler's solution consisted in a system of truss-framing adopted from the forms of contemporary railroad truss bridges."[44] Blueprints for Adler & Sullivan's Stock Exchange (1893–94) clearly indicate railroad bridge trusses.[45] The railroad bridge was a well-understood metal structure, with a broad range of specific designs for particular structural requirements. Its familiarity probably helped architects to persuade building owners of the soundness of their skeleton plans. Jenney related that he had to call in a seasoned engineer, General Arthur C. Ducat, to convince the Home Insurance executives that the design was sound and the calculations correct. Even after Jenney had successfully completed several skeletal structures, conservative owners could obstruct Chicago architects' plans for metal skeletons as having too much "risk and uncertainty" for the large financial investment of a multistory office block.[46] New forms of foundations, too, were the work of railroad bridge engineers.[47]

All the prominent Chicago architects employed railroad bridge engineers in some capacity. Jenney himself had served with General Ulysses S. Grant's engineering staff until Corinth, then through the end of the war with General William T. Sherman's staff, materially aiding his march to Atlanta. So proficient was Jenney at turning abandoned cotton gins, church pews, and chimneys into bridges that a southern newspaper stated that it was useless for the retreating armies to burn the railroad bridges because Sherman evidently carried with him a full line of ready-built bridges.[48] Jenney's construction superintendent for the pioneering Home Insurance Building was George B. Whitney, who had taken an engineering degree at the University of Michigan and then designed bridges for the Baltimore & Ohio Railroad. Before joining Jenney's office, he had worked some years for the Mississippi levee project of the Army Corps of Engineers; he later went to Burnham & Root's office as structural engineer for the Masonic Temple, the building Louis Sullivan hailed as the first "skyscraper." Afterward, with Theodore Starrett, his successor as engineer with Burnham & Root, he formed Whitney & Starrett Construction, a forerunner to the massive Thompson Starrett concern. Jenney & Mundie's chief engineer during 1893–98 was Louis E. Ritter, whose testimony in the Auditorium trial was quoted earlier; between graduating from the Case School of Applied Science in 1886 and joining Jenney's office in 1892, Ritter worked for the Erie Road and the Mississippi project. Holabird & Roche had the services of Henry J. Burt, who had technical training from the University of Illinois and bridge engineering experience with the American Bridge Company. From 1885 to 1893, Adler & Sullivan as well as Burnham & Root had the consulting services of Carnegie's Charles L. Strobel, who had experience with the Cincinnati Southern Railway. Adler later practiced with Elmer L. Corthell, who had been an engineer for the Hannibal & Naples Railroad, the Hannibal & Missouri Railroad, the Mississippi project, and the Merchants' Bridge at St. Louis.[49]

This influx of engineers was welcomed by those architects whose broad education, practical preparation, and professional outlook had led them to a new mode of architectural practice that embraced rather than denigrated technical details. Skeleton-frame buildings were "purely engineering," Jenney wrote. They "very materially changed an architect's duties" because the technical complexity of steel-framed buildings forced architects, draftsmen, and engineers to "work together" to devise the general plan, arrange the columns for tenant convenience and fireproofing, and make a sound foundation plan.[50] Dankmar Adler, an engineer and a pioneer in acoustical design, "was very apt to have almost an expert for every branch of the building that was built."[51] "Adler's brain was intensely active and ambitious," Sullivan recalled of their first meeting, "his mind open, broad, receptive, and of an unusually high order." Sullivan himself asserted that "the new form of engineering was revolutionary, demanding an equally revolutionary architectural mode."[52]

Significant personal linkages between architects and steelmakers provided the essential "contact" that Sullivan identified. William Jenney invited iron-fabricating companies into his working office. On 5 August 1879 Hough of Indianapolis, Younglove & Co. of Cleveland, and Union Iron Works of Chicago were busy completing their bids for the University of Michigan Museum. "As iron is so complicated," Jenney wrote, "they [come] to figure in [the] architect's office." Two days later, when Snead of Louisville arrived, Jenney's office was packed with four iron concerns.[53] Jenney's office practice was an especially significant model, considering that Martin Roche, William Holabird, Louis Sullivan, George Elmslie, and Daniel Burnham—two-thirds of the architects of the Chicago School—worked there during the 1870s and 1880s. Dankmar Adler did not, but in 1888 on board the Cunard Line steamship *Servia* Adler met Carnegie Steel's legendary general superintendent William R. Jones. Adler was beginning a whirlwind tour of European concert halls for the Auditorium project while Captain Jones, exhausted from his failure to install the eight-hour day at the Edgar Thomson steel mill, was en route to the Glasgow exposition. The two men had "great times" together, befriending the ship's chief engineer as well as several officers, and they continued their shipboard companionship onto land, comparing notes on the British railroad system and swapping travel stories. Jones promised to sponsor the architect for membership in the Iron and Steel Association when they rejoined in Glasgow.[54]

Perhaps the most significant linkage was the extensive consulting relationship that developed between Chicago architects and Pittsburgh steelmakers. Architects Jenney and Burnham met frequently with the Carnegie engineers to devise and develop the necessary properties of structural steel, "a trinity as it were, working together to advance this great architectural innovation."[55] In the New York Life and Fort Dearborn buildings of the mid-1890s, Jenney & Mundie developed a suggestion by a Carnegie-Keystone engineer into a new structural design for bracing tall buildings against wind pressures. Indeed, by about 1895 the pioneering phase of structural-steel construction was at a close, and a standardized method of making, rolling, and inspecting structural steel was emerging. In 1896 "American Standard" sections were adopted by the steel industry.[56]

Differing public policies in America's major cities helped shape distinct styles of structural engineering. The single most important public policy was the building code, which set the parameters of successful construction. Building codes reflected the intersecting demands of historical precedent, innovation from the building trades, and a city's political process. The New York building code, for example, was revised through negotiations among the Fire Department, Architectural Iron Association, American Institute of Architects, Mechanics and Traders Exchange, Real Estate Owners and Builders Association, and Real Estate Exchange, as well as fire insurers, fire engineers, building superintendents, theater consultants, and lawyers.[57] Thanks to varied policy constituencies, the building codes of the major cities differed significantly in the strengths they required of wrought iron and steel. As the Auditorium trial suggested, the choice of metal (wrought iron, steel, or cast iron) was not clearly determined by economic and technical considerations. Public policies therefore played a decisive role exactly when the first skyscrapers were being conceived and built between 1885 and 1895. During these years both steel and wrought iron were variously used in tall office buildings. Steel's advantage in strength was offset by its higher price. Comparison of New York, Chicago, and Boston shows that each city favored steel, but not to the same extent. Steel therefore gained a city-specific advantage over wrought iron to whatever extent one city's building law, compared to another's, permitted larger specified stresses for steel over wrought iron.

The details of city codes in force during 1895 reveal their subtle force. For extreme fiber stress (a severe index of tensile strength) in rolled beams and shapes, New York permitted steel a 25 percent advantage over wrought iron, while Chicago and Boston permitted steel a 33 percent advantage. For tension computations (which directly informed structural design) on beams and girders, the Chicago and Boston building codes gave steel a 33 percent advantage in allowed stresses, while oddly enough the New York code said nothing. For compression in the flanges of beams, Boston permitted steel a 20 percent advantage over wrought iron; New York, 25 percent; and Chicago, 35 percent.

The Chicago preference for steel in beams and girders vanished, however, for columns. Columns were almost entirely a compression application. While all three cities required structural computations using Gordon's formula, which related the shape and dimensions of a column to its permitted strength, each city specified a different strength constant.[58] For determining the strength of columns, New York and Boston permitted steel a 20 percent advantage over wrought iron, while Chicago permitted steel only a 12.5 percent advantage.[59] These calculations of (permitted) strength translated directly to differential economic incentives. Architects exploited the greater strength of a material by specifying components with thinner cross-sections and therefore lower cost. Consequently, Chicago architects could pay more per pound for steel beams and still have overall structural costs equivalent to a New York design, since they used less material. In contrast, depending on the dimensions of a column, Chicago architects might have to pay less per pound for steel to achieve equivalent structural costs, since they used more material.

More important, certain provisions of the New York building code nullified important advantages of the Chicago style of skeleton construction. For a Chicago skeleton design, the outer walls were nonbearing and could be very thin, indeed curtain walls; the thin walls were easy to suspend from the main structural system (creating a true skeletal construction) and also gave more floor space and admitted maximum natural light. The New York code, however, made only a slight difference in the thickness required of bearing versus nonbearing walls. This provision had large consequences. The code required that a tall building in New York have thick walls, whether bearing or not; the taller the building, the thicker the walls. The thick walls ate up floor space, hampered natural lighting and made it all the more difficult to suspend their increased weight from the internal structure of the building. Among the architects and engineers who railed against this restriction was William Birkmire, who argued in 1900 that "the thick and heavy walls required by the New York law are quite unnecessary." Although 12-inch walls were sufficient with skeleton designs, and walls of 16 or 20 inches provided adequate fire protection from adjoining buildings, New York nonetheless mandated walls up to 9 feet thick for the tallest structures.[60]

The timidity of New York architects when approaching skeleton construction can be traced not only to the New York building code but also to the hostile manner in which it was enforced. The building law of 1887 contained no reference to iron or steel skeleton construction or to thin curtain-type walls. These building types were in legal limbo until 1892, when, for the first time, the New York building law sanctioned "skeleton construction." How the code was applied in specific instances was equally important. In 1888, when architect Bradford L. Gilbert filed plans for the first skeleton building in New York, he proposed a bizarre hybrid structure with cast-iron columns from the basement through the seventh floor, wrought-iron columns on the eighth floor, and solid 20-inch-thick brick walls for the topmost three stories. One member of the Building Department board of examiners recalled "full well" that when these plans were submitted, several members of the board expressed a "strong preference . . . for solid masonry work" and a distinct "prejudice against iron work in general." Only after a long delay did the board pass on the design. The next two New York skeletons were hardly inspiring either. In 1889 the architects J. C. Cady & Company filed plans for a second skeleton building, at 25 Pine Street for the Lancashire Insurance Company. This was to be a ten-story structure using steel Z-bar columns. New York's third skeleton, the Columbia Building on Broadway, was twelve stories high, with steel columns and curtain walls 12 inches thick; plans for it were filed in 1890 by architects Youngs & Cable. The building was completed in 1891. These three narrow buildings (21, 24, and 30 feet wide, respectively) used skeleton construction to maximize the floor space available to front a popular commercial street. During this period, large buildings in New York were of a hybrid "cage" construction. George B. Post's Pulitzer Building (1889–90), at 375 feet, rested on load-bearing masonry walls 9 feet thick at the base; metal framing carried only the interior floors. Similar in construction were his New York World Building (1890) and Havenmeyer Building (1891–92). New York's first skeleton building of substantial size was Kimball & Thomson's 350-foot-tall Manhattan Life Insurance Company Building (1893–94), which featured cast-iron columns.[61]

By comparison the frontage of Jenney's Home Insurance Building (1884–85)—facing Adams and LaSalle streets for 98 and 138 feet, respectively—was larger than the frontage of the first three New York skeletons combined. Jenney's Ludington Building (1891) had a frontage of 120 feet on Wabash, four times that of the largest contemporary New York skeleton. By 1891 construction had begun on Jenney's mammoth nine-story Fair Store (1890–97), which fronted the north side of Adams Street for the entire block (730 feet) between State and Dearborn.[62] Besides Jenney's pioneering efforts, Chicago architects by 1891 had completed another six skeleton buildings between ten and fourteen stories high.

As a result of the interaction of its architects with its building code, New York effectively rejected the Chicago concept of skeleton construction and instead developed the concept of cage construction. In skeleton structures all the weight, including that of the outside walls, was transferred floor by floor to the metal frame. In cages the outside walls bore their own weight, while the frame supported only the insides of the building. The New York architect George B. Post was the most outspoken advocate of cages over skeletons.[63]

Moreover, anecdotal evidence suggests that New York financiers were hostile to Chicago architects, even after—and perhaps because—they developed steel-framed skeletons. Because large office buildings required large-scale financing, the chauvinism of a city's bankers could easily tilt the negotiations to favor local concerns. The president of the leading builder in New York, Thompson-Starrett, related that the banker Pliny Fisk, of Harvey Fisk & Sons, told him flatly, "We don't want to have any Chicago firm of architects." Chicago's bankers returned the favor, he reported. "Our rule was not to go out of Chicago for work," stated builder Henry Ericsson.[64] After about 1900 the largest builders, including Thompson-Starrett of New York and George A. Fuller of Chicago, expanded nationally and increasingly raised buildings in each other's city and indeed across the nation.[65]

Given the pervasive city chauvinism, it is all the more surprising that a Pittsburgh company was able to dominate structural steel in Chicago. "Always the competition among the big companies was extremely sharp," recalled one builder.[66] But what was the mode of competition? The price rivalry posited by economists was precluded by the "beam pool," a market-allocating cartel active continuously from 1885 to 1892—those crucial years for steel buildings—and intermittently until at least 1898.[67] Carnegie for his part resented paying $10,000 a month in "blackmail" to the beam pool, especially when its established allocation frustrated his company's plans for expansion. "[The threat of] low priced beams mean[s] ten fold to Jones and Laughlin or any of the others than to us. They all *need* the profit—need it badly and will agree to our getting what 3 years shows us entitled to," he stated in 1888.[68] Considerations like quality of product and prompt delivery were more important in mediating relations between users and producers in the structural trade. The worst nightmare was to save a dollar on steel but have its late arrival disrupt the intricate logistics of raising a building and filling it with paying tenants. The president of Thompson-Starrett, for instance, refused bids that were "too low" and instead favored certain steel companies for their track record. "If a subcontractor's figures were too low, we would throw them out, on the theory that the fellow must be either a fool or a knave; in either event he would be a poor person with whom to do business," he explained.[69] Andrew Carnegie's entry into this pell-mell world was no accident; neither was his success.

By the late 1880s, Carnegie clearly understood that "the outlook [was] . . . not bright for rails" and that a new product was urgently needed. Structural shapes, splice bars, and wire billets were the most promising new products, he believed, at least next to armor plate (see chapter 3). Carnegie relentlessly pushed his managers to line up structural contracts. For example, he responded to William Abbott's telegram announcing that Norcross Brothers had just placed a 4,000-ton order for the Exchange Building in Boston with a curt "Good boy—next!" As early as 1888, when the largest commercially available beam was 15 inches in height, he was formulating plans with Captain Jones for how to roll 20- and even 24-inch beams. Carnegie also understood that the shape of beams was as important as their height. "We should experiment on making a Beam for building purposes with narrow heavy flanges such as the Belgian in which we could do as they do—use all raw rails with a small piece of muck bar at each end of pile to act as flux," he wrote in 1889. "We could get all the business we could do in building beams if we put price down."[70] Apparently Jones's plans were well advanced, for even though he was killed in a blast furnace accident in September 1889, the company successfully made the leap to the large beams.

Indeed, the Carnegie concern's technical superiority can be identified with some precision by comparing its product line with those of its competitors. One appropriate measure is the *maximum depth* (the distance between the top and bottom flanges) of I-beam that a firm could produce. First manufactured by the Cooper-Hewitt and Phoenix firms in the 1850s, I-beams quickly became a fundamental element in the structural engineer's repertoire and were of comparable shapes even before they were standardized in 1896. Simply put, the deeper the I-beam, the stronger the structure would be. By this measure, Carnegie's largest beams were consistently in the lead (fig. 2.6). A second measure encompassing engineering and economic considerations is the *minimum weight* to which a standard-size beam could be rolled. Structural shapes gained strength principally by greater depth and only secondarily by heavier weight. For instance, a 15-inch I-beam weighing 41 pounds per foot was 20 percent stronger than a 12-inch I-beam of the same weight. Similarly, a standard 12-inch I-beam of 32 pounds per foot would, if rolled to a heavier section of 36 pounds, gain 6.28 percent in strength but double that (12.5 percent) in weight and cost.[71] Except when a large beam would not fit into the available space, what economy-minded structural engineers looked for was light sections of maximum depth. Here again, in a comparison of the minimum weight for standard 12-inch I-beams, Carnegie's were consistently at the top (fig. 2.6).

The Carnegie handbook for structural steel that advertised these technical advantages became a powerful competitive tool. "Architects, engineers and others who draw specifications seem to know of no book but Carnegie's," observed a rival's sales agent in 1897. "Ten years ago their book was no better than Phoenix, Pencoyd, Passaic or any of the others," he stated, "but in the last 5 years the others have been obliged to adopt Carnegie's sections right straight through." "No advertisement," observed the salesman, was as valuable as a handbook like Carnegie's that is "almost universally sought after" and gradually sets a standard that all competitors must meet or else lose business.[72] The ultimate compliment was paid earlier than the salesman might have known, when Passaic's 1886 handbook plagiarized an entire introductory section from Carnegie's 1881 edition.[73]

The transformation of the Carnegie handbook was launched by Charles L. Strobel. Strobel entered the Carnegie orbit in 1878 through the Keystone Bridge Company, where he served as engineer and special assistant to the president until 1885. During these years, Strobel originated a column design that used a dual line of rivets to connect four Z-bars to a central plate (fig. 2.7). This so-called Z-bar column soon became a structural signature of the company, which discontinued its two previous column designs. From 1885 to 1893, as noted earlier, Strobel worked closely with Chicago architects as a structural-steel consulting engineer and special agent for the Carnegie concern. After 1893 he built in Chicago several bridges of an innovative bascule design. The most innovative

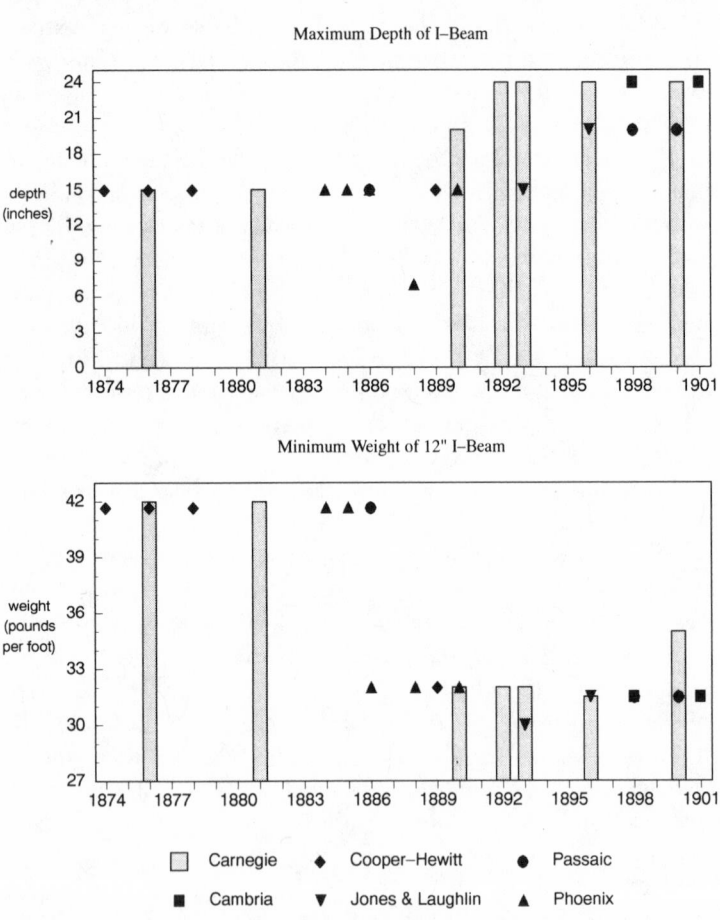

FIG. 2.6. Dimensions of structural beams, 1874–1901. Duplicates not shown are Passaic 1886 at 15″ and Cambria 1898 at 31.5 pounds. Steel I-beams from 1888; American Standard section adopted 1896.

Data from structural-steel handbooks for companies and years shown.

features of the Carnegie handbook appeared under his name.[74]

In 1881 the *Pocket Companion of Useful Information and Tables Appertaining to the Use of Wrought Iron as Manufactured by Carnegie Bros. & Co. Limited, Proprietors Union Iron Mills, Pittsburgh, Pa. for Engineers, Architects and Builders,* to give its full title, showed few signs, beyond its crisp electrotype print, of defining the new field. Its 177 pages began with two-color lithographed sections of the firm's structural products, including I-beams, deck beams, channels, angles, and segments for Keystone octagon columns and J. L. Piper's patented rivetless columns (both soon supplanted by Strobel's Z-bar columns). It continued with tables that explained the structural properties of the sections and gave safe loads, deflections, and proper spacings for beams. It concluded with a series of tables giving such information as shearing and bending values of rivets; decimal equivalents for fractional parts of inches and feet; comparisons of British, French, and American weights and measures; and trigonometry tables.

The feature that distinguished the 1881 edition from those of competing concerns—and which Strobel later expanded—was three pages of "General Notes on Floors and Roofs." Here Strobel went beyond the humdrum technical tables. He described, and illustrated with two dozen figures, the several methods available for connecting floor joists with beams, for connecting beams together, for making girders from beams and plates, and for constructing fireproof floors and ceilings. In the next decade, Strobel's textual notes were expanded and given even more illustrations. For an architect new to structural-steel construction, or for an experienced architect

FIG. 2.7. Z-bar column construction. The Z-bar columns are clearly visible at the bottom. The I- and channel-floor struts are visible at the top. Detail from the Venetian Building, Chicago, 1892.
From Joseph Kendall Freitag, Architectural Engineering *(New York: John Wiley, 1895), 146.*

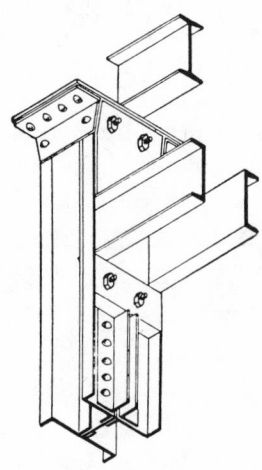

seeking to persuade an uncertain client, here was a handbook valuable beyond all others. Strobel's obituary identified these editions of the Carnegie Handbook as his "most important contribution . . . to Engineering Science."[75]

By 1890 Strobel provided a virtual textbook on structural steel. Besides the general notes on floors and girders, illustrated as before, he now included descriptive or analytical notes on iron roofs, the strength of riveted girders, the fireproofing of iron and steel columns, the strength of wooden pillars and beams (reporting a series of tests at Watertown Arsenal), and the use of steel I-beams in foundations. All of this, including the section profiles, load specifications, and trig tables as before, arrived in a pocket-sized package, 4 by 6½ inches, with a ten-page index for ready reference. The 1893 and 1896 editions, edited by F. H. Kindl, continued the Strobel tradition of information, clarification, and explanation. The 1896 edition added discussions on foundations, windbracing, and cantilevers. The obvious demand for a textbook in the mid-1890s prompted the publisher John Wiley to establish a pioneering series in architectural engineering.[76]

With its wide-ranging technical, promotional, and educational efforts Carnegie Steel simply outmaneuvered its competitors. The result, as Jenney put it, was that "the standard specifications of the Carnegie Steel Company are usually used in the West."[77] The Carnegie concern gained a vastly disproportionate number of structural-steel contracts in Chicago. No fewer than seventeen of the twenty-nine classic Chicago School buildings constructed from 1885 to 1895 used the Carnegie company's steel columns and beams.[78] No other steel company was even close. In the midst of technical uncertainty, Carnegie Steel had seized the moment and deployed (as Sullivan put it) "vision in salesmanship based upon engineering imagination." But innovative as the Carnegie company was in entering the structural trade, after 1895 it found itself playing catch-up with companies that had developed and perfected a new technique for making structural steel.

Technique

Engineers are all specifying for Open Hearth steel. It is impossible to sell Bessemer steel for bridges, boiler plates, ships, or even for these enormous 22-story steel structures which are going up throughout the country. —ANDREW CARNEGIE

Difficulties in obtaining a suitable structural steel first emerged in the mid-1870s as the transcontinental rail boom brought enormous demand for bridges. The first steel railroad bridges—the Chicago & Alton Railroad bridge spanning the Missouri River at Glasgow and the Chicago & Northwestern Railroad bridge at Kinzie Street in Chicago, both completed in the spring of 1879—were experimental structures using Bessemer steel, which the designer of the Kinzie Street bridge called "a rather unsatisfactory material." The Brooklyn Bridge (1878–83) was plagued by guide ropes of brittle Bessemer steel and suspension cables of fraudulent "crucible" steel.[79] Expansion of the production of structural wrought iron, steel's only real rival for structures with tension, faced certain limits, as the Phoenix case illustrated. Again, the gap between rising demand and existing production levels raised the question of whether a new production technique could be developed.

By 1887—the peak year for railroad construction—the problem of relying on Bessemer steel had become acute. "The Bessemer process . . . has also fallen into considerable disrepute for structural purposes," stated one engineer. "Not a few experienced engineers are now stipulating in their specifications that Bessemer steel will not be allowed to be used, especially for tension-members." Charles Strobel of the Carnegie concern argued that such cast steel was "not suitable" for bridges because "its strength is too uncertain." In 1890 two inspecting engineers wrote of "the general tendency among engineers to rule out Bessemer steel for tension-members and all of the more important structural purposes."[80]

From Chicago came a chilling story of a 20-foot-long, 12-inch I-beam that had snapped in two while being unloaded from a railcar. The beam was being shipped to the L. Z. Leiter Building (1889–91), and its architect William Jenney was naturally distressed. In one shipment, Jenney recalled, when teamsters making a delivery to the local fabricating company of Vierling, McDowell had reported an extra steel beam in the shipment, careful inspection showed that one of the long beams had broken squarely in two—solely from the jarring of the car. The beam, from the Carnegie company's Homestead mill, was of Bessemer steel. For its part, the steel company contended that the breakage could be traced to insufficient working in the rolling mill, and it asked Jenney to discontinue the use of these large beams until new rolls were installed at the mill. Others were unsure whether the problem was so simple. "We are at a loss to account for the breakage," wrote the fabricating company. "It shakes our confidence in the whole subject of steel beams."[81]

As the captive of railroads the Bessemer process had served their need for large output, but as it turned out this large output was "one of its inherent dangers." Bessemer mills in the United States could not properly manufacture structural steel for four related reasons. First, many companies had added structural mills alongside their rail mills. To keep both mills constantly busy, managers alternately charged their Bessemer-melting furnaces with pig iron suitable for making rail steel and with iron of "the higher qualities" suitable for structural shapes. When these two streams of metal inevitably got mixed up, the product was excellent rails but inferior structural shapes. Second, workers in the Bessemer casting pits, blooming mills, and bloom yards were "trained to get the steel out of the way" quickly, often regardless of its "precise quality." This also meant that not all defective steel could be identified and rejected. Third, in contrast to the Bessemer converter's 10-minute blow, the open hearth's leisurely pace allowed the operator to examine and test the metal, an "important and well-recognized advantage" in the making of uniformly good steel. An open-hearth heat could take from 4 to 24 hours; at intervals, the melter could dip out small ladles of steel and cast small test ingots. When these ingots were broken with a hammer, a knowledgeable person could identify when and if the steel had reached the desired quality. Finally, the mentality fostered by the Bessemer works made it impossible to manufacture high-grade structural steel there. "Too rapid rolling, too much draught between successive passages through the rolls, and finishing of the rolling at too high a heat are causes of failure in steel, which have proved much more likely to occur at Bessemer mills, where all hands are trained for output and tonnage, than at the open-hearth blooming-mills, where the men are more conservative and slower in their work and are less pushed."[82] The consequences of such reckless mass production continued to haunt the railroad industry into the twentieth century (see chapter 4).

Beginning in the 1880s U.S. steelmakers identified the lack of a suitable structural steel as a significant problem that deserved special attention. Together, the community of mechanics, metallurgists, and managers enacted Alexander Holley's directive that steelmakers must "take the initiative in practicing, perfecting and disseminating the new art of working steel for structural purposes."[83] As they had done with the Bessemer process—indeed, as they did for each major process technology—Americans turned to European precedent.

As Henry Bessemer's famous unmelted "shell" demonstrated, early open-hearth furnaces melted iron but failed to reach heats high enough to melt steel. Less than a decade after Bessemer's first iron experiments, open-hearth furnaces finally achieved high heats through the principle of regeneration. Charles William Siemens, a member of the prolific German family of engineers, first proposed the melting of steel in a regenerative furnace in 1861, and in that same year he erected a regenerative furnace to melt glass at a works in Birmingham, England. His licensees, Pierre and Emile Martin, made steel on a commercial scale three years later. The furnace at their Sireuil Works near Paris, according to Siemens, was "chiefly intended" for heating, but it was "constructed of such material (Dinas brick) and in such a form as to be also applicable for melting steel."[84]

A regenerative open-hearth furnace achieved extremely high temperatures by recycling the heat of spent exhaust gases, much as Holley's heat exchanger had done for the Edgar Thomson Bessemer works. Each regenerative furnace featured two parallel inlet-and-exhaust passages consisting of brick checkerwork, one of which was at any time being heated by the hot exhaust gases (fig. 2.8). When the operator reversed the direction in which the air flowed through the furnace, making the exhaust passage the inlet passage and vice versa, the hot checkerwork heated the incoming fuel and air (separately) to perhaps 1,200°C. When the fuel and air mixed their combustion raised the temperature above 2,000°C, and this very hot flame swept onto the open hearth of the furnace.[85] While a non-regenerative furnace required a worker who stirred, or "puddled," the molten iron, the regenerative open-hearth furnace used the high temperature alone to remove carbon and other impurities from the charge of pig iron, ore, and scrap that filled its hearth.

The early history of open-hearth steelmaking in the United States is littered with false starts, melted furnaces, and sundry explosions.[86] The early open-hearth steelmakers included Bay State Iron Works, the Nashua Iron Company, Singer-Nimick & Company, Otis Iron and Steel Company, and others. According to one observer, these companies "kept fast hold of their original domain of nearly the entire amount of sheared plates, of locomotive tires, of large forgings, spring-steel, agricultural steel, steel castings and the like."[87] This pattern of use persisted until scientific and technological developments linked open-hearth steel to the expanding structural trade. Just as U.S. Bessemer output had surpassed British output because of the unprecedented demand of railroad building, U.S. open-hearth output also exceeded British output because of the unprecedented demand of city building.

Some technological problems of the open-hearth furnace were easily addressed. One early problem was that the interior brick walls sagged because of the extreme heat of the open hearth. This danger disappeared when furnace designers enclosed the brick

hearth in an iron shell and used iron beams to carry its weight to
exterior walls or columns. Other incremental improvements sped
the charging of the furnace with raw materials and the changing of
its linings. Natural resources aided other problems. Siemens had
used an energy-poor artificial gas derived from coal, but the number of open hearths in the Pittsburgh area grew with the discovery
in the region of energy-rich natural gas.[88]

The principal scientific effort focused on the furnace linings and
the insoluble slags that they produced by chemically reacting with
the molten metal.[89] The open-hearth pioneers had used silica bricks
composed of ground quartz (silicon dioxide) and 3 percent or less
of lime (calcium oxide); the resulting silica-based slags were chemically acidic and so promoted certain sets of chemical reactions. The

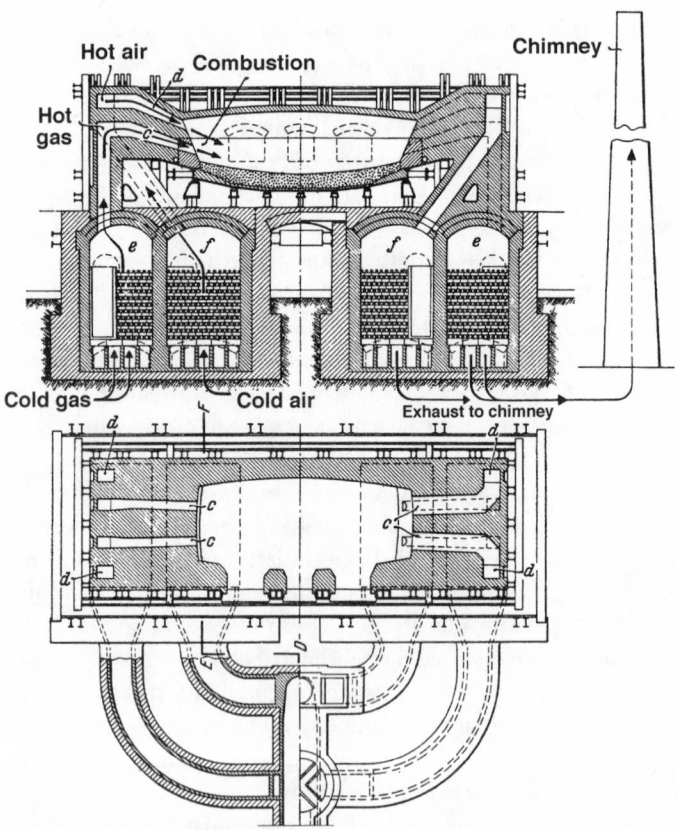

FIG. 2.8. Regenerative open-hearth furnace.
From Verein Deutscher Eisenhüttenleute, Gemeinfassliche Darstellung des Eisen-
hüttenwesens *(Düsseldorf: Verlag Stahleisen, 1949), 101–2.*

most important chemical consequence was that phosphorus stayed in the molten metal. In contrast, phosphorus was trapped and removed with a chemically basic process using dolomite or magnesite linings. The basic process was devised by the Welsh ironworks chemist Percy Gilchrist and his cousin Sidney Gilchrist Thomas, an amateur chemist and police court clerk. Because phosphorus made steel brittle, the development of the basic process was a major step in banishing the specter of broken Bessemer beams. The story of the Thomas-Gilchrist process has been told many times.[90] Less well known are the similar findings of Jacob Reese, a Welsh-born Pittsburgh ironmaster, accomplished engineer, and prolific inventor whose "basic" process patents were first purchased in 1879 by Andrew Carnegie. Two years later, after a patent interference case, Reese's patents were granted priority in the United States over those of Thomas and two other claimants. As Reese put it, he invented "the essential conditions" for the basic process. Like all good steel patents, those of Thomas and Reese were purchased by the Bessemer Association. After interminable litigation, Reese collected royalties until 1906 from the Bessemer Steel Company and its successor, the Steel Patents Company.[91]

The institution of yet another patent pool did not, of course, ensure the rapid dissemination of the technology. The license fee was $30,000, a figure beyond the reach of all but the largest companies. Further, the patent pool did not provide much help in locating the firebricks necessary to contain the white heat of the metal while promoting the proper basic chemistry. Dolomite bricks consisting mainly of lime (calcium oxide) and magnesia (magnesium oxide) were cheap and plentiful in the United States, while the preferred magnesite bricks, consisting of 80–90 percent magnesia, remained expensive imports. Domestic dolomite varied in chemical composition, and as one writer put it, "bricks made from it do not stand the weather well." Nor did they stand the furnace well. In the furnace magnesite lasted three times as long as dolomite.[92]

Thus a variety of problems challenged those who adopted the basic process. As early as 1879 Alexander Holley reported to the Bessemer Association on Thomas and Gilchrist's basic process experiments. Nonetheless, for almost ten more years American steelmakers simply exploited their access to inexpensive low-phosphorus ores that did not contain enough phosphorus for the basic process to work. Finally, in 1886 Samuel T. Wellman became the first steelmaker to employ the basic open hearth in the United States. After seeing the basic process in Europe, Wellman imported Styrian magnesite to line an experimental furnace at the

Otis Iron and Steel works in Cleveland. Wellman sent basic steel to the company's regular customers without event for four months, until pressure from the sales department to increase output forced him to return to the acid process. Other companies that erected basic open-hearth furnaces in the wake of this aborted trial included the Pennsylvania Steel Company (Steelton), the Cleveland Rolling Mill (Newburg, Ohio), the Shoenberger firm (Pittsburgh), and Carnegie, Phipps & Company (Homestead).[93] An unusually well-documented startup effort at Pennsylvania Steel illustrates not only the mechanical difficulties of working at extreme heats but also the critical role of science.

Pennsylvania Steel's experiments with the basic open-hearth process in 1889 were under the charge of Harry H. Campbell. After graduating from MIT in 1879, Campbell joined the company's main plant at Steelton, where he served in the Bessemer and open-hearth departments. In 1884, after the company had abandoned its brief experiment with the basic Bessemer process, he was named foreman of a new open-hearth department. As open-hearth superintendent, Campbell reported on 22 February 1889 to the general manager of the Steelton plant, Frederick W. Wood, concerning the difficulties he was having with the new process. Extreme heat had already ruined an arch of bricks protecting the roof of the melting chamber. "The heating up [after the arch was replaced] was carried on slowly and no trouble showed itself until we reached an extreme temperature; then the old trouble began; the joints, which up to this time had been tight began to open and the bricks which had been smooth and intact began to break and crumble, about two inches coming off at a time." Another problem was "setting" the magnesite firebricks. "The stuff remains slightly plastic and entirely unset at extreme temperatures," he observed, and when heated "the stuff" softened dangerously. "If our roof entirely fails us I shall put in a silica [acid] roof and magnesite [basic] bottom," Campbell wrote, identifying a popular solution. "If on this, we fail to set magnesite we can try dolomite."[94]

To guide his experiments Campbell turned to two recent papers that had just been translated by a colleague from the Swedish journal *Jern-Kontorets Annaler.*

> They advise that the [magnesite] bricks be laid very close—
> one works actually grinding the bricks together in the laying;
> the bricks are put in to the whole size of the bottom so that no
> bottom need be built of fine stuff "for this building requires
> hard work and careful watching." The fine stuff is put on in thin
> layers about half inch (in the hot furnace) and beaten down with

iron flails weighing 65 to 130 pounds—great stress being laid on this particular weight. It is described as very hard work.

In other papers they talk of filling holes in the bottom with a mixture of the magnesite with ore or sand—but in the description of building the bottom it is stated that unmixed magnesite is used—much stress seems to be laid on *dead burned* stuff.

At Campbell's urging, Superintendent Edgar C. Felton directed Wood, then traveling in England, "to find out what you can in relation to the use of these bricks."[95]

A week later Pennsylvania Steel made its first heat of basic open-hearth steel. Campbell replaced the basic magnesite roof with acid silica, and on 1 March when the furnace reached a melting heat, "everything stood up well." Four days later Felton wrote Wood: "We got the first heat out of Basic O.H. last night at about 7 o'clock and another today at 2.30 P.M. Blown metal was taken from [a Bessemer converter] in both cases and was in furnace 3½ to 4 hours. No attempt was made to work steel down very low [in carbon content] and no difficulty was experienced. Bottom is good and roof O.K. Have not had time to determine P [phosphorus] but test shows a very soft and ductile material." By 1895, on the strength of its success with open-hearth steel, Pennsylvania Steel diversified from Bessemer steel rails, which it had made for nearly three decades, by manufacturing cast-steel frogs and switches for railroad use as well as structural steel for its own bridge and construction department.[96]

Beyond this commercial success, Pennsylvania Steel's open-hearth effort also yielded an advance in scientific understanding. Campbell wrote a series of papers in the 1890s that demonstrated a deep understanding of the scientific principles of the open-hearth furnace. "The methods employed necessarily enter into the domain of what is called theory," Campbell explained, "but the results are eminently practical."[97] He distilled the operations of the open-hearth furnace at Steelton into a series of "universal laws." In 166 closely reasoned pages Campbell comprehensively analyzed open-hearth practice, including the economics of the open-hearth furnace, the regenerative furnace and its machinery, the various fuels, the regulation of temperature, the overall thermal equation of the furnace, and the chemistry of the acid and basic processes.

Specifically, Campbell demonstrated an understanding of equilibrium theory when he explained the chemical reactions that permitted the basic (but not the acidic) slag to remove phosphorus. For the basic process, as Campbell outlined the chemistry, "the silica finds bases in the lime, manganese and iron, or takes them from the hearth; the manganese burns to MnO and unites with the silica; the phosphorus changes to P_2O_5 and combines with the lime." The overall product, as Campbell recognized, resulted from "preferential relations." The acid process could not remove phosphorus because the high levels of silica reacted immediately with any basic elements in the slag; the intermediate form of phosphorus, phosphoric anhydride, could "obtain no foothold in the slag." In the basic process, on the other hand, phosphoric anhydride combined with the excess of lime to form phosphate of lime, a stable compound that stayed in the slag. "Thus the history of phosphorus, which metallurgists of the last generation could not write," concluded Campbell, "appears to be the record of a simple preferential relation."[98]

Finally, Campbell suggested the practical results of such scientific reasoning. To promote the removal of phosphorus, for example, he suggested including iron ore in the furnace charge to favor the oxidative reactions, adding sufficient lime to hold any P_2O_5 that formed and to prevent scoring of the hearth, and promoting a slag with high levels of iron oxide and low levels of silica. While it is difficult to prove that such a scientific analysis informed the practice of other steelmakers, the praise heaped on Campbell's papers ("by far the most valuable work on the open-hearth process that has ever been written") indicates that a sizable community appreciated his efforts.[99] Moreover, Campbell's book based on this work was titled, significantly, *Manufacture and Properties of Structural Steel* (1896), and it enjoyed at least four editions to 1907.

The connection between the basic open-hearth and structural steel, spotlighted in Campbell's book, oddly enough came late to the Carnegie concern. Although Andrew Carnegie had instantly relayed a siren call to Pittsburgh upon the announcement of the basic process in England in 1879, had personally steered the Thomas and Reese patents to the Steel Patents Company, and had promptly installed at the Homestead mill in 1888 one of the earliest basic open-hearth furnaces, he nevertheless failed to wean his principal structural mill from Bessemer steel for more than a decade.[100] The Homestead mill's broken beam for the Leiter Building in Chicago should have alerted the company that something was amiss; increasingly over the 1890s Phoenix and many other firms built basic open hearths for structural output. Toward the end of the decade, Carnegie saw the writing on the wall. "Engineers are all specifying for Open Hearth steel," he wrote in 1896. "It is impossible to sell Bessemer steel for bridges, boiler plates, ships, or even for these enormous 22-story steel structures which are going up throughout the country." In 1897 his board of managers learned from Charles Schwab that "we have the greatest difficulty in getting our customers to take Bessemer structural steel." Schwab indicated that fully 80 percent of the beams manufactured at Homestead were of Bessemer steel, while Illinois Steel, its principal competitor, used open-hearth steel to decided advantage. Worse, the Homestead Bessemer plant had just been rebuilt as "a modern one in every respect." To rescue its investment, the company diverted $375,000 from its $1 million plan for adding more open-hearth furnaces and instead renovated two rolling mills to absorb Homestead's Bessemer output. Funds remained for only ten new open hearths to make structural shapes. (The company's twenty existing open hearths were mostly used for armor plate, as related in chapter 3.) Compared with the heady days of the 1880s, when no technical change that promised to cut costs was overlooked, Carnegie's philosophy of innovation had undergone a sea change. "Nothing requires more conservatism than the manufacture of steel," Carnegie now declared.[101]

With the Carnegie concern lagging rather than leading, the production of basic open-hearth steel expanded handsomely. By 1900 the American Society for Testing Materials (ASTM) permitted *only* open-hearth steel for the structural shapes for bridges and ships. Iron and steel had contributed equally to structural shapes in 1889, but in 1909 there was one hundred times as much structural steel as structural iron, and twelve times as much basic open-hearth steel as acid open-hearth steel (table 2.1). Basic open-hearth structural steel was henceforth the material of choice for bridges, buildings, ships, and subways in the United States.[102]

Building Cities of Steel

The skyscraper is the most distinctively American thing in the world. —COLONEL W. A. STARRETT

After the rail market stagnated in the late 1880s, the mass production of steel in the United States became predicated on the mass consumption of steel for urban structures. To be sure, aggregate demand was increased by applications like agricultural machinery, itself the product of railroad-based agriculture in the Great Plains, and the entire metal infrastructure of a manufacturing economy. But the force that changed steelmakers' relations to users, and brought corresponding changes in the industry's technical abilities and scientific base, was the rapid growth of cities and the unprecedented demand for steel for urban structures. In this respect, the

TABLE 2.1. Production of open-hearth and structural steel, 1879–1909

	Open-hearth Ingots and Castings (weights in thousand gross tons)			Structural Shapes (weights in thousand gross tons)		
Year	Basic	Acid	% Basic	Steel	Iron	% Steel
1879				1	87	1.1
1886[a]				56	178	24.1
1889[b]				154	123	55.6
1899	2,080	867	68.5	830	27	96.8
1905				1,649	12	99.3
1909	13,418	1,076	92.6	2,102	21	99.0

Sources: AISA Annual Statistical Report (1886 and 1906); Peter Temin, Iron and Steel in Nineteenth-Century America (Cambridge: MIT Press, 1964), 270–71; Temin, "The Composition of Iron and Steel Products, 1869–1909," Journal of Economic History 23 (1963): 451.

[a] Structural shapes for 1886 are estimated.
[b] Total acid + basic for 1889 = 375.

rise of basic open-hearth steelmaking in the United States was ultimately a result of the nation's rapid urbanization.

"Construction and steel production are inseparably linked," stated the builder, Colonel W. A. Starrett. "The demands of the one furnished the incentive for the colossal scale upon which the other has been developed."[103] This was so because the city-building boom in the years around 1890 occurred just when the transcontinental railroad boom slackened. Yet as we have seen steelmakers found that Bessemer rail mills did not produce a suitable structural steel. Urban structures required a new steel, more ductile and above all more regular, consistent, and predictable than Bessemer rail steel. Beginning in the 1870s and accelerating in the 1880s, steelmakers collaborated with architects, builders, and scientific metallurgists to develop such a steel. This effort yielded a characteristic product: in the United States, if not elsewhere, the best structural steel was open-hearth steel made by the basic process.

The tall office building itself was a characteristic product of these changes. Americans accustomed to their own high-rise cities were surprised at how "short" other major cities were. From London in 1888, Dankmar Adler reported on the "rarity" of buildings over five stories in height.[104] A multiplicity of causes—ranging from new technologies, such as elevators and fireproofing, to shifts in business, financial, and demographic developments—combined to produce the distinctive form of the American city.

The impact of elevators on a building's rental profile suggests they were a precondition for tall buildings, a necessary if not sufficient cause. Without elevators there was little economic incentive to raise a building over five stories; desirable tenants simply would not pay to walk up so many stairs. A positive incentive to raise tall buildings, extracting the maximum rent from each plot of ground, appeared after Elisha Graves Otis introduced the passenger elevator in the five-story Haughwout Building in New York (1857) and then perfected a safety catch for electric hoists. So-called elevator buildings became commonplace in New York during the 1860s and in Chicago a decade later.[105]

Turning to the demand side, it is no coincidence that the tall office building came to maturity in the late nineteenth century. These same years witnessed unprecedented growth in new activities that demanded office space in urban commercial districts. "The arrival of the 'office building' was . . . an indication of and a contributing cause to basic changes then taking place in the internal organization and structure of commerce, industry and finance. An unduly large portion of our great combinations came out of offices in the skyscrapers of the 1880s," observed one Chicago builder. Indeed, skyscrapers were erected for the superstructure of industry: not meat packing or steelmaking but publishing, financing, retailing, and insuring (fig. 2.9). Life insurance companies especially

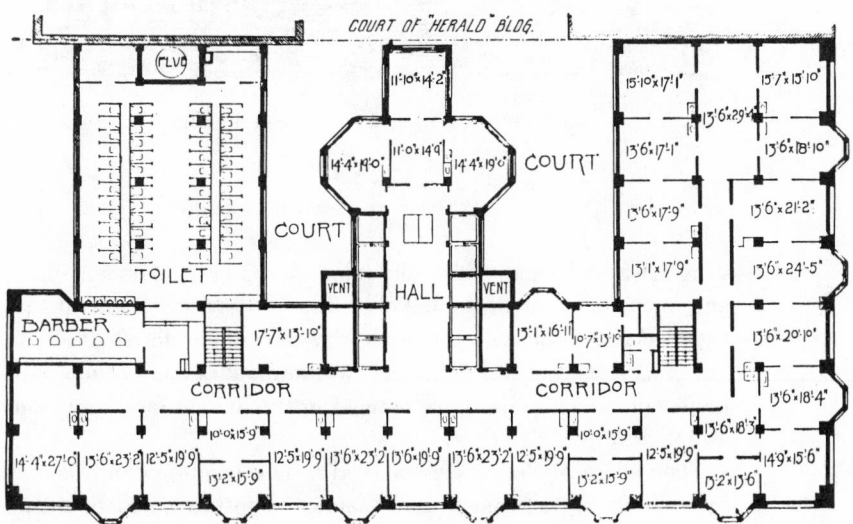

FIG. 2.9. Floor plan of Chicago Stock Exchange Building, 1893–94.
From Freitag, Architectural Engineering, 29.

FIG. 2.10. Entrance of New York Life Insurance Building, Chicago,
1894.
From Freitag, Architectural Engineering, *167.*

were flush with capital, and they, along with newspapers, retailers,
and financiers, saw in the tall office block a sound investment that
signaled prosperity and dignity to the community (fig. 2.10). "The
tall office building represents the working out of one of the dominant tendencies in the development of the human race," asserted
architect Adler.[106]

Technological changes also served as a precondition for this new
urban form by altering the economics of insuring and financing
tall office buildings. Fires and railroads were the key issues. The
so-called fireproof iron buildings that emerged in the 1850s were
in fact firetraps. Fires in several major cities, including the Chi-

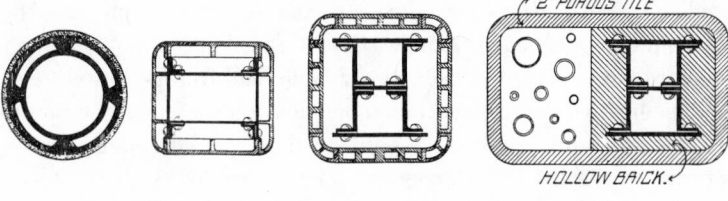

FIG. 2.11. Fireproofing of metal-framed skyscrapers, c. 1895.
From Freitag, Architectural Engineering, *134.*

cago fire of 1871, simply melted the exposed cast-iron structure of such buildings, which then collapsed suddenly and with a cascade of molten metal. Placing the iron columns inside brick piers, as Jenney had done with the Home Insurance Building, alleviated this problem; yet as late as 1894 insurance companies absorbed fire losses totaling $81 million—more than the value of the nation's production of pig iron. These unacceptable risks were cut dramatically by covering the load-bearing columns of buildings with casings of brick, hollow tile, or terra-cotta (fig. 2.11). Advocates of these fireproofing techniques pointed to the success of the ten-story Chicago Athletic Club against a disastrous fire during its construction in November 1892. Furthermore, working in concert with the reduced risk of fire losses, the falling yields on railroad bonds made investments in office buildings economically more attractive.[107]

The most impressive economic advantage of steel-skeleton buildings was how little time was needed to raise them. "Chicago is always *short* on time," Jenney explained to the National Brick Manufacturers Association. "We can not afford to put a lot of stonecutters at work picking away at big blocks of granite and spending two or three years in the erection of a building." With steel-skeleton construction it took only twelve months from turning out the old tenants and destroying the old building to installing new tenants in a new building. This speed of construction was easily appreciated by building owners, who faced heavy carrying charges, including insurance, taxes, and interest on millions of dollars, whether or not tenants paid them rent. Some sort of record was set in 1891 when, for Ernest J. Lehmann's Fair Store, Jenney put a space 95 by 180 feet into service within three and a half months of starting construction.[108]

Much of this speed was a result of the fact that the steel skeleton permitted the external covering of brick or terra-cotta on upper stories to be placed before the lower, and inevitably more ornamental, stories of granite or marble. By contrast, the load-bearing walls of New-York-style cage construction had to be built from the bottom up. In Chicago, Jenney's New York Life Building was under construction during the 1893 World's Fair, and European architects and engineers were at a certain moment astonished by a completed twelve-story steel skeleton, whose fourth, fifth, and uppermost two stories were clad with terra-cotta, but whose lower three stories and sixth through tenth stories were stunningly open. Work had proceeded on the terra-cotta for the upper stories and cornice even though the granite for the first three floors had not yet arrived. Construction photographs of Chicago-style buildings—including Adler & Sullivan's Guaranty Building in Buffalo as well as Daniel Burnham's Flatiron Building in New York—confirm the point (see the frontispiece to this chapter). Steel-skeleton construction avoided the logistical bottleneck of putting an entire building on hold because the fancy stonework for the lower floors was not ready.[109]

In addition to reducing downside carrying charges, the speed of construction also reinforced the increasing density and intensity of modern cities, a product of the technologies of skyscrapers and transit systems. Elevated railways, even more than the earlier trolleys, permanently extended the city by transforming its central residential areas into business (or warehouse) districts and turning its outlying areas into suburbs. Subways accelerated travel while sparing valuable air space. The steel-framed skyscrapers that piled thousands of office workers in huge blocks of standardized cubicles augmented these transit developments. In Manhattan the building frenzy reached its peak in 1899, just before construction on the subway system began in 1900; the next year came the magnificent Flatiron Building. Disentangling cause from effect in this rush of growth is futile, for the forces that drove cities upward and outward were interactive and systemic. Perhaps the result could have been different, for London, Paris, Vienna, and Berlin built outward but not upward. Nevertheless, by the first years of the twentieth century New York and Chicago would have been inconceivable without steel-lined transport systems and steel-framed buildings. They were the world's first cities of steel.

These developments were momentous for steelmakers. Success in structural steel required the ability to form new user-producer relationships, a necessary asset for technological innovation. Low prices in themselves were inconsequential. Price competition was rendered inoperative by the cartel active during most of this formative period, and high quality and timely delivery were the more pressing criteria for builders striving to raise tall office buildings and install paying tenants. In addition to contracts, specialized technical consulting and published technical handbooks helped create user-producer relationships. Companies entering the structural sector after 1900 faced established competitors—and slackening demand. Compared with the boom of the 1880s and 1890s, new building in Chicago grew very little in the following two decades, while new building volumes in Manhattan actually shrank; strong demand returned only with the skyscraper boom of the 1920s.[110]

Eighteen-inch-thick steel armor for the new navy's battleships, c. 1910: turret for 14-inch guns on battleship USS *Pennsylvania*, made at Bethlehem Steel.

Courtesy Bethlehem Steel Collection, Hugh Moore Historical Park and Museums, Easton, Pa.

CHAPTER THREE

The Politics of Armor

1885-1915

At noon on 31 May 1916, Admiral Horace Hood felt a familiar thrill and suspense, for his squadron of armored battle
cruisers was about to see action. A year ago at age forty-five, a dash-
ing figure, Hood had taken command of the Third Battle Cruiser
Squadron, a reflection of his impeccable Royal Navy lineage, his
battle experience in the Sudan (1898) and Somaliland (1903), and
his service as naval secretary to First Lord of the Admiralty Winston
Churchill (1911–15). Within six months Hood's squadron had sunk
four unlucky German commerce raiders in the Falkland Islands.
But that was small indeed in comparison to what they now awaited.
A creditable performance might launch him into the top echelon
of the Admiralty. After all, Hood had seen that Admiral Jellicoe,
commander-in-chief of the Grand Fleet, was at fifty-six years of age
mentally alert but physically drawn.

Shortly after midnight that morning, on intelligence obtained
from the compromised German code system, the warships of the
Grand Fleet had departed their harbors along the coast of Scotland to ready a surprise for the German High Seas Fleet steaming
out from its base at Wilhelmshaven. Hood knew that the Germans
were outnumbered and outgunned and could be outrun. If their
warships could be maneuvered into the terrible jaw of the Grand
Fleet's gunfire—the task of Admiral Beatty's battle-cruiser squadron, to the south of Jellicoe and Hood—the German Navy might
be demolished. Such a decisive victory, another Trafalgar, could
end the war in a matter of weeks.

. Hood's confidence rested ultimately on the principal vessels
under his command. *Invincible*, *Indomitable*, and *Inflexible* all resembled H.M.S. *Dreadnought*. At its launching in 1906 *Dread-*

nought had stunned the naval world with its astonishing speed and impressive firepower. All battleships worthy of the name had to follow in its likeness. Like *Dreadnought, Invincible* sported compact and powerful steam turbines; those of *Invincible* pumped out 45,000 horsepower and were the most powerful ever installed afloat. At full steam *Invincible* and its fellow ships could reach 28 knots, four knots faster even than *Dreadnought*. Hood's three warships featured eight 12-inch guns, whose effective range had been extended beyond 16,000 yards by centralized fire control and lengthy target practice. The latest 15-inch guns could score hits at 20,000 yards, nearly 4 miles. At ranges beyond the effective reach of torpedoes, then, the new vessels could drop high explosive, armor-piercing shells onto predreadnought vessels, whose 9-inch guns could not respond and whose 22 knots of speed would not allow them to escape the lethal danger. Indeed, gun turret crews could fire and reload in about twenty seconds—less time than it took for the shell to complete its airborne journey. Reloading was even quicker when the antiflash devices that hampered rapid vertical movement of shells from the magazines below were thrown aside. Hours of gunnery practice in which firing speed counted more than safety transformed this risky gambit into second nature.

While the two fleets were steaming toward one another, the architect of the dreadnought revolution was fuming on the sidelines. A decade earlier, Sir Jackie Fisher as first sea lord of the Admiralty (1904–15) had shrewdly assessed the generation's technical changes in steam, steel, and weapons, and how to integrate them into a coherent fighting system.[1] Modern offensive battleships, he concluded, should have guns—and speed. Achieving maximum speed with sufficient firepower meant cutting all other weight to the bone. Given the need to steam to all corners of Britain's global empire, ample space for crew and fuel was essential. The largest savings came from trimming the excessive thickness of armor, up to 22 inches thick, whose weight had consumed nearly half the displacement of a predreadnought warship and whose cost had drained government accounts. Such a reduction was feasible with a special surface-hardened armor perfected in the decade before the launching of *Dreadnought*. Those fantastic thicknesses of armor were no longer needed because the glass-hard face of the new armor was supposed to burst incoming shells harmlessly on contact. None of the great powers ignored this dramatic change in technology. For its part *Dreadnought* used 5,000 tons of the hardened armor to make room for some 7,000 tons of weapons, machinery, and fuel in its overall displacement of 18,000 tons. A similar logic shaped

Invincible's design at 17,250 tons.[2] Its four heavy gun turrets were protected by 10 inches of armor plate, the sides with 7 inches, the deck with a maximum of 2 inches.[3] Behind 10 inches of armor in his flagship's conning tower, Admiral Hood could reasonably believe himself to be invincible. He was not.

The Battle of Jutland began at about 3:45 P.M. when Beatty's cruisers closed to within 15,000 yards of their German counterparts. Within 20 minutes the Germans inflicted heavy losses. Beatty's flagship was hit by nine heavy shells, silencing its radio for several hours. More devastating, shortly after 4 P.M. the *von der Tann*, the oldest German dreadnought, trained its guns with terrible effect on *Indefatigable* at 17,500 yards. Two or three 11-inch shells penetrated the 7 inches of barbette armor beneath its X turret, then two more smashed into the A turret and forecastle. Although distant observers saw little immediate result, these shells caused havoc. They set off the turrets' complement of explosive cordite, which in turn blasted flame straight down into the magazine. Half a minute after *Indefatigable* was hit, a sheet of flame burst from the hull, followed by a dreadful explosion. Already listing to port, the ship capsized less than 17 minutes after the opening shots of the battle.[4]

"The way a ship seemed to blow up if you were far off her" fascinated a lieutenant interviewed later by Rudyard Kipling. "You'd see a glare, then a blaze, and then the smoke—miles high, lifting quite slowly. Then you'd get the row and the jar of it—just like bumping over submarines. Then, a long while after p'raps, you run through a regular rain of bits of burnt paper coming down on the decks—like showers of volcanic ash, you know." Others observed that only the explosion of an entire magazine, as in the case of *Indefatigable*, reached above the "ear-splitting, stupefying din" of battle. "It wasn't exactly noise," the lieutenant reflected. "Noise is what you take in from outside. This was *inside* you. It seemed to lift you right out of everything."[5]

The initial engagement between the battle cruisers quieted at about 4:30 P.M., and it had not gone well for either side. Besides *Indefatigable*, the *Queen Mary* had been rent by two internal explosions. Three German cruisers plus the battle cruiser *Seydlitz* were crippled. The full engagement between the great battle fleets themselves commenced at about 6 P.M. when Jellicoe's six columns of ships steaming southeast turned port to form a battle line against Scheer's ships steaming northward. Jellicoe momentarily gained an upper hand. Shot after shot hit home against the German warships, wrecking the cruiser *Wiesbaden* among others, and Hood on *Invin-*

cible's bridge called up to the gunnery officer directing fire from the forward topmast to "keep at it as quickly as you can." *Invincible* emerged briefly from a protective bank of mist, however, and the *Lützow* and *Derfflinger* trained deadly fire upon it at 9,500 yards. At 6:33 a full salvo, perhaps six shells, hit *Invincible* amidships. One shell pierced the Q turret, burst inside, and blew off its roof. Promptly this time the magazine below exploded, blowing the vessel in two. When the tremendous cloud of black smoke and debris cleared, observers saw both bow and stern projecting from the water, having come to rest on the shallow floor of the North Sea. The gunnery officer and five others lived to tell the harrowing tale; 1,026 men, including Hood, lost their lives. Wrote Corbett, "The mother of all battle cruisers had gone to join the other two that were no more."[6]

At nearly the moment *Invincible* sank, about 6:36 P.M., Scheer ordered a stunning "battle turn to starboard" that abruptly reversed his fleet's course. It was not a retreat, he later maintained. "While the battle is progressing," he wrote, "a leader cannot obtain a really clear picture, especially at long ranges. He acts and feels according to his impressions."[7] Since his impressions were of the "splendid effectiveness" of his fleet's gunfire, Scheer intended the battle turn to regroup his vessels for night battle. In the event, however, the turn brought Scheer's fleet into the cover of a fog bank and, despite one more brief and violent engagement of the fleets, toward the safety of the Danish coast. The ensuing night battle was madness: each navy proved more adept at damaging its own vessels through inadvertent collisions than at hitting the enemy. By the first light of dawn, twelve hours after hostilities had begun, the German High Seas Fleet was in safe waters. The Great Fleet returned to home port in Scotland.

The myth that defective armor was responsible for the poor British showing probably originated with Admiral Beatty. To him the indecisive and frustrating results were clear by the afternoon of 1 June. "There is something wrong with our ships," Beatty confided to his flagship's captain. "And something wrong with our system." Amplifying these remarks, Jellicoe wrote of the urgent need for "increased deck armour protection in large ships."[8] In contrast to the British vessels' propensity to blow up, German vessels, such as the shell-ridden cruiser *Wiesbaden,* sank without explosion. Although at least twenty-four heavy shells hit the capital ship *Lützow,* it had to be sunk intentionally by a German torpedo.

Ever since Jutland, experts have vigorously debated the relative merits of British and German armor.[9] The most thorough study of the battle presents data that are tantalizingly detailed but ultimately inconclusive without data on the critical damage to vessels that sank.[10] Most notable is the disparity in gun sizes and armor thicknesses. While armor on German warships measured up to 14 inches, armor on British warships, *Barham* excepted, was 9 inches or less. On the other hand, British 15-inch shells scored hits at nearly 20,000 yards, while German 12-inch shells had a maximum range of 17,000 yards. Based on these figures, Jellicoe's initial engagement of the High Seas Fleet—his actual range when he turned to form a battle line was about 15,000 yards, though he believed it to be longer—gave away the advantage of his long-range gunnery and exposed his thinly clad ships to German firepower. This miscalculation nullified the technical revolution ushered in by *Dreadnought* and *Invincible*.

Not defective armor but differences in ship and shell design, and especially in gun turret procedures, account for the differences between German and British losses.[11] Armor is not at issue, for German and British warships were protected by essentially the same armor. For this reason, in fact, American naval officers had good reason to scrutinize the Battle of Jutland. Not only was this the principal naval engagement of the entire war, the first and last battle between the great dreadnought battleships, it also afforded the best possible evidence concerning the potential performance of American armor. For the armor plate protecting the battleships of all the great powers was the product of a remarkable international technological convergence in the years before the war. The naval rearmament effort in the United States yielded the hardened armor that made the dreadnought revolution possible. In the years before World War I, the politics of armor transformed domestic steelmakers as well as altering their relations with government at home and their counterparts abroad.

Resurgence and Innovation

Railroads too poor . . . general trade may be better but there's no profit in that. Good thing we are to get more Armor this year.—ANDREW CARNEGIE

The year 1890, writes Walter Herrick, "marked the onset of a revolution in doctrine which transformed the United States Navy from a loosely organized array of small coast defenders and light cruisers into a unified battle fleet of offensive capability." Events such as the founding of the Naval War College (1885), the election of Benjamin Harrison as president (1888), his appointment of the activist Benjamin Franklin Tracy as navy secretary, and the publication of Alfred Thayer Mahan's *Influence of Sea Power upon History* (1890) capped two decades of naval resurgence and pointed to supreme opportunities for American steelmakers.[12]

Following the Civil War the navy remained moribund until a constituency for innovation overcame budgetary constraint and bureaucratic inertia. The war had actually served to deter change because the majority of sea battles—including that between the ironclads *Monitor* and *Merrimack*—reinforced the traditional view that blockade running and coastal defense by light vessels were the essence of naval strategy. Also, at a time when many European navies were switching to rifled guns made of wrought iron or steel, the U.S. Navy remained loyal to smooth-bore, cast-iron guns.[13] The impetus for innovation began with an energetic group of technically minded naval officers in the navy ordnance bureau. These men helped catalyze heavy industry just as army ordnance officers had done for light arms manufacture a generation earlier. In the course of several ambitious projects, ordnance bureau chief William N. Jeffers (1873–81) and his assistants found that the new rifled, breechloading guns favored by European navies required steel that American makers were unable to produce. Determined to resist imports, Jeffers was a persistent advocate of spending "our money encouraging our own steel makers," and to this end he worked closely with the Nashua, Naylor, and Midvale steel companies. By 1880 Midvale Steel could forge and oil-temper steel tubes suitable for 6-inch guns, and it became the nation's leading producer of ordnance steel.[14] At the same time Midvale supported young Frederick W. Taylor's first experiments in the art of cutting metals (see chapter 5).

Navy leadership and congressional action brought closer relations between the navy and steelmakers in the mid-1880s. During his tenure as ordnance bureau chief (1881–90) Montgomery Sicard expanded the procurement of steel guns and began building a modern naval gun factory at the Washington Navy Yard. During 1882–83, an influential board chaired by Commodore Robert W. Schufeldt planned two steel warships, while Congress authorized $1.3 million for the effort; the funding bill also forcibly retired the navy's wooden ships by limiting total repairs to 20 percent of the cost of a comparable new vessel. Congressional appropriations ratified steel as the preferred material for warships, including two cruisers in 1885, two battleships and a cruiser in 1886, two cruisers in 1887, and seven cruisers and a gunboat in 1888.[15]

For this unprecedented effort, the government devised several means to build up the nation's ability to produce high-quality steel. For manufacturing heavy forgings, the Gun Foundry Board recommended and Congress authorized in August 1886 an arrangement modeled on the French ordnance establishment: the navy would build an ordnance factory but purchase unfinished steel from domestic suppliers at attractively high prices. In May 1887 construction began on the Washington Navy Yard's Naval Gun Factory, and the navy soon lined up firms for the steel contract. Remarkably, not the established Midvale but Bethlehem, hitherto an iron rail maker, soon secured contracts exceeding $4 million.[16] Behind this accomplishment lay shrewd entrepreneurship and political savvy—characteristics of paramount importance in gaining lucrative government contracts.

The Bethlehem Iron Company had shown few signs of technological progressiveness since its organization in 1861. Like Pennsylvania Steel, Bethlehem was for many years the captive of a railroad. The company's organizer and primary customer was the Lehigh Valley Railroad, with which it shared directors and managers including Robert H. Sayre, who served the railroad as chief engineer and the iron company as general manager and director. Bethlehem erected its first blast furnace and puddling mill in the remote community of South Bethlehem in the Lehigh valley in northeastern Pennsylvania early in 1863, and in September of that year the rolling mill turned out its first batch of iron rails. In 1873 the company joined the Bessemer Steel Association and in October became the tenth U.S. concern to produce Bessemer steel, also rolled into rails. The move proved to be ill timed, however, as it coincided with the 1873 panic and the ensuing recession. When business improved in the late 1870s Bethlehem's directors, led by the imperious railroad man Asa Packer, reaffirmed their policy of making rails and nothing else.[17]

The directors' conservatism irritated John Fritz, whom Frederick Taylor recalled as "one of the greatest and one of the last" of the "school of empirical engineers." Best known as the inventor of the three-high rail mill, whose labor-saving potential was first revealed at Cambria Iron in the late 1850s, Fritz was recruited in 1860 by Sayre to come to South Bethlehem as general superintendent and chief engineer. In the next decade he worked closely with Alexander Holley to perfect the Bessemer process for rail manufacture. By the early 1880s Fritz became convinced that Bethlehem needed to cut its dependency on the rail market. The declining profits on rails—the result of a flood of cheap rails from Carnegie's Edgar Thomson mill and its counterparts—made a decision on diversification critical. He formulated plans for manufacturing structural shapes, plates for ships, and heavy forgings, but company directors rejected every one of them.[18]

Fritz gained a powerful ally in Joseph Wharton. A member of a distinguished Philadelphia Quaker mercantile family, Wharton made a fortune in zinc and nickel, and in 1881 founded the Wharton School of Finance and Economy in the University of Pennsylvania. Beginning in the 1860s he had invested profits from his mineral ventures in a growing block of Bethlehem Iron stock. In 1879 the Lehigh Valley Railroad's stranglehold on the iron company was loosened with the death of Asa Packer and then broken by the ineffective management of Packer's two sons. In 1885, with the passing of the last of Bethlehem's old-guard directors, Wharton gained effective control of the iron company and squarely faced its shrinking rail profits by taking steps to diversify. Rather than compete with Pittsburgh's high-volume rail mills, Wharton sided with Fritz and took Bethlehem out of rails altogether. Anticipating the navy's first large ordnance contract, Wharton personally conducted a series of negotiations at home and abroad that brought his company a state-of-the-art armor and heavy-forging plant. This effectively nullified competition from Midvale or any other potential rival in heavy-ordnance steel.[19]

Wharton's effort at technology transfer was expedited by the navy's policy of placing ordnance officers with European ordnance companies. With this policy the navy sought to retain career officers in a time of budgetary constraint (they were officially on leave of absence) while expanding their experience. The policy also created a path for transferring European technology to U.S. companies. Two such naval officers were especially important to Wharton. Both Francis M. Barber and William H. Jaques took leaves from the navy, became agents for leading ordnance manufacturers, and served as intermediaries during Wharton's negotiations with those companies. Barber had already distinguished himself as one of the navy's leading torpedo experts. In 1885, after a two-year stint on the Naval Advisory Board, he took a leave of absence and worked for four years as the U.S. agent of Henri Schneider & Compagnie, the armor and gun manufacturer of Le Creusot, France. Jaques, as secretary of the Gun Foundry Board, helped draft, and later implement, U.S. policy for obtaining heavy ordnance. Members of this board inspected British and European ordnance establishments in the summer and fall of 1883. It was during this trip that Jaques negotiated with Joseph Whitworth of Manchester, a leading manufacturer of heavy forgings, to become that firm's agent in the

United States. "To Lieutenant Jaques is due the main credit," stated Fritz, "for our subsequent acquisition of the Whitworth system of forging."[20]

From Wharton's perspective, Barber and Jaques were especially well placed because the navy's preference for the armor of Henri Schneider and the gun forgings of Joseph Whitworth had been well publicized.[21] Given that the navy was the only possible consumer of such products, its preference amounted to a forced technology choice for Bethlehem. Schneider had perfected the manufacture of all-steel, hammer-forged armor, while Whitworth excelled in the manufacture of heavy gun forgings using "fluid compression." With this technique steel ingots, immediately after casting, were subjected to heavy hydraulic pressure that minimized the blow holes, cracks, piping, and segregation that typically weakened large masses of steel.

Wharton's strategy got moving after Jaques approached Fritz with an attractive offer for access to Whitworth's heavy-forging technology. Already enthusiastic about diversification, Fritz used his approach to press for the directors' approval. Wharton and other Bethlehem officers were impressed with Jaques's formal presentation in October 1885; a month later, when a congressional delegation made a site visit to its plant, Bethlehem had already committed itself to securing a contract with Whitworth along the lines Jaques had suggested. The rail company would never be the same. In an agreement signed with the Manchester concern on 18 January 1886, Bethlehem gained patent rights plus the ability to purchase large amounts of machinery—an opportunity it took advantage of forthwith by purchasing an entire Whitworth plant. This included two complete hydraulic forging presses of 2,000- and 5,000-ton capacity; a complete fluid compression forging plant, including a press of 7,000-ton capacity and a 125-ton hydraulic crane; several large lathes and boring mills; and designs for open-hearth furnaces and special tools. The two companies also agreed to exchange engineers.[22]

Schneider too had approached Bethlehem seeking an entry into the American market, and Wharton initiated negotiations through its agent Lieutenant Barber. In Europe during May and June 1886 Wharton visited the Schneider and Whitworth plants, hoping to secure from Schneider similar access to patent rights, technical assistance, and machinery. Wharton's efforts received official sanction that August when Navy Secretary William C. Whitney (1885–89) issued a preliminary solicitation for bids on 1,310 tons of gun forgings and 4,500 tons of armor steel. Six months later the navy

FIG. 3.1. "The 125-Ton Bethlehem Hammer," 1893. A public view into the armor business was provided by this full-scale model of its oversize hammer that Bethlehem constructed for the 1893 World's Fair in Chicago.

From "The Bethlehem Hammer," Iron Age 52 *(13 July 1893): 60–63.*

secretary set the minimum figure for bidding at 300 tons of steel per month and expressed a preference for a bid that would provide both armor and gun forgings. Wharton and Schneider reached a final understanding in mid-March 1887, a few days before the bid opening.[23]

According to the terms of the contract, Schneider agreed to provide patent rights, drawings of machinery, and information on manufacturing methods and shop practice. At last setting aside the rail mill, Fritz turned his efforts to enlarging and enhancing the basic Creusot design by increasing the falling weight of the hammer from 100 to 125 tons, boosting the steam pressure from 75 to 125 pounds, improving the valve motion, and substituting hydraulic traveling cranes for power swinging cranes (fig. 3.1). For Fritz, as for Holley, European precedents in technology were inspirations and models rather than finished packages. Except for the machine tools for trimming heavy plates, Bethlehem constructed the whole plant, thereby avoiding the stiff import duties as well as the shipping difficulties that impeded its technology transfer from Whitworth.[24]

FIG. 3.2. Bethlehem armor shop, external view, c. 1893. Bethlehem's armor shop, constructed in 1886, had by 1893 been extended to 1,250 feet in length. The curved barbette armor being shipped was for the gun turrets of the USS *Monterey*.
From "The Bethlehem Hammer," 60–63.

Anticipating arrival of the Schneider and Whitworth machinery, Bethlehem constructed two new buildings of "ample proportions." A new steel plant with four open-hearth furnaces of a total capacity of 110 tons fit comfortably in one new building (1,155 by 111 feet) along with the forging presses, fluid compression apparatus, and plate mill. The other new building was a machine shop of "truly grand dimensions" (641 by 116 feet) that featured oversized machine tools, including a mammoth planer with a working area 13 by 50 feet to handle the large armor plates (figs. 3.2, 3.3). The first machines from Whitworth began arriving in September 1886, a month after Whitney's initial solicitation, and meanwhile Bethlehem's officers kept the navy fully informed of their extensive preparations for the armor and forging contract.[25]

The official opening of bids in March 1887 occasioned few surprises. Bethlehem tendered the only (officially preferred) bid for both armor and gun forgings. Cambria Iron and Midvale Steel both bid on forgings alone, while Cleveland Rolling Mills bid on armor alone. The sole twist occurred when the navy enforced its contracting policy that called for Bethlehem to drop its price on forgings to the lower figure quoted by Cambria. In June Bethlehem and the navy reached agreement that the steel company would deliver some 6,700 tons of armor and 1,300 tons of gun forgings, at prices ranging from $500 to $650 per ton. The contract was worth nearly $4.5 million.[26]

Bethlehem's 1887 contract with the navy obliged the company to begin monthly deliveries of 300 tons of armor by December 1889, but this date came and went with no delivery. The foremost problem was delay in receiving from Whitworth the heavy castings and forging presses, which did not arrive until three years after the 1886 contract with the British firm. The absence of armor at shipyards soon delayed construction on the armored cruiser *New York* and the ill-fated battleship *Maine*. In response to these frustrating delays, Navy Secretary Benjamin Tracy (1889–93) moved to secure a second domestic source of armor: the new Homestead mill of Carnegie, Phipps & Company.[27]

The Carnegie concern's entry into military contracting was much modified by Andrew Carnegie's anti-imperialist sympathies. In his best selling *Triumphant Democracy* (1886) Carnegie had written, "The Republic wants neither standing army nor navy. In this lies her chief glory and her strength." He went on to say, "Her Navy, thank God! is as nothing." Although Carnegie dismissed outright the navy secretary's first request for bidding on armor and gun forgings, his company made preparations. In 1886, three years after coming into the Carnegie orbit (see chapter 4), the Homestead mill erected its first open-hearth furnaces as well as a 119-inch plate mill and 32-inch universal mill that could handle the massive 25-ton armor ingots. Within a year of the navy's first contact, Carnegie permitted his company to complete an impressive entry into armor plate, a "defensive" technology that, as Carnegie helpfully assured the navy secretary, the government would not have to purchase from abroad. In 1890 the Carnegie concern signed its first contract with the navy for 6,000 tons of nickel-steel armor. To augment its armor capacity Homestead installed in 1894 an entire heavy-forging plant from Whitworth, and the next year it added a second open-hearth plant with a deep casting pit to accommodate the oversize ingots. Carnegie's company would manufacture armor—but not guns. Ever ambivalent about his beloved republic's military buildup ("international murder . . . still passes by the name of war"), Carnegie bankrolled the anti-imperialist movement,

FIG. 3.3. Bethlehem armor shop, internal view, c. 1900. Visible in the foreground are the side, or casement, armor plates for the battleships *Florida*, *Wyoming*, and *Alabama*.
Courtesy Hagley Museum and Library; Acc. 80.300.

in part with profits from armor. After the 1898 war with Spain, Carnegie tried to purchase independence for the Philippines from the United States for $20 million.[28]

Even as the Carnegie company ended Bethlehem's armor monopoly, the two concerns found much in common. Both tried to strengthen their contacts inside government, and legislators from Pennsylvania helpfully obliged to keep an enterprise that employed so many workers from falling to another state. Perhaps the chief political player was Sen. Boies Penrose (1897–1921), whose political career was entwined with Henry Clay Frick, the Connellsville "coke king" and Carnegie's chief lieutenant after 1889 (see chapter 4). Penrose was said to have "learn[ed] to speed to heel when Henry Clay Frick whistled. To Mr. Frick went the considerable distinction of Penrose's blind and unquestioning loyalty." Penrose repeatedly quashed legislation that threatened the steelmakers' positions and otherwise managed the Washington scene. Bethlehem's correspondence with Penrose details the quid pro quo: campaign contributions for legislative influence.[29] Carnegie's personal friends in government, such as Secretary of State James G. Blaine (1889–92) and the ambassador to Russia, Andrew D. White (1892–94), intervened periodically on his company's behalf.[30] Less exalted persons were no less useful. The U.S. Navy attaché in London, W. H. Emory, wrote confidentially to Carnegie in June 1890, just before Carnegie's first bid for armor plate, detailing the thickness of the armor on the upcoming contract and the number of tons expected, and providing a complete list of the hull components of the vessel. Navy officers who had been posted to the companies as governmental inspectors were recruited by both. Lieutenant Jaques (the erstwhile Whitworth agent) resigned his navy commission and came to Bethlehem in November 1887 to help build its new ordnance plant, and Lieut. C. A. Stone came to Carnegie about 1894. These naval technocrats proved invaluable for the steel companies, securing inside information on upcoming contracts, negotiating the maze of naval bureaucracy, and forestalling the entry of rivals into the armor market.[31]

Both companies used foreign orders to keep their armor shops full of orders in lull years. Through the 1890s the government provided a large armor contract every two or three years with nothing in the intervals (fig. 3.4). Moreover, the two companies shared a vigorous attachment to their armor profits, for this business was lucrative beyond all others. In 1895 Andrew Carnegie wrote privately to his cousin that while the railroads were in depression and ordering no rails, and the general trade was better but unprofitable, it was a "good thing" that an armor contract was coming up. A decade later, the chairman of Bethlehem Steel (a former Carnegie executive) stated, "Very little money is being made on anything outside of armor plate and Government gun forgings." More than half of Bethlehem's profits came from armor plate alone.[32] A series of public scandals, congressional inquiries, and incessant criticism did little to dissuade the companies from continuing this rewarding trade.

THE POLITICS
OF ARMOR

105

Above all, both companies recognized the danger of competing against each other. "There is a great deal [of money] in the armor-making plants working in perfect unison," Carnegie wrote

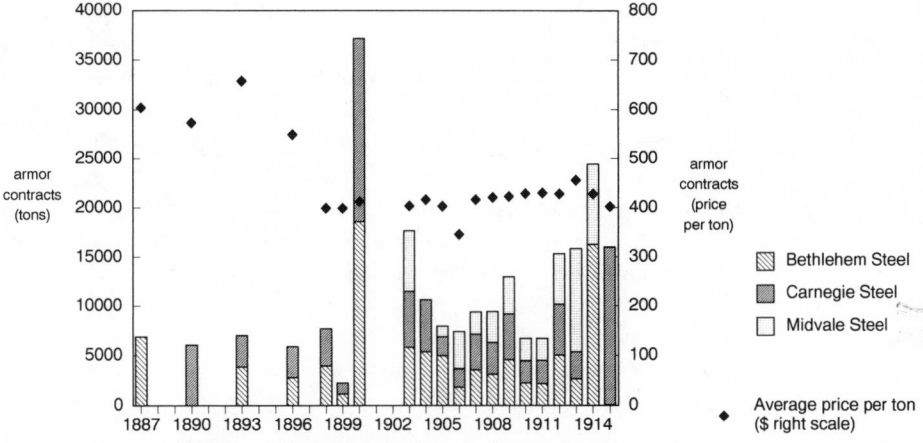

FIG. 3.4. Armor contracts (by firm and price, 1887–1920). For the graph, 1904 = 31 December 1903 for Bethlehem; 1908 = January 1909 for all three; 1910 = February 1911 for all three; 1915 = authorized 3 March 1915, but contracted April 1916. Not shown is Carbon Steel's 1913 contract for 61 tons. For the table provided here, the contract dates are approximations because the source statistics spanned several ships authorized as well as contracted over several years; the totals across 1916–20 are complete.

	1916	1917	1918	1919	1920	Total
Bethlehem	16,021	8,841	——	21,010	——	45,872
Carnegie	——	15,901	——	17,910	11,788	45,599
Midvale	——	29,909	——	——	——	29,909
Pittsburgh Screw	——	123	——	——	——	123

Data from E. A. Silsby, Navy Yearbook 1920–21 (Washington, D.C.: Government Printing Office, 1922), 916–19; Benjamin F. Cooling, Gray Steel and Blue Water Navy: The Formative Years of America's Military Industrial Complex, 1881–1917 (Hamden, Conn.: Archon, 1979), 224–29.

Bethlehem's president in 1895, and the two men affirmed an understanding of "equal division" in armor contracts that was achieved
by continually coordinating both bidding and pricing.[33] As figure
3.4 shows, such coordination divided armor contracts with arith-
metic precision between Carnegie and Bethlehem from 1893 until
1903, and thereafter until the war among Carnegie, Bethlehem,
and Midvale.

Divergence and Accommodation

*The tests were very ingeniously planned to give the plate the
victory over the gun.*—BENJAMIN FRANKLIN TRACY

From the datebooks of diplomats it appears that the decade after
1890 witnessed the escalation of Anglo-German antagonism that led
to war. While military planning on both sides of the English Channel proceeded apace, geopolitical rivalry spread from Europe to
Africa. In these years the American naval theorist Alfred T. Mahan
reached a worldwide audience with his thesis that great powers
needed great navies to secure and protect overseas trading, and that
great navies provided the necessary geopolitical might for would-
be great powers. "I am just now not reading but devouring Captain

FIG. 3.5. Belt armor of
Indiana-class battleship,
1892.

*Courtesy Hart Nautical Col-
lections, MIT Museum: Union
Iron Works, Proposed Sec-
tions through Belt Armor
(approved), USS Oregon
(Indiana class), Drawing 1322-
39528 Sheet 63 (4 April 1892);
William A. Baker Collection;
MITHNC.*

Mahan's book and am trying to learn it by heart," wrote Kaiser Wilhelm II. "It is on board all my ships and constantly quoted by my captains and officers." Diplomats around the world agreed that great navies meant large armored warships, the most complex machines in the world.[34]

From the letterbooks of steel companies and the reports of naval officials, however, an even more complex picture emerges. The U.S. Navy's shift to an offensive, "blue water" strategy squarely engaged the United States in the naval arms race between England and Germany. The navy bill passed in June 1890 allotted $9 million to build three "sea-going coast-line battleships designed to carry the heaviest armor and most powerful ordnance" and to compare favorably with the best European warships. These new, Indiana-class battleships were to have 3-inch nickel-steel decks, 5-inch internal armor, 18-inch belts of nickel-steel armor along the hull, and 17-inch armor around the main gun turrets. Armor producers viewed them as a stiff challenge and a supreme opportunity (fig. 3.5). British battleships remained the most serious enemy four years later, according to Lewis Nixon, a Naval Academy graduate (1882) and one of the nation's leading naval architects. He wrote of the English vessels Blake and Blenheim, protected cruisers of 9,000 tons, that their "object . . . in war would have been to destroy American commerce destroyers." In response the United States built the New York, an armored cruiser that was faster, lighter, and better protected than the Blake and "could outfight" it as well. Britain began building—as a counterresponse, according to Nixon—eight enormous battleships of 14,000 tons displacement.[35] American entries in the accelerating naval arms race were watched closely by German and British naval officials.[36]

In this context of geopolitical rivalry and technical change, the naval superiority of the great powers rested ultimately on the ability of warship armor to withstand enemy projectiles. In the absence of effective protection, warships were simply large targets; for this reason armor plate became the critical and strategic technology required for great-power status. But given the condition and evolution of armor-plate manufacture in the 1880s and early 1890s, naval officials faced a curious predicament, which could be turned to advantage. None of the several armor types—all-steel, compound iron-steel, chilled cast-iron, and nickel-steel—could be proven superior to the others. This indeterminate state of the art permitted each navy to adopt or retain a particular armor type. By staging supposedly objective firing trials, each could claim that its battleship armor was superior to the next and therefore that its navy was superior to all rivals. In effect, the technical divergence allowed (and was a precondition for) the accommodation of armor tests to whatever results suited a nation's procurement patterns. This indecisive "battle of the armor" persisted into the early 1890s, when a truly superior technology permanently altered the international politics of armor.

In theory, testing armor plates was a simple matter. A superior armor plate withstood the impact of a series of test shots; an inferior one did not. Before 1880 the world's leading artillerists had agreed on standard test conditions for wrought-iron armor, making one test comparable with another. After 1880 this consensus collapsed with the rapid spread of several new armor types.[37] Consequently, the widest variety of test conditions obtained, so that no test could be fairly compared with another. The ballistics of the shot (projectile weight, charge of gunpowder, velocity, and energy on impact), the type of projectile (standard, armor-piercing, or capped) and its angle of attack, and the nature and thickness of the test plate's support all affected the outcome of a trial. Altering any of these factors could skew the test results. The "racking" versus "punching" controversy, never really resolved, illustrates these issues. Hard armor plates were most vulnerable to heavy, slow-moving shells ("racking"), while soft armor plates were most vulnerable to light, fast-moving shells ("punching"). Racking shots that would destroy hard plates might not damage soft plates, while punching shots that would pierce soft plates might not damage hard plates.[38]

Given that no armor type was clearly superior, and that each had its weaknesses, the setting of test conditions decisively influenced the outcome of ballistic tests. The all-important perception of military superiority hung in the balance. The steelmakers and the navies in each rival country had the strongest possible incentive to arrange successful tests, to demonstrate the superiority of their armor and hence of their battle fleet. It is no surprise that rival navies disagreed. American and English tests yielded flatly contradictory results during the 1880s. Tests in the United States, according to Navy Secretary Benjamin Franklin Tracy, "were very ingeniously planned to give the plate the victory over the gun."[39]

The "battle of armor" was set up after English armor makers first began using steel. Two Sheffield companies developed and patented the compound iron-steel plates adopted by the British Admiralty in 1878. The "Wilson system" produced armor plate at Charles Cammell's Cyclops Works. In this method a hard face was formed with a charge of molten steel poured onto a hot wrought-iron backing plate; later, an even harder face was achieved with a second charge of steel high in carbon. The "Ellis system" prevailed at John Brown's Atlas Works. This method used a charge of high-carbon steel to bond a 5-inch steel face to a 10-inch wrought-iron back. The compound plates from either method, after being rolled down to the desired thickness, planed flat, and trimmed to size, featured a hard steel face that blunted incoming projectiles plus a resilient iron back that flexed on impact (fig. 3.6). Cammell and Brown, the only British steelmakers equipped for making compound armor through 1890, informally divided each Admiralty contract.[40]

A more competitive situation obtained in Germany through the 1880s. In this period German warships and coastal fortifications were protected by compound armor patterned on the Sheffield practice and by Gruson chilled cast-iron armor. Gruson armor, like the compound armor, combined a hard outside with a flexible inside, and it could be cast into a variety of complex curved shapes. These armor types were produced by Dillenger Hüttenwerke, Bochumer Verein, Hörder Verein, Gruson-Magdeburg, and Fried. Krupp.[41] Through the 1880s Krupp was a strong mili-

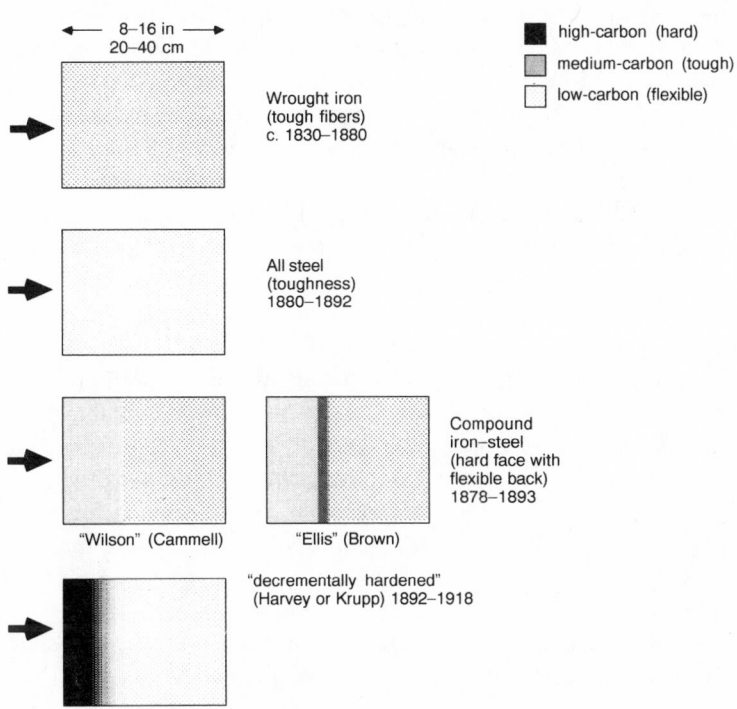

FIG. 3.6. Evolution of armor types, 1878–1918.

tary contractor, and in the mid-1890s it achieved preeminence by mastering high technology.

Compound armor did not find much favor outside Germany and England. All-steel armor plates from the Schneider works protected the navies of France and Italy after 1880, and that of the United States after 1887. Unlike the compound plates, Schneider's all-steel plates were not prone to cracking upon a shell impact, although they had other weaknesses, such as being "soft" on impact. They began as large castings of open-hearth steel whose chemical composition had been carefully monitored during the 12–24 hours that a heat required. At Schneider's Le Creusot works the huge castings were forged down to the desired thickness by a 100-ton hammer. The U.S. Navy decided to adopt Schneider all-steel armor without comparing it against any of its numerous rivals. In fact the navy's preference for Schneider armor flew in the face of a well-known armor test conducted in 1885 at the Italian Navy's Spezia proving ground. That test suggested that Schneider armor was inferior to the compound armor of Cammell and Brown. "The Schneider plate exhibited comparative softness," reported a U.S. Navy observer. "An examination of the fragments of the projectiles showed that the one which passed through the Schneider target had been less damaged, though much distorted, by the blow. It was observed, also, that the head of the shell showed only meridional cracks, while in the case of the other two [the Cammell and Brown plates] the cracks were of a spiral character, demonstrating that the resistance opposed had been so great that the shot had been conspicuously twisted while passing through the plate."[42] These results—specifically the wrenching twist of the projectiles caused by the Cammell and Brown compound armor—suggested to this American observer that Schneider all-steel was inferior to compound armor.

The Spezia test, conducted in October 1885, was reported to the navy in 1886, but it failed to halt or modify the March 1887 contract that formalized the navy's tie to Schneider through Bethlehem. The strong working relationship among Bethlehem, the Navy Department, and Schneider proved invincible. The navy's disregard of these results illustrates how armor tests that yielded inconvenient results could simply be ignored. Moreover, when the test location shifted from Spezia to Annapolis, and once a domestic rival to Schneider existed, entirely new results emerged.

At the Annapolis proving ground in September 1890 the navy conducted a test of its own to compare the two leading types of armor (all-steel and compound) and a promising new contender (nickel-steel). Previous accounts have mistakenly assumed that the *results* of this test persuaded the navy to adopt nickel-steel armor.[43] In fact, Navy Secretary Tracy had committed the navy to invest in nickel months *before* the test, and he carefully stage-managed the affair to ensure that the nickel-steel plate was declared superior. Tracy then used the test results to rush emergency legislation through Congress allotting $1 million for the purchase of nickel, preempting the rival efforts of a British nickel syndicate to corner the rich North American market in this metal. With these actions Tracy effectively and imaginatively served as an entrepreneur—a role that frequently requires principal actors to adopt stances that may not reflect a narrow objective reading of the evidence at hand and whose consistency is best explained by the overall goals being sought.[44] In this case, Tracy was determined to have the next 12,000 tons of battleship armor be of nickel steel.

The remarkable properties that nickel imparted to steel were something of a metallurgical sensation beginning in 1889. In July of that year S. J. Ritchie approached Tracy with news of unusual international activity in nickel. The owner of several Canadian nickel ore mines, Ritchie also reportedly controlled three-fourths of the world's nickel supply. Tracy received from Ritchie the scientific paper that fanned interest in the metal across the English-speaking world. From the naval attaché in London whom Tracy assigned to the nickel campaign, the Navy Department learned that several European companies had been experimenting with nickel armor. That autumn Tracy ordered a nickel-steel plate from Schneider, and in December he formally asked other foreign and domestic steel manufacturers to submit armor plates for a forthcoming competitive test at Annapolis. In February 1890 Tracy took the nickel campaign to Congress as part of his overall plan to secure nickel armor. Ritchie testified to the House Ways and Means Committee in favor of reduced duties on nickel ore, while Tracy lobbied the House Naval Affairs Committee for an appropriation to purchase nickel. Tracy activated another portion of the plan in March when he requested Rep. William McKinley, then drafting the protectionist tariff bill that would bear his name, to drop the stiff import duties on nickel ore. In July, as the final piece in his plan, Tracy scheduled a competitive armor test for September.[45]

Meanwhile, Tracy persuaded Carnegie to look into the matter. Carnegie told his chairman, W. L. Abbott, that the navy secretary had big plans for armor: "Hopes we will 'read up' on it and be prepared to offer Armor. Wants to ask bids as soon as tests are made—Ritchie, our Akron, O. friend, has the big nickel mine and is selling to Krupps & c. . . . There may be millions for us in armor. To one man should be assigned 'Armor' and he should read up and keep up on the subject."[46] The armor assignment fell to Millard Hunsiker, who had joined Carnegie on 1 May 1890 as engineer of tests and had begun an improbable ascent toward the highest reaches of London business, politics, and finance. Early that June—one hundred days before the Annapolis trial—Tracy called Abbott to Washington and expressed to him a "strong inclination for nickel." Pending the trial, Tracy was even delaying the opening of bids on 3,000 tons of armor. With 9,000 more tons in the appropriations pipeline, Tracy's insistence on nickel ("we need it at once for ships, contracts for which were recently awarded") left little doubt.[47] On 23 June, Abbott secured 2,000 tons of nickel matte (the first product of smelting nickel ore) from Ritchie's Canada Copper nickel mine at Sudbury, Ontario. Three days later the Carnegie company secured a license for the manufacture of patented ferro-nickel compound armor plate from the Abel Rey Company of London, agents for the French syndicate Le Ferro Nickel. During ongoing negotiations with Carnegie, who pragmatically favored using whatever armor seemed best, Tracy indicated a clear preference for nickel steel; by August the two men had verbally agreed to an armor contract.[48] Tracy was about to transform a potentially promising technology into an unmistakably superior one.

All these considerations weighed on the eleven members of the naval board convened on Thursday 18 September at the Annapolis proving ground. Chaired by Rear Admiral L. A. Kimberly, the board was to test three armor plates: one of Schneider steel, one of Schneider nickel steel, and one of Cammell compound. The three plates, each 6 feet wide, 8 feet tall, and 10 inches thick, were carefully mounted on 36-inch-thick oak backings positioned in an arc 28 feet distant from the 6-inch rifled breechloading gun. To provide a striking velocity of 2,075 feet per second and a striking energy of 2,988 foot-tons, the officers carefully set the ballistic conditions: the Holtzer armor-piercing shells were carefully adjusted to the standard weight of 100 pounds, and each round was propelled by 44½ pounds of Du Pont brown prismatic ("cocoa") powder. The naval officers, safe behind "bomb-proofs" located 50 feet from the gun, began the firing at 11 A.M. To document the results, a board member meticulously described each plate after each of the four shells had been fired. Photographs were taken of each plate after each shot, and after each plate had received four shots the plates were photographed as a group.[49]

After a full day of firing the board prepared to adjourn, but Tracy, giving the test "his personal attention," ordered the firing to resume after the weekend—with changes in certain of the crucial test conditions.[50] From Tracy's pronickel perspective, Thursday's trial had failed: the nickel-steel plate had not been proven superior to the all-steel plate. "The all-steel certainly showed the least penetration and showed that it could be relied on up to that limit more certainly to keep an enemy's shots out of the ship," reported the *New York Times*.[51] The compound plate fared worst of all.

The pressure to ratify nickel steel intensified that weekend when Sir James Kitson telegraphed his aim to secure all possible sources of nickel in North America.[52] Kitson was a prominent steel manufacturer from Leeds, the president of the British Iron and Steel Institute (the audience for the paper that had ignited the nickel craze the year before), and the head of an English nickel syndicate. The official purpose of his trip to New York was to present the Bessemer medal to Abram Hewitt, founder of the venerable Trenton Iron. From Tracy's perspective, however, Kitson's visit spelled danger. Kitson sent his message no later than Sunday, 21 September, when his ship, the *Servia* of the Cunard Line, weighed in at the Irish port of Queenstown before steaming on to New York.[53]

Already chagrined by nickel's poor showing, and now impelled by Kitson's threat, Tracy literally ordered up the big gun. On Monday 22 September an 8-inch gun replaced the 6-incher. Now 85 pounds of brown prismatic Du Pont powder propelled a Firth armor-piercing shell weighted to 210 pounds. Compared with Thursday's test these new ballistic conditions yielded a similar striking *velocity*, but the much heavier shell boosted the striking *energy* by 67 percent. This would be a "racking" shot. Each plate received just one—square in the center. For the all-steel plate, as the *New York Times* reported, "the projectile penetrated through the 10½ inches of steel and 4½ inches into the oak backing and rebounded," while the plate suffered four cracks extending to its edge. For the nickel-steel plate, "the projectile pierced clear through it and into the oak backing, a total of 21½ inches, being 6½ inches more than the penetration in the all-steel plate and its backing," but no cracks appeared. While the large shell clearly demolished the compound plate, "between the all-steel and the nickel steel, however, there was a question. The penetration was greater in the nickel-steel, while the all-steel was cracked."[54] The newspaper's accurate report that there was a "question" about which plate was superior did not survive the mounting pressure for a clear victory for nickel.

The next day Tracy launched his campaign for an emergency appropriation for nickel. He and the new ordnance chief, William M. Folger, visited the Capitol that afternoon to meet with the House

Committee on Naval Affairs, and its chairman Charles A. Boutelle agreed to introduce a joint resolution authorizing $1 million for purchasing nickel. In reporting the story, the *New York Times* reversed its previously balanced assessment. Now, the "remarkable results" of the Annapolis tests "signally developed the superiority of the nickel plates."[55] The newspaper's more enthusiastic account doubtless pleased Tracy.

Boutelle on Wednesday introduced the $1 million resolution as promised. When he reported the measure back from his committee to the full House the next day, he affirmed the "superiority" of the nickel steel as "incontestable and remarkable." Boutelle argued that it was "imperative" that the House act favorably, which it did the same day. That morning's *New York Times* obliged the nickel campaign with an article praising the nickel-steel plate "for enduring almost perfectly . . . this fifth shot from the eight-inch gun."[56] (No mention of the 21½-inch-deep hole!) After a hurried debate on Friday, the Senate passed the appropriation on Saturday, 27 September. The swift action of Congress was "something phenomenal," according to *Scientific American. Iron Age* observed that "somebody has evidently frightened the authorities."[57]

When that somebody, Sir James Kitson, stepped off the *Servia* on Monday morning, 29 September, the front-page headline of the *New York Times* dashed his hopes of cornering the North American nickel market: "Nickel Steel Armor Plates: An Appropriation which gives this Government great Advantages." The Navy Department could now secure, the paper continued, "a virtual monopoly of the use of nickel steel for armor plates and projectiles." Ending its session on the last day of September, Congress approved the protectionist McKinley Tariff, which as Tracy had requested nonetheless dropped all import duties on nickel ore. Tracy immediately sent a $50,000 order to Canadian Copper. When on 2 October he opened bids for the three "coast-line" battleships that marked the emergence of the United States as a major naval power, he had achieved his goal of providing the new warships with nickel-steel armor. In November Tracy and Carnegie signed a written contract for 6,000 tons of armor, with the Navy Department to pay the two-cent-per-pound royalty demanded by Le Ferro Nickel.[58]

The report of the Kimberly Board, issued on 11 October, dutifully pronounced the victory of nickel steel. The report noted that the 8-inch shell "badly cracked" the all-steel plate, while the nickel-steel plate "remained uncracked." Only a determined reader would find that the nickel-steel plate, like the all-steel plate, had been penetrated by the large shell.[59] The board rated the three plates "in the following order of 'relative merit': 1. Nickel Steel. 2. All Steel. 3. Compound." Skillfully managing the rush of events, Navy Secretary Tracy had succeeded in arranging armor test results that ratified his decision to adopt nickel steel. Tracy's victory paved the way for both the Bethlehem and Carnegie companies to produce nickel-steel armor for the navy. Indeed, nickel remained the U.S. armor of choice until the adoption of a new, hardened armor in the mid-1890s.

As these episodes through 1890 show, armor tests were not an independent source of hard data that informed naval policy. Rather, the decision on armor policy came first (for Schneider and then for nickel), and the tests accommodated and ratified the policy. Results of tests could be wholly ignored, as illustrated by the navy's adoption of all-steel armor despite the apparent superiority of compound. These examples suggest that armor tests should be analyzed as cultural artifacts rather than as scientifically objective findings.[60]

As mentioned earlier, the firing ranges of the United States and Great Britain witnessed a great if indecisive "battle of the armor" throughout the 1880s and early 1890s. American artillerists demonstrated the superiority of all-steel over compound, while (remarkably enough) their British counterparts demonstrated precisely the reverse. Naval experts in the two countries also disagreed on the merits of nickel steel, especially after Kitson failed to secure a nickel supply for Britain. The British naval establishment vigorously advocated compound armor from its adoption in 1878 until the adoption of (nonnickel) hardened Harvey armor in 1894. The members of the quasi-official Institution of Naval Architects, for example, roundly savaged a Schneider representative who dared to assert the superiority of his company's all-steel plates.[61] Indeed, British armor experts could and did cite tests supporting their stance. At this time compound armor "held the field" in Great Britain.[62] Conversely, Americans never tired of recalling the 1890 Annapolis test that destroyed the compound plate. It can hardly be a coincidence that English tests ratified the Admiralty's commitment to compound plates, while American tests ratified the Navy Department's adoption of Schneider all-steel plates.[63]

Since the Navy obliged with hospitable test conditions, American steelmakers had no need to influence the tests themselves. Occasionally they expressed concern about the tests. "It would have been *extremely dangerous* to have allowed an *increase* in ballistic test," observed the head of Bethlehem's armor department in 1899. "While it may be well for us to test plates *beyond* the requirements, it is likely to prove *detrimental* to our interest in making *future* contracts if this data obtains too much *publicity*."[64] Indeed, whatever the navy desired, the consensus on "domestic only" contracting determined that U.S. vessels would be protected by the armor that the Bethlehem and Carnegie armor plants could produce. Neither steelmakers nor naval officers showed any inclination to challenge the secure assumption that U.S. steelmakers produced superior armor fit for a world-class navy.

Finally, armor "experts" claimed a central role in interpreting test results. "No report conveyed to my mind all the facts," noted one naval architect. "One wants to be there, to see the whole thing, to see the plate taken down, to see the condition of the target behind the plate."[65] In fact, it was the pronouncements of naval experts, not the reportage of newspapers, that resulted in "definitive" statements, as the 1890 Annapolis trial demonstrated. In interpreting this test, the ordnance bureau technocrat Edward W. Very upheld the expert's role. "A knowledge of this method and of the results obtained does not make one an expert judge of armor," Very asserted. "It requires an expert to properly weigh all the cir-

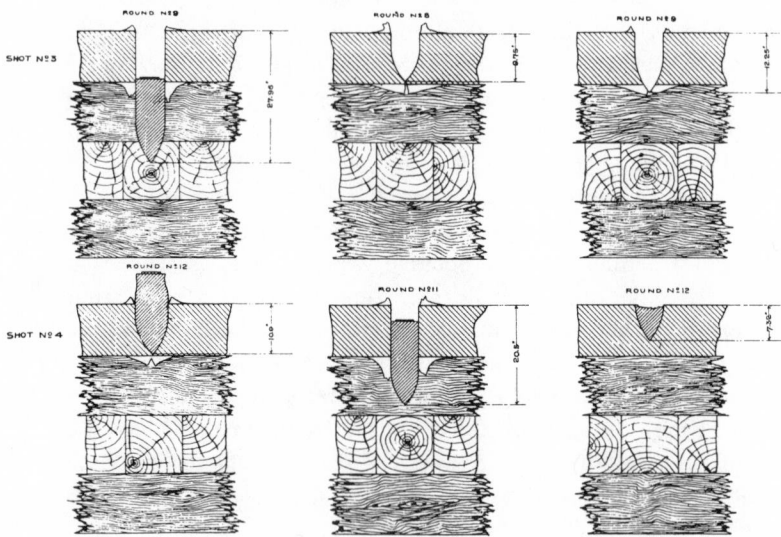

FIG. 3.7. Armor trial results, 1892. Armor trials rarely yielded clear-cut results. The left column is Bethlehem low-carbon steel, the middle column is Carnegie low-carbon nickel steel, the right column is Bethlehem high-carbon nickel steel. Later rounds of this trial, using special armor-piercing shells, decisively penetrated the low-carbon steel plate.
From F. Lynwood Garrison, "The Development of American Armor Plate," JFI 133 (June 1892).

cumstances of any test." As figure 3.7 suggests, the results were rarely clear-cut. To devise a quantitative comparison of armor tests, Very and others devised formulas relating the striking energy of the projectile to the thickness of the plate under trial. Earlier, the so-called de Marre and Gâvre formulas had been devised to compute the energy required to perforate a wrought-iron plate of given thickness, and the new generation of naval artillerists tried adapting these formulas to steel. The formulas reflected the fact that through the early 1890s the best armor-piercing projectiles were able to pierce the thickest armor plates. In the event of battle, the big battleships were vulnerable. But the balance between offensive and defensive technologies was reversed with the development of a new type of hardened steel armor. The new armor shattered the assumptions of quantitative formulas. Battleships, steel companies, and governments on both sides of the Atlantic felt the repercussions.[66]

Convergence and Conflict

The striking ends of the projectiles appear to have been splashed on the face of the plate.—BENJAMIN FRANKLIN TRACY

A remarkable technological convergence in the ten years after 1895 propelled the navies of the great powers in a single direction that no one could have foreseen or intended. Such a convergence illustrates how technologies can appear to determine the course of history.[67] Before the 1890s, as we have seen, technical uncertainty about armor permitted substantial diversity, a precondition for the interminable "battle of the armor." Beginning in the mid-1890s, however, no battle fleet with aspirations to great power could fail to adopt a new type of "decrementally hardened" armor plate. The technology was invented in the United States, improved in Germany, and ultimately controlled in London through a web of international patents and cross-licensing agreements that linked the most diverse armor makers and navies. The new armor plate served to erode the political control of the great powers over their own rearmament efforts. The essential technology for building world-class battleships was now held by a group of London capitalists fully loyal to no one country, not even England. In addition, the rearmament effort brought about sources of private economic power that the governments could not control. The "whole history" of the hardened Harvey-Krupp armor, as one participant remarked, "is rather a remarkable one."[68]

The loss of domestic political control attending this technological convergence was especially troubling because it occurred during the acceleration of the Anglo-German-American naval arms race. Radically expanded navy bills during 1898–1900 in each of the three countries signaled the onset of intense naval rearmament and direct geopolitical competition. From 1902 to 1908 England added sixty-nine warships to its navy, Germany added thirty-eight, and the United States added fifty-one. Pressures for rearmament in the United States included the jingoist fervor surrounding the Spanish-American war, which resulted in an 1898 navy appropriation of staggering size (fig. 3.8). "The pressure to build ships is something enormous," complained one member of Congress. "Public opinion is manufactured, and the pressure seems irresistible." Two years later, when the new warships were ready, armor contracts increased fivefold. As one Bethlehem officer put it, "We were specially requested—almost commanded—to increase, or to double, rather, our producing capacity" for armor.[69]

The technological convergence brought about a transformation of naval-industrial relations in each of the three major rivals. In England, Brown and Cammell, which had jointly monopolized the armor trade since 1878, were displaced by Vickers and Armstrong. Vickers and Armstrong fully mastered the new technology, and directors of these companies also took a leading role in the international armor-patent cartel. In Germany the competitive situation of the 1880s dissolved with the ascendancy of Krupp. In 1893 Krupp formed an alliance with its principal rival, Dillenger, which had

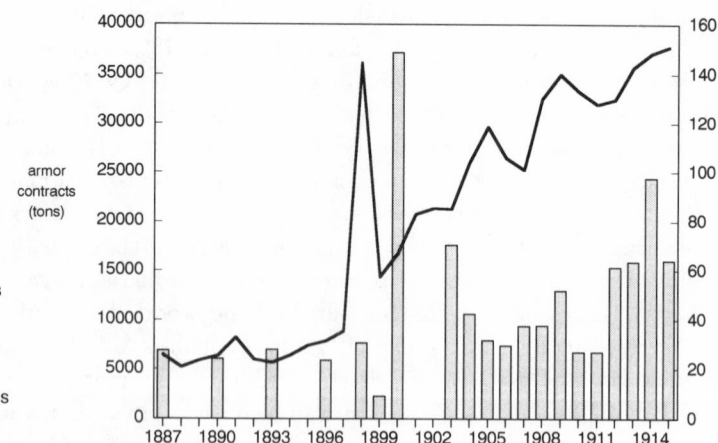

FIG. 3.8. Navy budget and armor contracts, 1887–1915.

From Silsby, Navy Yearbook 1920–21, *805; Cooling,* Gray Steel and Blue Water Navy, *222.*

supplied the German Navy with armor plate for nearly two decades, and simultaneously bought out the Gruson concern. Perfecting hardened nickel-steel armor in 1894 gave the alliance, now dominated by Krupp, a corner in high-quality armor. Five years later the German Imperial Navy Office specifically blamed advanced technology for the armor monopoly. Only a few companies, it noted, successfully employed the Harvey-Krupp process for manufacturing armor: Krupp and Dillenger in Germany, Vickers and Armstrong in Britain, Schneider in France, and Carnegie and Bethlehem in the United States.[70] The politics of armor, once shaped by nation-specific procurement patterns, now exerted a force for change on these same institutional structures.

The technology that provided the impetus for these realignments originated with an unlikely New Jersey inventor, Hayward Augustus Harvey. Harvey had collected some seventy patents, beginning in the 1850s, but only his screw-making machines proved commercially viable. In 1886, in an attempt to manufacture railroad hardware, Harvey conceived a process to improve the properties of cheap, low-quality, open-hearth or Bessemer metal. To do this, he "steelified" the low-carbon metal by baking it in charcoal. At low cost, so Harvey claimed in his patent, "the characteristic qualities of refined crucible-steel are imparted to steels of low grades." (Sheffield steelmakers noticed that his process resembled the age-old technique of cementation, the first step in making crucible steel.) In 1886 he organized, with two business partners, the Harvey Steel Company of New Jersey, his final and by far most successful commercial venture. Beginning in February 1887, Harvey used his process at a small works in Jersey City to make steel suitable for axes, files, and cutlery as well as cutting and boring tools. In June 1888 the company increased its capitalization to $600,000 and built a plant in Newark capable of producing 10 tons per day of the special steel. Nine months later the new works stood ready for an inspection by stockholders.[71]

Sometime in 1888 Harvey realized that his process could be adapted for armor plate, and early the next year he made up a small inch-thick demonstration block, which was "so hardened . . . that it could not be indented by a centre punch driven by either a heavy sledge hammer or by a steam hammer." In May, after preliminary contacts suggested the navy had a "vast amount of material" suited for the new process, Harvey took his demonstration block to Washington. After a successful performance, Harvey wrote that William M. Folger, then ordnance inspector at the Washington Navy Yard, was positively enthusiastic about the prospects for hardened armor. "He says it's a big thing, and very important to the Government, bigger than I can know."[72]

Folger indeed had spotted a winner: better even than the nickel steel he was championing for Secretary Tracy. With his elevation in 1890 to ordnance chief, he was well placed to promote the new process. It produced stunning results. Test projectiles were "completely shattered" during preliminary ballistic trials in June 1890 of a 6-inch plate and in February 1891 of a 10½-inch Schneider nickel-steel plate that Harvey himself had treated at the Washington Navy Yard, the first time the two armor technologies were combined. "These results are remarkable," reported Folger. The Harvey process appeared to yield "the ideal armor plate, a hard front combined with a tough back without any weld or other line of demarcation between the two."[73]

In light of these favorable results, the navy readily shouldered the burden of developing the Harvey process for armor making. In March 1891 the navy acquired an option on the right to use the Harvey process for treating armor for warships; it formally accepted the option a year later. The navy would pay a royalty of half a cent per pound of the finished plate, up to a maximum payment of $75,000, and it specifically agreed to "bear all the expense of the experimental development of the process as applied to armor plates" as well as to reimburse the Harvey Steel Company for installing an armor furnace at the Washington Navy Yard. When the patent office initially dismissed Harvey's claim on "decrementally hardened" armor plate, the navy intervened on his behalf and quickly secured the patent office's approval.[74]

Success depended on making the Harvey process work on the mammoth 10-ton blocks of steel, for the basic theory behind the process—that carbon permitted steel to harden, and that hardening was achieved by rapidly cooling a hot piece of steel—was long received wisdom. The process that Harvey brought to Bethlehem during the summer of 1891 was followed in general terms wherever Harvey armor was made. The all-important step of "decremental carburizing" was conducted in a firebrick compartment erected inside an oversize heating furnace. The face to be hardened was embedded in 10 inches of carburizing compound, a patented mixture of finely powdered wood charcoal and coarsely ground animal (or bone) charcoal, while the other face was protected by a thick layer of sand. The furnace was fired up. When iron test rods extending into the treating compartment showed a "sparkling white heat," above the melting point of cast iron, or approximately 2000°F, the proper heat had been achieved. Heating up the furnace could take seven days or more, and carburizing the plates at high heat took equally long. After the carburized plates had been removed from the furnace, and had cooled slowly to a cherry-red heat, they were cooled quickly by a strong spray of cold water. With this treatment the depth of "supercarburization" extended 1–3 inches into the plate. Carnegie's armor plant at Homestead initially tried to circumvent Harvey's special techniques, but the alternative technique devised by Superintendent Charles M. Schwab proved unsuccessful: in August 1891 one all-steel plate was heated up and cooled down eleven times without hardening satisfactorily. The Carnegie company too was forced to adopt the entire Harvey process. From Harvey's personal perspective, the navy's aggressive sponsorship of his process rankled, but there was no gainsaying the navy's success at developing the process for military purposes.[75]

Armor-piercing shells, once the superior technology, now took a fearful beating in full-scale trials against Harvey plates. At the Annapolis and Indian Head proving grounds during 1891 and 1892 the incoming shells were variously "smashed," "pulverized," "broken up," and otherwise destroyed without causing significant damage to the hardened plate. At Bethlehem's proving ground in July 1892, "the striking ends of the projectiles appear[ed] to have been splashed on the face of the plate." Previously the shell's ballistic energy had been absorbed by the plate itself (sometimes liberating enough heat to set fire to the wooden backing); now the energy was dissipated by the shell bursting into fragments. The consequences were not lost on observers. "The new ships will be rendered practically impervious to the guns of the average battery of any of the ships of the European service," stated one newspaper. "Let Italy take notice," stated another. "These results," stated the navy with supreme confidence, "speak for themselves." The navy formally adopted Harvey armor in October 1892 for all its new warships and made arrangements to Harveyize the armor of vessels under construction, including the Indiana-class battleships.[76]

Faced with these alarming results, British armor makers initially fought Harvey armor in the patent courts. Early in 1892 they dismissed the overtures of Harvey's European agent, E. M. Fox, by observing that the carburizing step had been practiced for two centuries in Sheffield as cementation and that the hardening step was already protected by an 1891 English patent owned by the Brown concern. During the summer and fall of 1892, while Brown and Cammell fought the Harvey process in the patent court, Vickers made two trial plates under a license from the Improved Steel Syndicate. This was a group of twenty-one aristocrats, military officers, and industrialists (among them Earl Cairns, a director of the Armstrong concern), which Fox had organized to promote the Harvey process in England. These plates withstood a formidable ballistic trial upon HMS *Nettle* in Portsmouth harbor on 1 November 1892 (the projectiles were completely "destroyed" and "pulverized"). "It is a great—a vast triumph," stated Vice Admiral P. H. Colomb, vice chairman of the syndicate. In another test two months later more Admiralty shells met spectacular ends (being "shattered" or "smashed" into "fragments"). After Harvey's English patents were sustained in a series of patent-law decisions completed by December 1892, the British armor makers realized that contesting the Harvey process was futile. "We are glad to see," commented the English *Marine Engineer*, "that the authorities were not too proud and conservative to be taught any wrinkle that our cute Trans-Atlantic cousins could put them up to." Indeed, the authorities had begun negotiations with Harvey Steel of New Jersey.[77]

The Harvey process began spreading nearly as rapidly as the news of its phenomenal success against armor-piercing projectiles. In April 1893 the German marine minister among others had seen a Krupp steel shell "smashed" by a Harvey plate at the Meppen proving ground. The consequences for global naval power were unmistakable, and eager licensees began lining up. Harvey had assigned his three key armor patents as well as his foreign rights to Harvey Steel (N.J.), which after his death that August served more as a patent-holding than a manufacturing concern. By September Harvey furnaces had been erected under a direct license from the New Jersey firm in no fewer than four European countries: at the works of Vickers, John Brown, and Charles Cammell (all in Sheffield, England); William Beardmore (Glasgow, Scotland); Witkowitz (Moravia); Terni (Italy); and St. Chamond and Montluçon (France).[78]

But 1893 was the last year in which the Harvey process was controlled from New Jersey. A family of Harvey companies headquartered in London mobilized capital to a total of £630,000 (by comparison the initial Improved Steel Syndicate had been capitalized at just £7,000). The larger figure was nearly two-thirds of the Bethlehem company's total capitalization, and a significant if smaller fraction of the Carnegie company's.[79] For armor no less than railroads, London was still the place to achieve such large-scale financing; for this reason alone British influence over the Harvey patent cartel was substantial if not absolute. When the Harvey Steel Company of Great Britain, Limited, issued a prospectus in 1893, representatives of Brown, Cammell, and Vickers were among the directors.[80] The next year, when the Admiralty (still smarting from Tracy's nickel campaign) adopted nonnickel Harvey armor,[81] prospectuses appeared for separate Harvey companies covering France, Italy, and Germany (the last of which was styled as covering the rest of the world exclusive of North America). Again the big three British armor makers, as well as representatives from French, Italian, and German armor makers, were the directors. For each of the four new companies, the original New Jersey–based Harvey Steel held less than one-third of the shares, and presently it too came to be controlled by London-based Harvey Steel.[82] Henceforth, Harvey armor was controlled from London.

In the United States, an odd situation developed when the two domestic steelmakers contested the domestic Harvey patents. Neither Bethlehem nor Carnegie was ever a Harvey licensee; rather, the Navy Department licensed the process and permitted— or more accurately, required—the two steelmakers to make Harvey armor. Manufacturing Harvey armor for sale abroad was forbidden under the navy's license, however. This issue came to the fore in 1894–95, when Bethlehem won a sizable Russian contract for just over 1,000 tons of Harvey and nickel-steel armor. The American Harvey company flatly refused Bethlehem's request for a license, fearful of throwing the international family of Harvey companies into disarray. Faced with the Russian contract, and no Harvey license, Bethlehem adopted the stance that the Harvey patent was "invalid" and proceeded with the Russian order. The Harvey company promptly launched a suit for patent infringement.[83]

What forced Bethlehem and Carnegie into the fold was a new Krupp patent acquired by the omnipresent London Harvey concern. By 1893 Krupp had begun carburizing armor using methane-rich illuminating gas produced from coal (rather than Harvey's powdered charcoal). Still, perhaps for promotional purposes, the German outfit was still keen to produce "Harvey" plates.[84] Within two years, however, Krupp patented and licensed its process to a syndicate formed by the six largest European armor-plate works outside Germany, whose worldwide rights to the Krupp technology were purchased by the London Harvey company. When the London company began cross-licensing the Krupp armor process, the Bethlehem and Carnegie companies were among its licensees.[85] This cross-licensing arrangement brought about the global technological convergence on Harvey-Krupp armor.

The essential similarity of the two processes was revealed in the transfer of Krupp technology to the United States. In October 1897 the Bethlehem and Carnegie companies sent two men each to Krupp's factory at Essen, where they jointly extracted precious technical information from their recalcitrant hosts. "The Krupp people," complained Archibald Johnston, Bethlehem's armor superintendent, "own the very air hereabouts" and "[are] afraid of our getting too much information." Krupp's famous gun shop, containing numerous special machines, was an especially sensitive site. What impressed the Americans most about the armor works was the German's obsession with controlling each step of the armor manufacture. "The principal feature of the process," Johnston observed, "is the accurate and regular manipulation of the steel." From initial melting to final trimming, all plates were manufactured under identical conditions. The heating furnaces bristled with pyrometers to monitor the temperature; carburizing was conducted at 950°C and "never above 1000°C" as this produced a brittle metal. (By comparison Harvey's patent stated vaguely that "the more intense the heat the better.") The armor shop was packed with machine tools. Most important, a small army of well-trained workers, supervised by competent, educated engineers, continually monitored the furnaces to achieve the necessary uniformity. "They have worked up the armor business in a very systematic and thoughtful manner," Johnston concluded.[86]

Much of the specific technology was familiar to the Americans. Krupp's regenerative furnace used for heating, cementing, and tempering was "similar" to Bethlehem's original Harvey cementing furnace; both featured a rail buggy to move the heavy plates in and out of the furnace. Although Bethlehem's other armor furnaces needed some alterations, Carnegie's twelve armor furnaces were "almost identical" to Krupp's and needed only hookups for the carburizing gas. Indeed, the only points of departure were the use of methane-rich illuminating gas to carburize the plates, the making and handling of nickel-chromium steel, and the precise control of manufacturing. Johnston detailed a seventeen-step process for manufacturing control, which yielded a tough and fibrous structure throughout the plate (to prevent it from cracking under the shock of impact) and a face that was as hard as glass (the secret to shattering the incoming shells).[87]

Given the technical brilliance of Krupp's manufacturing and the cross-licensing through the London Harvey company, the navies of the great powers had to adopt the Harvey-Krupp armor. The lone holdout in England, the Armstrong concern, already committed to the Harvey process, signed up for a Krupp license in 1898.[88] By the turn of the century the technological convergence was complete. The new Harvey-Krupp regime coupled the domestic dynamics of armor procurement in each country to international dynamics involving careful patent management, cooperative technology transfer, and competitive naval rearmament. The strength of these relationships were soon put to the test by a newcomer to the American armor field.

Midvale Steel mounted a concerted effort in 1900 to discover the Krupp and Schneider methods of producing armor, just as Midvale's president, Charles J. Harrah, Jr., made his first bid for armor. Naturally, Midvale investigated the leading European armor plants of Krupp at Essen and Schneider at Le Creusot. "The chief engineer at Creusot told me that they began to consider a week incomplete if a representative from Midvale did not visit their works," Bethlehem's metallurgical engineer reported from Paris that September, "and at Krupp's Works they have a record of over 20 different Midvale men who have visited them during the past year." The resurgent Philadelphia company installed a Whitworth heavy forging press, but managed to evade international patent entanglements by using its own patented process that varied in minor details from the Harvey-Krupp process. In December 1904 the navy tested a pair of Midvale 7-inch carburized armor plates at the Indian Head (Virginia) proving ground. The two established American armor makers watched the results carefully, and Bethlehem president E. M. McIlvain was forced to admit that Midvale had "succeeded, in making plates of more or less merit." In fact, it appeared that Midvale had simply replicated the efforts of the Bethlehem and Carnegie companies and slapped its own name on the process. Later that month McIlvain wrote Charles Schwab, Carnegie's former armor-shop superintendent and Bethlehem's new owner and chairman, about getting "the Harvey people to begin suit against [Midvale] for infringement of the patents."[89]

Nevertheless, the two established American armor makers soon found that the time-honored strategy of suing a competitor for patent infringement was especially risky in the armor field. One problem was that the Harvey and Krupp *patents* did not disclose the information necessary to conduct the Harvey-Krupp *process*. Both sets of patents, for instance, failed to disclose the crucial steps of tempering the plates before carburizing and hardening them as well as of annealing the plates afterward. Tempering and annealing were crucial because they gave the armor a fibrous and tough structure that prevented cracking upon the impact of a shell. In the absence of this deep structure, hardened plates were little better than the outdated compound iron-steel plates. Detailed letters between the Bethlehem and Krupp armor works are compelling testimony that much additional information beyond the patents was needed to succeed with the armor process.[90] A related point is that although patent lawyers did their best to create detailed technical arguments, patent suits in this context were only incidentally about such technical points. As Archibald Johnston, Bethlehem's onetime armor-shop superintendent and then vice president, put the matter, "Our idea of this whole armor situation is to use this [patent suit] as a lever for making better terms abroad."[91]

The more serious problem, as the Bethlehem and Carnegie companies discovered to their dismay, was that vigorously prosecuting a patent lawsuit could have disastrous consequences for their own contracts with the government. Working in concert they first set the redoubtable Senator Penrose to "stir up" the Navy Department about Midvale's tardy armor deliveries. Then they turned to secure the legal approval and practical assistance of the Harvey directors abroad. Harrah of Midvale reacted to this legal-political assault by viciously cutting prices, and in 1906 domestic armor prices hit an all-time low of $345 per ton. As customary, Navy Secretary Charles J. Bonaparte dropped the price to the lowest bid ($345 including royalties) and divided the contracts between the companies. Bethlehem and Carnegie split the armor for the *South Carolina* (1,824 and 1,865 tons, respectively), while Midvale received the entire lot of armor for the *Michigan* (3,690 tons).[92] The confrontational legal campaign against Midvale had hurt rather than improved the domestic position of the two established American firms.

Smarting from these ruinously low prices, executives at Carnegie and Bethlehem saw the wisdom of bringing Midvale into the fold, and they changed tactics. Much to the dismay of the Europeans, who had reluctantly agreed to sue Midvale, the two steel companies now obstructed the preparation of that patent suit and instead began negotiating with Harrah for domestic concord. All three U.S. companies shared an interest in increasing the price of armor, and with another round of armor bids due in June 1907 the trio agreed to allot a fixed amount of just over 2,000 tons to Midvale. "Probably the least suspicious procedure," stated a Bethlehem memo, "would be if Carnegie and Bethlehem were to follow the general practice of each bidding the same price for the entire tonnage, and letting

[Midvale] cut under to the extent of a few dollars per ton to secure the 2060 tons referred to above." True to plan, when bids were opened that summer Bethlehem and Carnegie each received just over 3,500 tons for the battleships *Delaware* and *North Dakota*, while Midvale received a 2,200-ton lot for the *North Dakota*. Most striking, the three steelmakers boosted the average price per ton back up to $420. "Our armor bids in America, you must know, were satisfactorily arranged and at no small effort," Archibald Johnston of Bethlehem wrote Millard Hunsiker, Carnegie's former armor expert who was then serving as the two companies' joint representative to the London Harvey concern. "Thank goodness, affairs have been conducted in harmony in America which is a thing that had never before been accomplished."[93] Given the Harvey-Krupp entanglements and Midvale's impetuosity, Carnegie and Bethlehem could have either patent protection or high contract prices, but not both.

Technological change promised an escape from these entanglements. "The foreign people handed us a lemon" with the Krupp process, one Carnegie executive stated in 1907, and "they will hand us another one next year if we have nothing." Since cemented armor was locked up by the Harvey-Krupp patents, Carnegie and Bethlehem experimented with various alloying elements, including vanadium, as well as with noncemented processes that relied on forging and heat treating alone to harden and strengthen the plate. They cooperated in this development effort by exchanging the details of manufacture as well as results of ballistics tests.[94] Bethlehem executives meticulously designed its new "noncemented" armor process to avoid infringing the Krupp patents. Despite the navy's support once again with the patent office, the companies failed to circumvent the cemented-armor patents—that is, the Harvey-Krupp process.[95]

By 1907 the Navy Department too desired relief from these foreign entanglements. It forced Bethlehem to investigate armor made by James Churchward (who was later revealed as an "imposter" who "knows absolutely nothing concerning the manufacture of armor plate"). Of more immediate promise was the Navy Department's artful interpretation of evidence gathered for the ongoing patent-royalty case of Harvey Steel against the United States (to collect royalties from all the American companies, including Midvale, that used the Harvey process). The evidence seemed to suggest that the Carnegie and Bethlehem companies had abandoned the Harvey process in 1904. If so, the navy was no longer liable for the Harvey royalties.[96] The result would be a permanent savings of staggering proportions as well as a retroactive savings of $336,000 (considering the Harvey royalties of $11.20 per ton on the 30,000 tons manufactured by Bethlehem and Carnegie during 1904–7). Nonetheless, so long as the lawsuit dragged through the courts, the navy was still not off the hook.

Though harmlessly coopted in the domestic armor market, Midvale's aggressive attempt to seek armor contracts in Europe caused havoc there. Italy was its chief target. In 1906 Midvale submitted a tender to the Italian marine minister to produce face-hardened Krupp armor, an offer that caused consternation in the Italian armor company, Terni. Formed in 1884 by direct state intervention to produce steel and armor for warships, Terni had installed experimental Harvey armor furnaces in 1892. It expanded its Schneider-built plant rapidly after 1895 with capital from the Banca Commerciale Italiana. In 1905 Terni constructed a Krupp-process plant, and the next year it launched an ambitious joint venture, Vickers-Terni Società Italiana di Artiglierie e Armamenti, to construct a large ordnance plant near the Italian proving ground at Spezia. The Terni concern already faced a rival Italian-English joint venture, Ansaldo-Armstrong, and as a Krupp licensee Terni expected to be protected against yet another encroachment on its home market. Condemning the "Midvale effect" on its armor prices, Terni wanted revenge. To protest the ineffective protection offered by the Harvey-Krupp concern, Terni obstructed an accommodating out-of-court settlement in the Krupp patent suit against Midvale. Although Italy was probably the greatest battlefield, Midvale's price cutting at home and aggressive stance abroad also depressed armor prices in Germany. All told, Midvale caused more damage to the established armor makers in Europe than to those in the United States.[97]

The tangle of legal proceedings against Midvale wound to a close in the years immediately before the war. In the Harvey company's suit seeking patent royalties, the U.S. Supreme Court decided that the U.S. government was indeed liable for royalties on the armor plate that all three domestic steel companies had produced for the navy. But the Krupp suit against Midvale for patent infringement was unsuccessful. After so many years, the Krupp lawyers could not produce sufficient evidence to prove that Midvale had infringed Krupp's process. Moreover, on an appeal the circuit court even struck down several of the Krupp patents themselves. This was a severe blow to the Harvey-Krupp concern, and the coincident expiration of some of its core patents as well as the gathering diplomatic crisis was too much. On 31 July 1912, the Harvey United Steel Company of London was liquidated.[98]

The Political Economy of Armor

The liberality of the government . . . has advanced the steel-trade interest far beyond any point it could have attained by reason of purely commercial influence.—AMERICAN MACHINIST

The breakup of the Harvey-Krupp patent cartel did not initiate any breakthrough in armor, for by this time the three leading powers had largely completed their naval rearmament. In the United States, navy appropriations and armor contracts actually shrank during 1909–11 (fig. 3.8). Similarly, in Germany Admiral Tirpitz's grand plan for naval rearmament was completed in 1909–12, and from then on the army received top priority. In Britain, too, naval budgets were falling, even though the naval scare of 1909 helped ensure the supremacy of naval interests. Germany's approach to numerical parity in dreadnoughts during that year, and trumped-up estimates of dreadnoughts to come, overthrew all caution about building new warships. As Winston Churchill, home secretary at the time, later quipped, "The Admiralty had demanded six ships; the economists offered four; and we finally compromised on eight."[99]

For all the great powers, the naval arms race leading to World War I transformed their political economies. The rise of technology-based monopolies in the armor sector ended the competitive heavy-industrial economy in Germany and brought different companies to prominence in the United States and Great Britain. Never before had so many government officials interacted so intimately with so many managers and executives from private industry. Probably it could not have been otherwise, for in this sector the user-producer interactions necessary for vigorous and sustained change inevitably included the government. Seen in this light, the Harvey-Krupp patent cartel only intensified and internationalized the characteristic dynamic of this sector. The consequences of naval rearmament persisted to the Battle of Jutland and beyond. In the United States, steelmakers proved adept at fixing prices during the war largely because of practice gained in the armor sector.[100]

Naval rearmament brought direct and obvious changes in the steel companies that came inside the government orbit. Profits from military contracting bankrolled Bethlehem's exit from the rail market in the 1880s, its diversification into structural steel after 1905, and its acquisitions and expansion in the 1920s.[101] Midvale never regained the prominence it had enjoyed in the 1880s. By the early 1920s it—along with Lackawanna Steel, Pennsylvania Steel, Cambria Steel, Fore River Shipbuilding, and a dozen lesser concerns— had been acquired by Bethlehem, investing its wartime profits to become the nation's second-largest steel producer (fig. 7.2).[102] The changes were just as momentous at Carnegie, where military contracting yielded a flow of retained profits that were a necessary condition for the company's self-financed transformation into the industry's most fully integrated producer. Seen in this light, Andrew Carnegie's remark in 1895 ("Good thing we are to get more Armor this year" because "there's no profit" in "general trade") underscores how the multimillion-dollar armor profits[103] provided the Carnegie concern with the means for purchasing in 1896–97 the iron ore lands and associated transportation that made it self-sufficient in this vital raw material. As chapter 4 recounts, these moves destabilized the railroad-and-steel system that J. P. Morgan had just salvaged from competitive disarray and economic depression, triggering the events behind the formation in 1901 of the U.S. Steel Corporation.

Other companies besides these three felt the government's influence indirectly. As a writer in *American Machinist* put it, political machinations by Pennsylvania's legislators kept other states out of the military market, limiting its direct effects. But when the three Pennsylvania armor producers deployed the military-funded techniques in the civilian market, they put unprecedented competitive pressures on all manufacturers. In this way, the "liberality" of the government advanced the steel industry as a whole "by leaps and strides far beyond any point it could have obtained by reason of purely commercial interest." The military-funded technical advances included domestication of heavy forging for guns and armor and large-scale engineering for steel warships, expansion of open-hearth furnaces, and experimentation with exotic alloys of nickel and chromium. The persistent pressure of government inspectors was no less important. These inspections, admitted Charles Schwab, who had suffered through them in the 1890s as Homestead's armor superintendent, "were the real means of producing the quality of material now so universally used in the industry." For *American Machinist*, however, "the most valuable and important result of the Government exactions for armor and projectiles" was the development of high-speed-tool steels, which had begun at Midvale in the 1880s and continued at Bethlehem in the 1890s. "Causing a demand for such steels is what our Government has done for our steel industry. [Such] indirect benefits . . . are actual and of incomparably more value than just the making of guns and the forging of armor."[104]

The rail-centered mentality of U.S. Steel, 1901: dinner celebrating corporate organization of U.S. Steel.

Courtesy Carnegie Library of Pittsburgh; negative A-146.

CHAPTER FOUR

The Merger of Steel

1890–1910

On Friday, 10 July 1885, two of Andrew Carnegie's oldest sources of capital hung in the balance, along with his philosophy of business and his ambition for inexpensive transport. Carnegie was singularly oblivious to the risk. Bitten with the bug of political journalism, he was in London supervising his stable of seventeen money-losing Radical newspapers. But the action that day was in New York.[1]

Ever since Carnegie's "retirement" from railroading in 1865, capital for his ventures had flowed abundantly and repeatedly from the House of Morgan and the Pennsylvania Railroad. Indeed, Carnegie's reputation as a super-salesman of railroad and bridge bonds resulted largely from his congenial relations with Junius S. Morgan, the international banker of London, and his contacts in the railroad world. Profitable schemes had regularly begun with Thomas A. Scott and J. Edgar Thomson, Carnegie's former mentors at the Pennsylvania Railroad. The Keystone Telegraph Company, organized by Carnegie in 1867 to compete with Western Union, flourished largely because the Pennsylvania Railroad had allowed the upstart company to string its wires on the railroad's poles for a nominal fee. Similarly, Carnegie's venture into railroad sleeping cars had depended on his mentors' financial backing and especially on their granting him exclusive rights to serve the railroad and its subsidiaries. Construction of bridges for the evolving transcontinental railroad system had resulted in the most byzantine arrangements. The classic Carnegie-Pennsylvania promotion was the Eads Bridge, part of the road's plans to cross the Mississippi at St. Louis (fig. 4.1). With the Pennsylvania promising to provide the traffic, Carnegie had concurrently organized the bridge-building company, run the bridge-fabricating firm, made the steel, and sold the bonds to finance the whole—including $1 million worth of bonds that he had sold to Junius Morgan in March 1870. Eighteen months later, Carnegie had placed with Morgan railroad bonds totaling $10 million, the product of a refinancing scheme again involving the Pennsylvania. In the summer of 1874, Carnegie had traveled to London with Alexander Holley to secure a $400,000 bond issue with Morgan to complete the Edgar Thomson steel mill. Carnegie's desire to stay in the good graces of Junius Morgan had even contributed to his painful fallout with Scott, mired in a doomed railroad promotion in Texas.[2]

By the early 1880s, however, the Carnegie-Pennsylvania concordat had crumbled. Thomson had died, Scott had left the road, and Carnegie, fully committed to his Pittsburgh steel mills, now saw the Pennsylvania as a principal threat to their profitability. With the Baltimore & Ohio, the Pennsylvania represented Pittsburgh's only connection to the Atlantic coast. The Pennsylvania alone was the steel city's link south to the Connellsville coal fields. And both roads, according to Carnegie, exploited their position to exact monopoly profits from a pitiable Pittsburgh steelmaker. When his private crusade to lower their rates went nowhere, Carnegie

FIG. 4.1. Eads Bridge, St. Louis, under construction, c. 1871.
From Thomas M. Cooley et al., The Railways of America: Their Construction, Development, Management, and Appliances *(London: John Murray, 1890), 95.*

threatened a public offensive. "I was so indignant at the extortions practiced upon us that I was almost ready to incite public indignation against the Pennsylvania Road," he admitted.[3] Achieving the lowest imaginable railroad freight rates became something of an obsession.

Carnegie's dream for cheap transportation materialized in the person of William H. Vanderbilt. The leading figure of the New York Central Railroad, Vanderbilt had had his eye on southwestern Pennsylvania for some years. The region's mineral resources were rich but untapped, and developing this traffic would extend the reach of his rail system. In 1879 the Philadelphia & Reading Railroad had surveyed a route that would connect Pittsburgh with the Reading's line at Harrisburg. If built, such a road would compete directly with the Pennsylvania—a consideration that appealed to Vanderbilt, who had reason to believe his rival was taking certain steps to threaten the New York Central main line running up the Hudson Valley. Working through the Reading, he secured control of the paper road in 1883 and began turning it into steel. Pittsburgh manufacturers, including Carnegie, embraced the new lifeline to the East; soon the project had $15 million of paid-in capital, enough to complete two hundred miles of single-track line between Harrisburg and Pittsburgh.[4]

Bids for the ill-starred South Pennsylvania Railroad were let in August 1883, and construction began the next summer. Believing he would soon be able to play his trump card, Carnegie set aside his plans for a public campaign against the Pennsylvania Railroad. By midsummer 1885 nine separate tunnels were proceeding well, steel rails were being rolled with dispatch at Carnegie's Edgar Thomson mill, and the road was planning full operation within fifteen months. But in late July Carnegie was notified without explanation that construction was being terminated; when he learned the details he was furious. "The first blow the South Pennsylvania Railroad received was the blow of an opponent that openly sought its capture," he told the Pennsylvania Legislature, referring to Vanderbilt. "The fatal blow was the act of the assassin, who stabbed secretly from behind and in the dark."[5]

At 10 A.M. on that fateful Friday in mid-July, a sleek motor yacht pulled up to the docks at Jersey City and took aboard two well-dressed gentlemen just in from Philadelphia. The two men were accustomed to luxury, for they were George Roberts and Frank Thomson (a nephew of Edgar Thomson), president and vice president of the Pennsylvania Railroad. Even so, the 165-foot motor yacht, with its sharply pointed prow and raked-back smokestack, must have impressed them. The invitation to spend a day on the water was welcome, given the summer heat, but considering that the vessel was the *Corsair*, owned by J. Pierpont Morgan, the son of Junius and a rising star in railroad finance himself, the two men knew that more than pleasantries were in store. Morgan welcomed them aboard at Jersey City. The fourth member of the party, who had joined Morgan dockside in Manhattan, was Chauncey Depew, president of the New York Central. The assassins had gathered.[6]

Morgan had been watching these two roads with increasing alarm. For years the Morgan and Drexel partnerships had been lead financial houses for the Pennsylvania Railroad; Pierpont himself was a director of the New York Central. Now the two roads were on opposite sides of the South Pennsylvania gambit, threatening a fresh round of ruinous competition. For also in peril was the New York Central main line running north from central Manhattan up the east bank of the Hudson River to Albany, where it turned west for the Great Lakes. Some years earlier a party of railroad promoters had duplicated this route with the New York, West Shore & Buffalo Railway, which shadowed the New York Central all the way to the Great Lakes. Construction during the summers of 1882 and 1883 had disrupted the Morgan family's retreat at Cragston, overlooking the West Shore line, but the echo of pickaxes and dynamite was the least of Morgan's worries that July morning. The Pennsylvania was quietly buying up the West Shore's mortgage bonds, then available at a bankruptcy-induced bargain. If revived, the West Shore line could connect to the Pennsylvania system at Jersey City and bring mayhem to the New York Central. A rate war was already on among the eastern trunk lines, and Morgan knew their finances were suffering. By November 1884 clients were "constantly" calling on the Morgans' senior partner in London to inquire whether the New York Central would pay its next bond dividend.[7]

Extraordinary measures were necessary to head off this destructive competition between the nation's two largest railroads, or so the Morgans concluded. Together father and son met several times with the roads' officers to urge some amicable settlement. In March 1885 Pierpont traveled to London for discussions with Vanderbilt, who was then vacationing in Europe; on the return ocean passage in early June he won Vanderbilt's go-ahead to settle with the Pennsylvania. By now, the railroad press was forecasting a titanic struggle between the two roads. Although Chauncey Depew and Pierpont Morgan separately held a series of meetings in Philadelphia with the Pennsylvania Railroad officers, a settlement remained elusive. "Negotiations still pending, feeling much more confident," cabled

Morgan to his father after returning from one five-hour meeting; the next day, he added, there would be an "interview" in New York.[8]

The interview began as the *Corsair* steamed slowly up the Hudson River, with the four men—Morgan, Depew, Roberts, and Thomson—seated comfortably under a large awning aft of the funnel. Depew began by outlining the case against rate wars, ruinous competition, and construction of parallel lines. He had been thoroughly briefed by Morgan, and he knew Vanderbilt's mind as well. Depew stressed the mutual profitability of an arrangement that would respect each company's "sphere of influence," ratifying the "community of interest" principle. The essential first step, he proposed, was to have the Pennsylvania take up the unfinished South Pennsylvania line and the New York Central take over the defunct West Shore line. Morgan for his part sat smoking a cigar, adding a word now and then for emphasis. "The question was discussed all day," Depew recalled. Thomson was soon won over to the cooperative plan, but Roberts, ever mistrustful of Vanderbilt, remained silent and uncommunicative. The *Corsair* turned about at West Point and steamed slowly down the Hudson, out into the New York harbor as far as Sandy Hook, then back to the Jersey City piers where the return train to Philadelphia awaited. It was by then 7 P.M. Roberts made no commitment until the moment he stepped off the yacht. Taking Morgan's hand he said, "I will agree to your plan and do my part."[9]

Morgan moved quickly to execute the fatal blow. By the end of July the New York Central had officially agreed to lease the West Shore "in perpetuity." Inside two weeks the West Shore was harmlessly brought within the New York Central system. Stopping the South Pennsylvania was more difficult because state law prohibited the Pennsylvania Railroad from buying it directly. Morgan personally took a 60 percent share in the road and then transferred his holdings to the Pennsylvania. The route across Pennsylvania remained disused for half a century until the Pennsylvania Turnpike cleaned out the tunnels and transformed the route into a four-lane highway.[10] Having averted ruinous competition, Morgan found his prestige rising in parallel with railroad securities, tariffs, and fares. A ticket from New York to Chicago, which had fallen as low as $7, now rose to $20. Beyond Chicago was another matter, however. A nasty rate war among western railroads demonstrated the basic instability of the industry, which the Interstate Commerce Act (1887) did little to address. Indeed, shaky railroad finances contributed to the economic panic of 1893 and the resulting depression.

What did stabilize the railroad industry was the reorganization of the mid-1890s, led by Morgan. Briefly, the consolidated railroad industry that emerged around 1900 resulted in unprecedented heavy loads on new heavy rails—which began breaking in odd circumstances. Addressing the ensuing "rail crisis" meant revamping the relations between rail producers and rail users, as new scientific expertise, new forms of standards, and new intermediary bodies replaced the old patterns of interaction that had failed so dismally. The associated scientific, technical, and industrial changes were as profound as those associated with the Bessemer steel rail. The compact reached aboard the *Corsair* had other repercussions as well. Mistrust between Carnegie and Morgan persisted to the point of spite, even as they built up the most impressive organizations in their respective realms. Carnegie Steel reached epic status with its achievement of vertical integration in the 1890s. At the turn of the century the two men's paths crossed again, and Carnegie and Morgan had one final transaction, which formed the core of U.S. Steel, the nation's first billion-dollar corporation.

Consolidating the Railroads

The rail question today . . . is a matter, you might say, of life and death. —W. C. CUSHING

The nation's railroads, as Albro Martin has emphasized, were built twice over—once in the nineteenth century and again in the early years of the twentieth. Both periods were crucial for the steel industry. The first, extensive phase of building culminated in three great spurts of new mileage as railroads rushed to cover the continent (see chapter 1). The second, intensive period of building featured smaller increases in mileage but involved colossal rail consumption. The all-time peaks for new rail mileage were in 1881 and 1887 (twelve thousand miles each year), while the all-time peak for total rail consumption was in 1906 (4 million tons). The two phases were separated by the 1893 panic, which hit railroads hard. "Never in the history of transportation in the United States has such a large percentage of railway mileage been under the control of receiverships," reported the Interstate Commerce Commission in June 1894. At that time 192 railroads were held by receivers, in default of their bonded debt and technically in bankruptcy. These defaulted railroads operated 22 percent (40,000 miles) of the nation's total mileage and represented 25 percent ($2.5 billion) of its total railway capitalization. By 1895 receivership had claimed three of the nation's five transcontinental systems, one large southern system, four eastern trunk lines, and a host of smaller roads.[11]

As in 1873 a fatal weakness in railroad finance had triggered the panic. Again the center was Philadelphia, but instead of Jay Cooke's financial house it was the Philadelphia & Reading Railroad in extremis. With pretensions of becoming a major eastern trunk line, that venerable road to the coal fields had amassed an unsustainable debt of more than $125 million on capitalization of $40 million. In February 1893 the Reading declared itself insolvent, tilting the country toward financial calamity, which hit three months later with the "spectacular failure" of the National Cordage Company. The immediate toll of 573 closed banks and 8,105 commercial failures was bad enough. The tumultuous events of the next year—including the march of Coxey's ragtag "army" on Washington and the unrest in Chicago accompanying the Pullman strike—revealed strain in the country's social and economic fabric. Railroad construction fell drastically; fewer than two thousand miles of new tracks were built in each year from 1893 to 1897. In 1895 a pitiful 1,420 miles were built, the lowest figure for three decades.[12]

Severe as it was, the panic was not the only cause for the situation of the eight largest defaulted railroads. Failures of the Norfolk & Western and the Northern Pacific were due to financial weakness resulting from overexpansion. In the years just before 1893, the Norfolk had embarked on a massive expansion plan that sent lines into Ohio and to undeveloped mineral areas in and beyond West Virginia; while its tonnage and mileage increased smartly, its net revenues languished with the low rates obtaining on its shipments of coal, iron, and cotton. When the depression hit, fully 28 percent of its mileage lay in promising but undeveloped regions. Similarly, the Northern Pacific's problems owed much to overextension and bad timing. The financing of feeder lines for its main line from Lake Superior to Puget Sound was suspended in 1890 by the failure of its London bond merchant, Baring Brothers, and the railroad opted to take out massive short-term debt rather than cancel construction contracts at loss. In contrast, the failures of the Erie, the Union Pacific, and the Atchison, Topeka & Santa Fe resulted mainly from management mistakes that had been public knowledge. The Erie, now in its fourth period of receivership, was no stranger to financial distress. One of its subsidiary roads had first, second, and third mortgages, and a prior lien that preceded them all; accordingly, all unpaid mortgage interest was added to capitalization, which spiraled out of control. The Erie itself was capitalized at $165,000 per mile, a dangerously high figure, yet the infamous subsidiary's capitalization per mile was 240 percent greater, beyond hope of ever being paid down. A series of bumbled reorganizations in the

1880s crippled the Atchison; a delayed but massive payment to the federal government of $52 million hung over the Union Pacific. Finally, the failures of the Baltimore & Ohio, the Richmond Terminal, and the Philadelphia & Reading became full-blown scandals when these roads' financial legerdemain became public. The books of one of the Richmond Terminal's subsidiaries, itself crippled by nearly 800 miles of light iron rails and Civil War–era locomotives, listed such "assets" as property destroyed by fires or lost in accidents, losses on traffic contracts, and worthless accounts receivable. Likewise, the Baltimore & Ohio's books were "in a hopeless tangle."[13] Lawsuits and recriminations, as well as prolonged financial disorder, were preempted when stringent reorganization plans were devised and implemented.

One financial firm beyond all others dominated the reorganization process. The House of Morgan's ability to reorganize railroads rested on financial engineering as impressive as the mechanical engineering of the transcontinental railroads had been. Indeed, J. Pierpont Morgan was described by Lincoln Steffens in the 1890s as the greatest of "the private bankers who are the constructive engineering financiers." Inheriting his father's international banking house in London, J. P. Morgan himself developed subsidiaries in Paris and Philadelphia and built up the New York branch into a major entrepôt for European capital, much of which flowed into American railroads. Morgan's philosophy of industrial finance was forged during his service as director of the New York Central and the New York, New Haven & Hartford roads; his participation in the failed 1885 reorganization of the Philadelphia & Reading; and his efforts to establish pooling arrangements between competing roads (before these were banned by the Interstate Commerce Act). Even the notorious gentlemen's agreements between competing roads to divide traffic or otherwise damp ruinous competition, such as the *Corsair* compact, were not a strong enough bond to realize Morgan's grand objectives. He wanted cooperation between rival concerns, stability of the industrial system, and profitability for investors at home and abroad. With the depression-given opportunity to devise the reorganization of four major roads, and to implement the reorganization of several others, Morgan managed to achieve these goals and more.[14]

Morgan was personally occupied with massive gold-bond issues for the U.S. Treasury during 1894–96, and he could never have reorganized so many roads so effectively without his leading partners. Chief among them for railroads was Charles H. Coster, who was described as "a kind of rare genius, a sort of financial chemist." Coster had the gift of grasping the proper value of each of the innumerable bonds, shares, and securities under scrutiny and also seeing, where many were baffled, how to implement the specific goals of reorganization. At his death the *New York Times* included him "among the four or five great organizing minds" of the city. Like Coster, Samuel Spencer, an outside expert for most of the railroad reorganizations, had tremendous capacity to master detail. "There was no man in the country," reported the *Times*, "so thoroughly well posted on every detail of a railroad from the cost of an air brake to the estimate for a terminal." Morgan's leading attorney, Francis L. Stetson, was chiefly responsible for the remarkably small number of lawsuits ever brought against a Morgan reorganization; he is also credited with inventing no-par-value stock and a new form of railroad mortgage debenture.[15]

As effective as the new financial techniques was the simple premise of a Morgan reorganization: all stakeholders made a sacrifice. If a simple majority of stockholders and a sizable majority of bondholders agreed, the railroad was in effect turned over to Morgan's experts for surgery and reconstruction. It was no quick fix, requiring from fifteen to forty-eight months of intensive negotiations, complex legal proceedings, and careful underwriting of new securities. A Morgan reorganization began with accurate calculation of the road's minimum income and corresponding adjustment of all financial parameters. The goal was to ensure viability in bad economic times, a test the defaulted roads had conspicuously failed. Excessive fixed charges—which for the defaulted roads had averaged 110 percent of net income—were typically reduced 30 percent by slashing the interest due bondholders. New working capital, as well as resources to retire short-term debt, was raised by assessing stockholders for additional funds or forcing them to accept new issues at heavy discount.

A Morgan reorganization then turned to management. If a road was deemed to require improvements and construction, the plan provided additional working capital and specified that it be properly employed. Unprofitable subsidiaries were excised, and the remaining subsidiaries, which had often persisted in the rivalries they had pursued as independents, were brought under effective central control. Until a road was completely rehabilitated and once again paying dividends on common stock, a three- to five-man voting trust nominated by Morgan assumed the shareholders' powers and assured that the stringent goals of reorganization were implemented. In this fashion, Morgan and his experts reconstructed four major railroads: the Erie, the Philadelphia & Reading, the Northern Pacific, and the Southern (made from pieces of the Richmond Terminal). They also participated in the financing of four more reorganizations. The effort, writes Morgan's biographer, was an "extraordinary accomplishment" that brought the firm "unprecedented power and influence."[16]

The first major railroad reorganization of the 1890s in which Morgan did *not* take the leading role was that of the Atchison, Topeka & Santa Fe. In the years before the depression, its managers had tried without success to patch together a series of regional railroads to form a transcontinental line serving the great Southwest. The Atchison emerged from receivership in 1896, under new management, and embarked on a successful campaign to realize its transcontinental promise. The following year it replaced nearly four hundred wooden bridges and trestles with steel ones to prepare for heavy locomotives, began double-tracking from Kansas toward Chicago, and concluded a swap of lines with the Southern Pacific that gave the Atchison a through line from Chicago to Los Angeles. By 1900 traffic was $5.6 million higher than the preceding year, while total operating expenses actually fell by $85,000.[17] Similar improvements in financial structure and technical infrastructure followed the rehabilitations orchestrated by James J. Hill, with Morgan's assistance, for the Northern Pacific and the Baltimore & Ohio as well as that conducted by the New York bankers Kuhn, Loeb for the Union Pacific. By the first years of the twentieth century Edward H. Harriman had shaped that road, along with the Illinois Central, the Central Pacific, and the Southern Pacific, into a transcontinental empire that was Morgan's only serious rival.[18]

The railroads that avoided financial crisis took steps during the 1890s to rebuild themselves. The Pennsylvania Railroad, the nation's largest in capitalization, already stretched from Jersey City in the east through Pennsylvania to western terminals at Chicago and St. Louis. In the first years of the depression, the Pennsylvania elevated its tracks in Elizabeth, New Jersey, and North Philadelphia, to avoid grade-level cross-traffic; linked Philadelphia more effectively with the popular Jersey coast by bridging the Delaware River; and committed $2 million to regrading grades and curves. In the late 1890s the Pennsylvania, already double-tracked from Jersey City to Chicago, began quadruple-tracking its most heavily used lines. By 1900 it was quadruple-tracked from Jersey City to Harrisburg and pressing on to Pittsburgh. In the next decade it spent $500 million in a vast reconstruction project that virtually rebuilt its entire freight system east of Pittsburgh and culminated in tunneling under the Hudson and constructing the magnificent Pennsylvania Station (1902–10). Similarly, the New York Central went from strength to strength. Its principal line, up the Hudson and Mohawk valleys and along Lake Erie to Chicago, carried a mix of freight that the depression scarcely reduced. Chicago's vigorous growth ensured profitable traffic for its well-run subsidiary, the Lake Shore & Michigan Southern. In 1898 it solidified its position in the Midwest with a substantial purchase in the Cleveland, Cincinnati, Chicago & St. Louis Railroad. Increased traffic and deadly congestion compelled the rebuilding of its Manhattan terminal, which resulted in the new Grand Central Station (1899–1901). By 1898, when the cooperative "community of interest" model had been established among the New York Central, the Pennsylvania, and the Morgan-Hill roads, the nation's railroad system had been effectively stabilized and substantially rebuilt.[19] Restoring the railroads to financial health set the stage for the wave of mergers that transformed American industry in the ten years after 1895, including the steel merger detailed later in this chapter.[20]

The consolidated systems resulted in exceptional physical burdens on the track, especially when traffic picked up in the late 1890s. Greater passenger and freight traffic was now met not by building more lines or buying out a competing line, but by making more intensive use of existing lines by running heavier engines and rolling stock and longer trains. While the nation's overall system mileage grew steadily from 1890 to 1910, the intensity of system use grew even more (fig. 4.2). Impressive enough in the passenger system, these changes were especially pronounced in the freight system that accounted for around three-fourths of total traffic. In these twenty years, the weight of an average freight train (tons per train) more than doubled. Heavier trains demanded stronger, heavier locomotives, and consequently the maximum axle loads on rails, which had already jumped 25 percent from 1885 to 1890, rose ever higher. Given these strains railroads had to adopt rails of unprecedented weight.

Meeting these traffic loads might have been a simple matter of getting the rail mills to adjust their rolling machinery. At first glance, prospects for the railroad industry had never looked better. Receivership or rationalization had returned the industry to financial health, and the demand for freight and passenger service was encouraging. Federal regulation was little more than an administrative nuisance, at least until the Hepburn Act of 1906 gave the Interstate Commerce Commission rate-setting powers.[21] But in the first years of the twentieth century, steel rails unaccountably began chipping, cracking, splitting, mashing, breaking, and generally wearing out.

The dismal performance of the heaviest rails especially frustrated railroad managers. Rails weighing 60–70 pounds per yard had sufficed even for main-line use throughout the boom of transcontinental railroading in the 1880s. The extraordinarily heavy traffic after 1895, however, required rails weighing 85 and even 100 pounds per yard.[22] The shift was especially pronounced after 1903, as figure 4.3 shows. Curiously, these heavier rails seemed to be less durable than the lighter rails rolled twenty years earlier. Beginning around 1900 a litany of complaints against soft rails, piped rails, poorly wear-

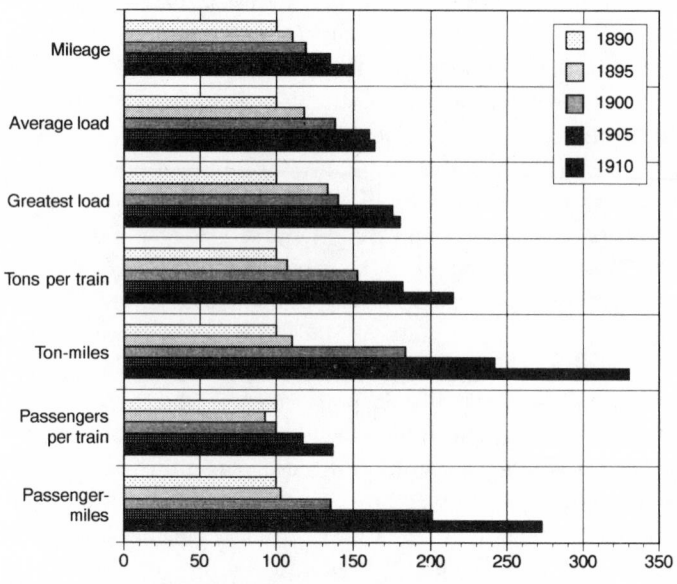

Total operated mileage (100 = 167,000 miles)

Average load per axle (100 = 137,500 pounds)
Greatest load per axle (100 = 150,000 pounds)

Tons per train (100 = 177 tons)
Ton-miles (100 = 77.2 billion)

Passengers per train (100 = 41 passengers)
Passenger-miles (100 = 11.8 billion)

FIG. 4.2. Railroad system mileage and loads, 1890–1910. For each data series, the data have been indexed so that the point for 1890 equals 100. *Data from William H. Sellew, Steel Rails: Their History, Properties, Strength and Manufacture (New York: D. Van Nostrand, 1913), 15; Department of Commerce and Labor, Statistical Abstract of the United States (Washington, D.C.: Government Printing Office, 1920), 337.*

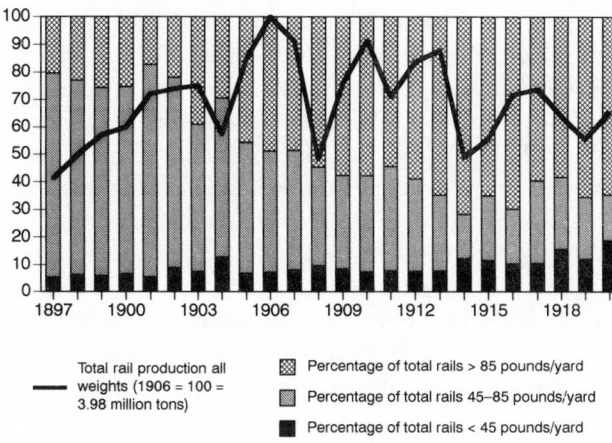

FIG. 4.3. Steel rail production by weight per yard, 1897–1920.
Data from AISI, Annual Statistical Report *(1912 and 1920)*.

ing rails, and generally unsatisfactory rails filled the engineering
literature. That year W. C. Cushing of the Pennsylvania Railroad
claimed that his road had never received "worse rails in our history." The manufacturing of rails was obviously at issue, and the
key questions were how to make changes in the rail mills themselves, and how to revamp the relationship between rail users and
producers.[23]

A cadre of activist railroad engineers spearheaded a successful
approach to the rail crisis through the newly formed American Railway Engineering and Maintenance of Way Association (AREMWA).
Its membership of railway engineers, railroad managers, and rail
mill inspectors constituted the country's leading reservoir of technical knowledge about rails. Formed in 1900, the association saw
in the rail crisis a supreme opportunity to address a difficult problem beyond the scope of existing engineering societies. The railway engineers worked closely with the American Society of Civil
Engineers (ASCE), whose standard rail section of 1893 had been
adopted by 84 of the nation's 128 railroads, providing a focus for
technical debate on rails. Both railway and civil engineers helped
write specifications for steel rails through the American Society
for Testing Materials (ASTM), the U.S. offshoot of an international
standards-setting body. All three of these societies formally backed
the ASCE sections, which some railroads had blamed for the rail
crisis. These engineers understood that the problem was not in
the rail's *section* but rather in the *metal*. For expertise in metals,

the railroad engineers turned to the American Institute of Mining Engineers (AIME).[24]

For the metallurgists, the rail crisis of the 1900s was a timely stroke of luck. It was a high-profile problem for which physical metallurgy could offer pertinent, and sometimes counterintuitive, insights. First, unlike the Bessemer process, the rail crisis was not primarily a question of correct chemistry. Robert W. Hunt, one of Alexander Holley's first assistants and himself an early advocate of chemical metallurgy, argued repeatedly that chemical composition was now secondary to how the metal was physically and mechanically treated. As an inspector of steel rails—an intermediary between producers and consumers—Hunt could rise above commercial partisanship. To understand why rails gave good or poor service, he observed, "we have to look beyond the chemical constituents." Hunt and others drew on metallurgical insights that suggested straightforward remedies for the defective heavy rails. Second, the metallurgists understood that the problem resulted from a revolution in rolling-mill practice achieved during the preceding two decades. By a host of ingenious tactics and techniques, rail mills had cut the total time needed to form a rail from a moderate 10 minutes down to a slapdash 2 minutes.[25] The tyranny of Bessemer technique, already the scourge of structural steel, was revealing its consequences.

Bessemer rail-mill practice around 1880 had been a comparatively unhurried affair. At the blast furnace, ores naturally low in phosphorus and sulfur were used, and the crude iron was tapped out and cast into pigs at a reasonable pace, then allowed to cool completely—to "come to rest molecularly" as one engineer put it. A mix of pig iron selected for its desirable chemical composition was remelted in cupolas and blown into steel using Bessemer converters of around 5-ton capacity. The resulting ingots, 12–14 inches square, weighed 1,400 pounds—enough metal for two 30-foot rails of 60 pounds per yard. The ingots were again allowed to cool before being stripped of their molds and charged into a reheating furnace. When an ingot reached the appropriate heat, it was rolled down to a bloom 7 inches square, using a dozen or more passes through the rolls. Any cracks appearing in the metal were chipped out with a hammer. The bloom was once again cooled down completely. The mystique of steel, at that time, carried some "controlling influence on the worker," who knew intuitively that steel would not stand high heat, and so before the steel blooms were charged into it the final reheating furnace was partially cooled off. The blooms received up to 90 minutes of thorough heating, then went to the

FIG. 4.4. "Reckless" mass production of steel rails, c. 1890.
From Cooley et al., Railways of America, 39.

final rolling. For making an average 65-pound rail, the final rolling
was done ("very slowly, very carefully") in thirteen to fifteen passes
through a rolling train running at 400 feet per minute. "More or
less care," wrote Robert Hunt, "was exercised on each and every
rail." As a result of this "deliberate and thorough" working, wrote
one rail-mill manager, the steel was never "subjected to violence"
and the treatment produced "a most salutary effect on its molecular
structure."[26]

By 1900 Bessemer rail-mill practice did violence to the metal.
In response to the depression of the 1890s, in which railroads had
cut orders to the bone and rail mills had competed mercilessly,
the mills had elevated reckless mass production to a dangerous
art (fig. 4.4). The steel was worked at temperatures as hot as it
would stand; the single-minded goal of "volume of production was
the thing most considered." From the blast furnace that smelted
the ore to the saws that trimmed rails to their proper length, the
metal was continually in motion and constantly hot. Both high-
and low-phosphorus ores were used in the blast furnaces, where
"hard driving" techniques made for prodigious output. Charges of
raw iron as large as 20 tons were tapped as needed from the Jones

mixer into a waiting converter and there blown hard and rapidly into steel, in a scant 10 minutes. The white-hot liquid steel was cast into mammoth 4,000-pound ingots, 6 feet long and up to 22 inches square. Their outsides barely solidified before being mechanically stripped of their molds, and their insides remained fluid for some time thereafter. Such ingots were either charged into heating furnaces (soaking pits) or sent directly to the blooming mill to be reduced to a bloom 8 inches square in a dozen fast passes, with no time wasted by chipping out defects. The 8-inch bloom was again not allowed to "readjust its outraged and crushed particles" by cooling or resting, and was either kept hot in a furnace or stuffed directly into the finishing rolls, where it was reduced to a 70-pound rail in just nine passes by rolls running at 900 feet per minute. Heavier rail sections required even less rolling. A mere 2 minutes separated the rail trimmed to length from the bloom heated nearly to melting. According to some descriptions, it never really solidified at all. "The metal," as one engineer put it, "is squirted through the rolls."[27]

To combat the rail mills' habit of reckless mass production, the railroad engineers needed the insights of physical metallurgy, and the metallurgists made the most of this new audience. The railroad engineers learned that two key lessons were "of the greatest importance" to rail-mill practice. First was the "heat and work" principle that centered on sustained mechanical treatment. The most persuasive evidence here was the improbable success of E. W. McKenna's rail-rerolling process. McKenna had rented space at the North Chicago Rolling Mill in 1895 and gone into production on his own two years later; by 1901 his factories in Joliet, Kansas City, and New York had processed 100,000 tons of rails with impressive results. His secret was simple: an old rail was heated in a special furnace to 1700°F, not even as hot as the finishing pass in the fast practice. It was then given three sets of passes through rolling trains: an initial pass that closed up the rail and produced something like a long bloom, a roughing pass that impressed the outline of a rail, and a finishing pass that yielded the final form. The finishing temperature was a most modest 1480°F. The rails from McKenna's process, up to 80 pounds per yard, were reduced in section by 6–10 percent and were correspondingly longer.[28]

The second lesson was that the ever higher temperatures used by the rail mills had been a grave mistake and that substantially *lower* rolling temperatures were urgently needed. Lower temperatures were cogently justified by the metallurgists' physical theory of hardening: that steel possessed a complex microstructure whose nature varied greatly with mechanical and thermal treatment, and was not merely a homogeneous mass of "chemical constituents." Furthermore, the metallurgists' instrumentation had visible impact. From behind every rolling train, it seemed, sprouted a pyrometer to measure the rail's temperature or a microscope to reveal its microstructure. In the 1880s rails had been finished at a cherry-red heat (now measured at 1,400–1,600°F) that yielded a fine granular structure and durable rails. By 1900 rails were finished as hot as an orange heat (2,000–2,200°F), which yielded a coarse crystalline structure and inferior rails. Investigations by Albert Sauveur at Harvard and the Carnegie, Maryland, and Illinois steel companies, as well as by the New York Central and the Philadelphia & Reading railroads, all showed that lower rolling temperatures, especially during finishing, were necessary to achieve the desirable fine-grained structure.

Engineers turned to realizing the benefits of lower heat in the rail mill. An early effort took place at Carnegie Steel's Edgar Thomson works, where Superintendent Thomas Morrison and consulting engineer Julian Kennedy devised a patented process that extended and rearranged the rail-finishing mill. "A heavy outlay of money" and three weeks of downtime for construction were required before the mill restarted in December 1900. The process featured a special 90-foot-long cooling table that held back the flow of partially rolled rails for some seventy seconds just before the final rolling pass. The configuration of the table permitted the hot rails to cool down uniformly (with the thick head of one rail transferring its heat to the thin flange of the next). Still continuous, the Kennedy-Morrison process dropped the final rolling to a desirable 1,580°F— down sharply from the previous 1,800°F.[29]

Working from a more explicitly scientific base, Simon S. Martin of Maryland Steel achieved results "nearly identical" to the new Carnegie practice. Martin reasoned directly from the theoretical insights of physical metallurgy to the practical characteristics of rail rolling. He began with a twenty-year-old catechism of physical metallurgy. When steel cools from a high heat, the cooling momentarily halts at certain temperatures; at these "points of recalescence," or "critical points," molecular rearrangement occurs. Low-carbon steel had two or three such points, but rail steel had a single critical point around 1,300°F. The microstructure of rail steel depended on the rate of cooling to the critical point: the higher the initial temperature, as well as the slower the cooling, the coarser the grain (and the weaker the material). Below the critical point the structure of steel did not change, and above the critical point mechanical working could break up a coarse granular structure into a desirable fine-grained structure, so Martin reasoned that to get the best structure a rail must be rolled as near the critical temperature as possible. Moreover, to obtain the desirable microstructure throughout the rail, and not just at its surface, the temperature must be reduced early in the rolling process. His reasoning suggested that rails should be rolled direct from hot ingots that were already heated uniformly throughout their interior and at temperatures of 1,550°F or less. By contrast a reheated ingot was inevitably overheated on the outside, yielding rolling temperatures of 1,850°F or more. Conveniently enough, Maryland Steel, as well as other direct-rollers like Lackawanna Steel, already had rail mills that could easily be modified to conform to these guidelines.[30]

Visual evidence from the microscope graphically illustrated these points. Articles on the rail crisis in *Railroad Gazette, Iron Age, The Metallographist,* and other engineering journals featured photomicrographs showing the desirable fine-grained structure achieved with low rolling temperatures. For decades, a crude form of physical metallurgy had been practiced every time a worker or foreman took a bar of metal, broke it, and examined the fracture by eye. Microscopes added a compelling visual picture of the microstructure of the metal.[31] Moreover, physical metallurgists could use iron-carbon-temperature diagrams to predict for each level of carbon the temperature that would yield the most suitable microstructure and hence the most satisfactory rail. Such evidence was cited directly in AREMWA discussions.[32]

The best evidence of structural metallurgy at the rail mill was in the "shrinkage clause" that became part of standard contracts after around 1905. Once excessive rolling temperatures had been defined as the problem, the solution required devising a way to ensure lower rolling temperatures. The shrinkage clause relied on the simple fact that steel expands when heated and shrinks when cooled. Just seconds after passing through the finishing rolls, rails were cut—or "hot sawed"—into standard lengths (28, 30, or 33 feet). To allow for contraction upon cooling, rails were cut somewhat long; their measured contraction after hot sawing was an *indirect* practical measure of their final rolling temperature. For instance, a 30-foot rail finish-rolled at an undesirable orange heat (2,000–2,200°F) might shrink by as much as 10 inches; the same rail finished at a desirable dull-red heat (1,300°F) might shrink only 5 inches.

The shrinkage clause, affirmed William R. Webster, the rail-inspecting engineer from Philadelphia who chaired AREMWA's key rail committee, "is an absolute check" on the rail's finishing temperature. Beginning in October 1901 Pennsylvania Railroad contracts stipulated that rails should be finished at a dull red heat. But since a "dull red" heat varied somewhat from person to person (and dramatically from day to night), the Pennsylvania was forced to specify that, with no artificial cooling between the finishing pass and hot saws, 30-foot rails of 100 pounds should shrink by no more than 5⅝ inches after hot sawing, and 85-pound rails should shrink by no more than 5½ inches. Rail mills thus set their hot saws at 30 feet plus the allowable shrinkage—and adjusted their rolling procedure to meet the shrinkage clause. The Pennsylvania's shrinkage clause was quickly adopted by the Norfolk & Western, which reported excellent deliveries from National Steel, Maryland Steel, and Carnegie Steel at "no variation in price," and formally endorsed by AREMWA. The shrinkage-clause concept was widely discussed by railroad and steel-mill engineers, and by 1905 the shrinkage clause—or explicit temperature requirements—was a common feature of rail contracts and steel rail specifications.[33]

Even with the lowering of rolling temperatures, rail failures were still troublesome. Two additional defenses against brittle or defective steel rails were devised, discussed, and included in standard specifications. The first was the "drop test," in which a 3-foot rail segment had to withstand the impact of a 2,000-pound weight dropping, say, 20 feet. Excessive bending could be measured; breaking was obvious. Manufacturing representatives favored the status quo of testing one in five Bessemer heats, but railroad officials successfully pressed to drop-test every single Bessemer heat. The second defense was the "shearing" of ingots—a certain if costly method of improving rails by discarding as scrap a measured portion of the steel ingot's top, which was often spongy, piped, or otherwise inferior.[34] While armor mills discarded up to half of their oversize 50-ton ingots, rail mills rightly feared that railroads would never pay the necessary premium prices (armor prices were fifteen times higher than rail prices). Where rail mills had typically sheared 10 percent off the top of the ingot, the railroads pushed up the standard contract figure to 25 percent. A third defense, using special alloy steels, was widely discussed. Several railroads, including the Pennsylvania and the New York Central, experimented with special rails containing nickel, but although results were encouraging the high cost prevented widespread use of these rails.[35]

Persistent penny pinching by the railroads gave rise to a complex series of developments. During the depression of the 1890s, when the price of rails fell below $18 a ton, rail buyers had become hypersensitive to price. Price variations of as little as 1 percent had sent a contract to one rail mill or another. The resulting insecurity spurred the rail makers in May 1901 to form a producers' cartel that locked in the handsome price of $28 a ton. Standard steel rails remained at this level through 1915. Although railroads bitterly complained that the producers' cartel was compromising rail quality, they were notoriously unwilling to pay any premium for better rails. "The buyer of steel rails," Robert Hunt stated, "should . . . pay what they are worth, or . . . cease complaining."[36]

Just as it is easy to date the beginning of a political revolution and impossible to date its end, so too with the rail crisis. In 1906 railroads like the New York Central complained bitterly about having three times as many broken rails as in the previous year. Then in 1907, when rail orders collapsed with that year's financial panic, the perception of crisis in the engineering community vanished and did not reappear even after production picked up again around 1910.[37]

In a few years the tyranny of Bessemer technique was replaced by a more quality-conscious, continuous open-hearth process, but it is not clear (except to its self-interested promoters) that open-hearth steel was in itself a solution to the rail crisis. Rather, the pattern of development and diffusion suggests that the growing use of open-hearth rails was a *defensive* response to abrupt changes in the market for Bessemer-grade ores, detailed below. A continuous open-hearth process originated with work by Benjamin Talbot, an English metallurgical chemist recruited by the Southern Iron Company in 1889 to install open-hearth furnaces at its mill in Chattanooga. There Talbot devised a technically successful "duplex" process that removed excessive levels of silicon and phosphorus from the economical grades of southern iron before they were run into the open hearth for refining into steel. Financial difficulties, however, prevented further development. A decade later his process spread to northern steelmakers through Pencoyd Iron of Philadelphia, where Talbot devised a special 75-ton open-hearth furnace that tilted, permitting it to be charged with partially converted fluid Bessemer metal. Talbot's process was continuous because a portion of the steel remained in the hearth to help refine the next charge. What most impressed contemporaries was not the furnace's continuous aspect but rather its ability to pour off phosphorus-enriched slags and so remove the principal cause of brittleness in steel. Scrap-rich northerners lamented that Talbot's furnace could not use steel scrap, which by 1900 could often be obtained more cheaply than pig iron.[38] A fully continuous open hearth was realized with the addition of Samuel Wellman's mechanized charging machines, which facilitated the economical use of steel scrap. Thus began the industry's half-century quest for supplies of scrap (see chapter 7).

Firms that did not gain access in the 1890s to iron-rich, low-phosphorus Bessemer ores had little choice except to adopt open-hearth furnaces. In the six years after 1901, with the best Bessemer ores locked up by the U.S. Steel Corporation, the cost of Bessemer pig iron (the product of smelting with Bessemer-grade ores) increased 47 percent. "Today," wrote one sales manager in 1907, "there is but little available low phosphorus ore in the United States." [39] Because rail prices remained fixed at $28 per ton, it was precisely those mills with the worst access to high-grade Bessemer ore supplies—for example, Tennessee Coal and Iron—that led the switch to open-hearth rail steel. Jones & Laughlin, the most integrated of the "independent" mills in Pittsburgh, found itself excluded from the best grades of Bessemer ore and consequently erected thirteen Talbot-process open-hearth furnaces between 1901 and 1912.[40] It appears that the rapid increase in open-hearth rail production after mid-decade (fig. 4.5) represents a forced technology choice, at least for those firms excluded from Bessemer ores, as much as a truly superior product.

The issues behind open-hearth rails can be examined in some detail through Bethlehem's reentry into the rail sector. After entering the military market in the 1880s, the company had gradually phased out the manufacture of Bessemer rails. At the turn of the century it began planning a new open-hearth plant to expand its product line beyond armor and guns, to include merchant bars,

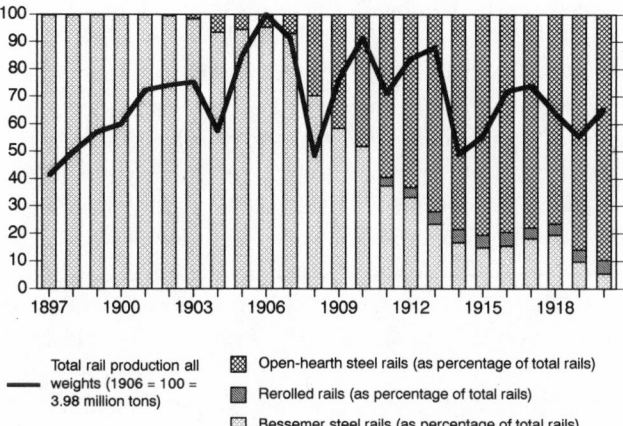

FIG. 4.5. Steel rail production by process, 1897–1920.
Data from AISI, Annual Statistical Report *(1912 and 1920).*

structural shapes, and rails. Because it too was excluded from U.S. Steel's high-grade Bessemer ores, Bethlehem opted to develop iron-ore mines in Cuba at considerable effort and expense. Despite an energetic campaign in 1907 to promote the new open-hearth rails as a "quality" product, the company found that these rails commanded a premium over Bessemer rails of only $2 a ton. And some of this modest differential must be attributed to the lower phosphorus content of open-hearth rails; without becoming brittle, low-phosphorus rails—whether open-hearth or Bessemer—could be higher in carbon and hence tougher and longer lasting. Judging from Bethlehem's sales figures, premium prices were commanded only by rails containing nickel, chromium, and other exotic alloys. Facing a temporary turndown in the armor market, and a financial crisis during the panic year of 1907, Bethlehem was all too willing to melt down its nickel-chromium armor scrap and convert it to cash. While its sales campaign asserted that the advantages of open-hearth rails were "too fully recognized . . . to require discussion," Bethlehem nonetheless mentioned that these rails would undergo more passes in the rolling mill and be finished at a proper temperature, so that they would possess a fine crystalline microstructure "insuring better wearing qualities."[41]

Ultimately, what resolved the rail crisis was the revamping of relations between the producers and users of rail steel. By around 1907 all of the technical aspects of making satisfactory heavy rails—adequate top discards, drop tests, and lower rolling temperatures—were well known. Discussion at that year's AREMWA meeting focused on improving interactions between the rail makers and rail users to deliver on the practical promise of this technical knowledge. In the words of P. H. Dudley, the designer of the New York Central's rail specifications, "Making satisfactory rails is only a question of getting together and doing it." Bethlehem vice president Archibald Johnston concurred that such cooperation was "the meat of the coconut." Joseph T. Richards, the Pennsylvania Railroad's chief engineer, concluded: "We ask the mill men to come in and help us make a good rail."[42] Constructively addressing the rail crisis required organizing well-known engineering knowledge in new ways. The railroad engineers as well as new independent intermediaries, such as the rail-inspection firms of Robert Hunt in Chicago and William Webster in Philadelphia, flourished because the preexisting ways of organizing and applying technical knowledge were not working.

The rail crisis also provoked the need to apply well-known scientific knowledge in new ways. Indeed, the application of microstructural metallurgy provides an apt illustration of how "industrial technologies have shaped the demand for scientific knowledge."[43] The puzzle is why microstructural metallurgy found its initial application in the United States thirty years after it was developed in England and some twenty years after it was first used in French and German industry. American metallurgists were not ignorant of European advances in physical metallurgy; indeed, Albert Sauveur at Harvard, as well as Henry Howe at MIT and then Columbia, contributed to these advances. But from the 1880s efforts to promote microstructural metallurgy in the United States were unavailing.[44] Such fundamental principles of steelmaking, as Robert Hunt put it, had been "completely ignored" owing to the stress of business competition. Sauveur himself relished telling of his firing from Illinois Steel in the early 1890s: rejecting his scientific investigations, management assumed that any step to lower the rolling temperatures would interfere with output, "and this was not to be considered."[45] What changed after 1900 was not the basic scientific insights of physical metallurgy, available since the 1880s. Rather, what changed was the market for scientific knowledge—a result of the consolidation of the railroads and the subsequent rail crisis for which the reigning chemical metallurgy could provide few insights. During these same years, the consolidation of the nation's railroads also linked the destinies of J. P. Morgan and Andrew Carnegie. In this way, the changes in the steel industry's largest user were organically linked to changes in its largest producer.

Integrating the Company

Fortunately, I woke up in time.—ANDREW CARNEGIE

Vertical integration at Carnegie Steel in the 1890s ensured that it, along with the other companies that made comparable moves to control raw materials, would dominate the maturing steel industry. It is no accident that the majority of companies that integrated vertically during the late 1890s wound up in the U.S. Steel combine of 1901. The structure of the mature steel industry took shape when Carnegie Steel, already in control of the limestone and coke needed for smelting iron ore, moved to gain control over iron ore as well, forced its competitors to do likewise, and ultimately triggered the most important consolidation in the industry's history, with Morgan again at the helm. For those companies with sufficient financial means, vertical integration reduced raw material costs and also erected barriers to entry that hampered would-be competitors.

The existence of some "irresistible process of vertical integration" was not clear, however, to Carnegie and the others until after the fact.[46] Vertical integration, defined as purchasing (or controlling) additional units "upstream" or "downstream" from the core manufacturing stage, began in 1871 when Andrew Carnegie formed a subsidiary blast-furnace company to supply pig iron for his Union Iron Mills. Six years later, when the energy and enthusiasm of three early partners were flagging, Carnegie confided in a younger, more aggressive partner that he intended forming a new partnership around his iron smelting, rolling, and steelmaking concerns, "and go it on that basis—the largest and strongest concern in the country." Although around this time Carnegie identified iron ore as an "all important" concern about which "we can afford to run no risk," oddly enough his solution was overseas. "The foreign ores are No. 1 no question about them. I hope you will make them your chief reliance."[47]

These plans found expression in the 1881 reorganization that founded Carnegie Brothers & Company. At this time the Edgar Thomson works (opened in 1875) lost its separate corporate identity and joined the Union Iron Mills, the Lucy Furnaces, the Scotia ore mines, and several Carnegie-owned coal mines and coke ovens in the Connellsville district in far southeastern Pennsylvania. The 1881 consolidation appears to have been motivated not by considerations of efficiency but by fallout from "the internal discord in which all the Carnegie enterprises were born and brought up."[48] Profitable investments in the securities of railroad, bridge, and iron-rolling ventures, as well as the commissions for selling these securities in New York and London, gave Andrew Carnegie the wherewithal to make major purchases. Yet it was the bitter infighting among his partners and among some of their competitors that gave him the opportunity to buy up shares in all these partnerships and consolidate them under his aegis. One insider observed that "it was other considerations than increased efficiency and economy that prompted the first imperfect combination of the Carnegie properties."[49] Moreover, vertical *dis*-integration occurred within two months when the Lucy Furnaces were sold off. "Thus the first venture into developing a truly vertical steel organization had been quickly dissolved," his biographer writes. "At this point neither Carnegie nor his associates saw the advantages of having within one company the entire process, from the mining of raw iron ore and coal through the manufacture of pig iron to the converting of pig iron into steel."[50]

Five years later, in 1886, the annual net profit of Carnegie's "other" company (Carnegie, Phipps & Company, founded a dozen years earlier) reached 60 percent of its total capitalization. At that point Carnegie reorganized it as a limited copartnership. Carnegie, Phipps & Company held the Homestead mill, purchased in October 1883, and the recently repurchased Lucy Furnaces.[51] Carnegie Brothers & Company remained a separate entity. Again, as far as can be ascertained, the motives for acquiring the Homestead mill—a state-of-the-art facility built in 1879 as the Pittsburgh Bessemer Steel Company to produce rails—were prescient opportunism and the desire to buy out "a dangerous rival" rather than any strategic vision of vertical integration. Again, too, the occasion of the purchase was infighting among the partners of the Homestead mill compounded by labor unrest that had culminated in a ten-week strike in the spring of 1882.

Once in the control of Carnegie, Phipps & Company, Homestead was converted to the production of structural steel shapes, an important diversification for the industry as a whole. The extent to which this new market for structural shapes required compensating organizational change remains, in the absence of a detailed account of this period, a matter of deduction and conjecture.[52] An odd consolidation it was. The anomaly of retaining two parallel companies in 1886 reflected Carnegie's fear of creating a single concern with heavy capitalization, as well as his desire to maintain the purely financial advantages of having two separate companies. For example, working capital was obtained by discounting notes given by one company to the other, as buyer and seller, thus providing the "two-name paper" required by banks. In fact, a three-cornered arrangement emerged among the twin Carnegie companies and the Frick Coke Company, and in 1890 $2,635,000 of such notes was issued.[53] The buildup of the Carnegie enterprises "was a natural and unconscious development growing out of trade conditions," wrote James Bridge. "There was at no time a well-defined plan or policy of expansion."[54]

Considerations of mortality were also a significant motivation for organizational change. During the fall of 1887, Andrew Carnegie's younger brother Tom died of pneumonia, while Andrew himself lay sick with typhoid fever. Tom Carnegie was the second largest shareholder in the two companies and also a capable manager and trusted adviser—roles that were especially important because Andrew had directed his Pittsburgh operations from his home in New York for twenty years. The twin companies barely recovered from the loss of Tom Carnegie, who held 17.5 percent and 16 percent of them; they could never have survived the loss of his brother, who held 54.5 percent and 52.5 percent. As privately held, limited partnerships, they were vulnerable to loss of partners. Andrew personally had the financial means to purchase his brother's share from his widow, but settling Andrew's much larger share of the companies would have bankrupted them. The 1887 crisis shocked the companies into creating the "Iron Clad" agreement, which provided that each company could purchase for itself at book value and across a duration of four months to fifteen years (depending on the size of the share) any partner's share in the case of the partner's death or (as later amended) in the case of a three-fourths vote that any partner's continued presence was detrimental.[55]

No special strategic vision seemed to motivate Carnegie's offer to Henry Clay Frick on 1 January 1887 of a 2 percent share in Carnegie Brothers for the book value of $184,000. Like other favored junior partners before him, Frick paid for his share over time through its declared dividends. Like others, Frick signed the Iron Clad agreement; unlike others, though, Frick was eventually forcibly ejected from the Carnegie empire through the Iron Clad. The partnership represented an alliance between the Pittsburgh district's most prominent steelmaker and its most aggressive coke producer. Through the 1880s the Carnegie companies provided handsome financing for Frick to expand his coking operation in the Connellsville region. They displaced the Pittsburgh banker, Judge Thomas Mellon, as Frick's chief source of capital and emerged as the majority owner of Frick's expanding coke empire. This partnership completed one phase of vertical integration (control of coke). Carnegie, for his part, praised Frick's talents as a hard-boiled, cost-cutting manager and in January 1889 elevated him to chairman of Carnegie Brothers & Company, increasing his share from 2 percent to 11 percent. However, there is no reliable evidence that Carnegie appreciated the strategic importance of vertical integration until nearly another ten years had passed. Indeed, for the second phase of vertical integration, "Carnegie's junior partners would have to prod him into risking capital for the acquisition of iron ore."[56]

The acquisition of the nation's richest iron-ore lands was effected by the Carnegie Steel Company, Limited, formed 1 July 1892. The new company's capitalization of $25 million—two and a half times the combined capitalization of Carnegie Brothers and Carnegie, Phipps—reflected remarkable growth since the 1886 reorganization. At last, the Carnegie enterprises were collected into one organization, presided over by Frick.[57] Before the privately held company was a week old, the bitter Homestead strike put it in the public eye and into the annals of labor history. Despite personal chagrin to Carnegie, who had professed respect for the workingman and then left the dirty work of strikebreaking to Frick, Homestead was worth considerable pain to the company. Retained profits from the armor shop at Homestead—which Carnegie identified as the company's singular source of profit in 1895—bankrolled the purchases of ore lands in 1896 and 1897.

With the Carnegie company's acquisition of iron-ore lands in the mid-1890s, one can for the first time identify actors motivated principally by the strategic advantages of controlling this essential raw material. Nonetheless, this outcome did not reflect the single-minded motivation of any of the Carnegie partners—not even Frick, who had done so much to build up an empire in coke and deliver it to the Carnegie realm. Carnegie had a saying that Fortune knocked once at every man's door, but in this case Fortune—the richest deposits of iron ore in the nation—knocked time and again before Carnegie permitted entry. "It is a story," Bridge observed, "that shatters all preconceptions of the genius necessary to achieve millionaireship."[58]

By the early 1890s the insatiable maw of the nation's blast furnaces had drawn down the iron-ore deposits of the Marquette, Menominee, and Gogebic ranges in Michigan's upper peninsula, and ore prospectors turned their attention to the newly opened Vermilion and Mesabi ranges. As early as 1865 Minnesota's geologist had reported finding "immense bodies of the ores of iron" on the eastern Mesabi, but this observation was not systematically followed up until the Merritt family explored this region beginning in 1888. In 1890, based on their mappings, the Merritt brothers obtained 155 leases on mineral rights of up to 160 acres each. By the spring of 1891, when Lon Merritt set out to rally support in Pittsburgh, the brothers had joined with Duluth investors to establish two companies with a total capitalization of $4 million. Frick's reception of Merritt was cool in the extreme, though Merritt did persuade Frick to dispatch his leading expert to look over the range. The expert knew the hard ores of the established ranges; seeing instead the soft granulated substance characteristic of the Mesabi range, he "swore it was not iron ore" and declined to have a sample assayed in Duluth. Had he done so he would have discovered pure Bessemer-grade ore, 50–65 percent iron with scarcely a trace of harmful phosphorus.[59]

Fortune knocked again, insistently, on Carnegie's door in the person of Henry Oliver. Oliver was a childhood friend of Carnegie who had entered the steel industry in the early 1880s after nearly two decades of promoting various Pittsburgh railroads and iron manufactures. He first visited the Mesabi range during the summer of 1892 after attending the Republican national convention in Minneapolis. He went north, saw the efforts of the fifty-one mining companies then sending out iron ore from the Mesabi to the harbor at Duluth, convinced the Merritt brothers by means of a $5,000 check that the Carnegie steel companies would make a large investment, then hastened back to Pittsburgh to make good his promise. Already party to one of Oliver's railroad ventures, Frick favored the deal Oliver now proposed: a one-half interest in Oliver's mining company (which he had organized with the Merritts) for a loan of $500,000, secured by a mortgage on the ore properties. But Carnegie, writing from Scotland on 29 August 1892, demurred: "Oliver's ore bargain is just like him—nothing in it. If there is any department of business which offers no inducement, it is ore. It never has been very profitable, and the Massaba [that is, Mesabi] is not the last great deposit that Lake Superior is to reveal." After the settlement of the Homestead strike, Frick successfully concluded the deal with Oliver, and the Carnegie company took up its half-share in the Oliver enterprise. Twenty months later, in a letter to his board of managers dated 18 April 1894, Carnegie still denied the fortune that Oliver offered: "The Oliver bargain I do not regard as very valuable. You will find that this ore venture, like all our other ventures in ore, will result in more trouble and less profit than almost any branch of our business. If any of our brilliant and talented young partners have more time, or attention, than is required for their present duties, they will find sources of much greater profit right at home. I hope you will make a note of this prophecy."[60] Many, indeed, delighted in Carnegie's stunningly mistaken prophecy.

By this time, however, forces as powerful as Carnegie's company and twice as determined had moved onto the Mesabi range. Indeed, the transformation of the Mesabi in these years from an entrepreneurial venture run by families to a rationalized empire controlled by corporations nicely encapsulates the merger movement that transformed U.S. industry between 1895 and 1904. For the Mesabi entrepreneurs, the 1893 panic and the depression that followed had been a disaster. Orders for iron ore from steel mills vanished just as the Merritts' credit collapsed. Into this situation stepped John D. Rockefeller, who in autumn 1893 took the major share in the newly formed Lake Superior Consolidated Iron Mines Company, which included the six mines that the Merritts held independently from Oliver as well as Rockefeller's other ore holdings. "I was astonished," Rockefeller later stated, "that the steelmakers had not seen the necessity of controlling their ore supply."[61]

Although contemporaneous speculation had it otherwise, Rockefeller saw the Mesabi mostly as a source of freight for his transportation network, and by December 1896 he was willing to strike a mammoth deal with Carnegie. For fifty years the Oliver Mining Company and Carnegie Steel were to lease all the ore property of Lake Superior Consolidated Iron Mines, paying a royalty of 25¢ per ton extracted and guaranteeing a minimum purchase of 600,000 tons of ore each year. Rockefeller was to gain the traffic for his railroad lines and lake steamships of this 600,000 tons plus an additional 600,000 tons from the Oliver properties, and each side agreed not to enter the other's domain (thus protecting Rockefeller's Mesabi ore holdings and Carnegie's steelmaking empire from each other). The agreement in effect slashed the going ore royalty from 65¢ per ton to 25¢ per ton; its consequences swept across the entire ore-producing region. "This alliance with the Rockefellers . . . produced a panic among the other mine owners; and stockholders in Boston, Chicago, Cleveland, and the Northwest hastened to get rid of their ore properties at almost any price. The demoralization extended to the ore markets; and Norrie [ore from the largest mine on the Gogebic Range], which sold at $6 a ton in 1891, dropped to $2.65 on the docks at Cleveland."[62]

With ore properties in a doubly depressed condition, Oliver labored mightily to deliver Fortune to Carnegie's door. In January 1897, a month after the deal with Rockefeller, Carnegie increased his share in Oliver's mining company from one-half to five-sixths.[63] Carnegie then directed Oliver to make preparations to deliver a full 2 million tons of iron ore during 1897, and Oliver soon interpreted this as a mandate for getting options on ore properties, whose owners, numbering in the hundreds, were all too willing to sell. With the 1896 agreement barring the Carnegie concern from Rockefeller's Mesabi domain, Oliver turned to the other principal ore regions.

Oliver focused on three important mining areas in the Gogebic and Vermilion ranges. On 27 July 1897 he reported to Frick that they could acquire by leasing or purchase the Norrie, Tilden, and Pioneer mines. These new mines together would add 1.6 million tons of ore to the 1.2 million tons shipped each year from the Mesabi, and Oliver urged Frick to acquire all three properties. Now, for the first time, Oliver saw the strategic advantages that the company stood to gain and explained them to Frick. The Oliver and Carnegie companies, he maintained, had paid enough "tribute" to the northwestern miners, who had retained handsome profits and had unavoidably wasted resources in exploration and development, as well as to the Cleveland middlemen who charged excessive freight rates. "All this should stop," he said. The three mines he had selected comprised over 80 percent of the region's developed or "in sight" ore. In addition, the companies held a 60 percent share in the yearly allotment of the region's ore-producers' cartel, by which the established concerns hoped to maintain high ore prices, and so could exert considerable influence on the cartel's policies. Oliver had considered all the important ore properties, except those of the Minnesota Iron Company ("the banns have been published and union with Illinois Steel is only a matter of time"). Only these three could mine ore at present prices and make money. An "important point" was that the possession of a large body of ore would "strengthen our position" in holding the Rockefeller people down to low freight rates from the Mesabi range.[64]

Carnegie accepted Oliver's plans to lease the Tilden and Pioneer mines, but he flatly rejected his recommendation to purchase the Norrie mines, the largest of the three properties. Although Oliver's strategic vision should have warmed Carnegie's heart, he seemed unimpressed. "We are entitled to less than average freights as Rockafelow [sic] saves 10 [percent] Commission," Carnegie scrawled on a follow-up letter.[65] In part, Carnegie was worried about "litigation, censure, etc." from the Norrie's minority shareholders, a concern that reflected his desire to control his companies absolutely through closed, limited partnerships. But fundamentally Carnegie was fixated on short-term financial calculations suggesting that the ownership of ore properties would not be as profitable as manufacturing. "Can make three times any probable profit out of 3 000 000 in *manfg*," he asserted. Specifically, he had in mind expanding the company's open-hearth furnaces, building a bar mill to augment the output of finished steel, and constructing a rail link to Frick's coke fields—a direct blow to the Pennsylvania Railroad, which had secured this valuable traffic through an 1896 agreement with the steel company.[66] "Policy of firm not to invest Capital in mining, but to lease mines upon favorable terms," he concluded on 18 August.[67]

Although Frick wanted more ore, he sided with Carnegie against purchasing the Norrie mines.[68] It was Frick and not Carnegie who cabled to Pittsburgh the oft-quoted message of 17 August: "Norrie rejected owing minority stockholders."[69] At a board of managers' meeting on 31 August, Frick reviewed the ore situation. Oliver was now likely to get all the necessary Norrie options in two weeks, and they would be good for another two weeks, that is, until 30 September. Independently from Oliver, and without his knowledge, the company was getting options to lease two other desirable ore properties. Carnegie, Henry Phipps, and George Lauder were still "strongly opposed" to purchasing ore properties although "anxious" that good properties be leased. Frick could easily have identified himself among this group, for that same day he wrote Phipps that while "we should not buy ore property," the Norrie bargain "would be a great thing for us" at Oliver's knock-down price.[70] Not even Rockefeller, who Frick hoped would purchase the mines in question so the steel company could lease them, was in the market for additional ore properties.[71]

Besides Oliver, the only backer of the Norrie purchase within the Carnegie orbit was Charles Schwab, the company's president since April 1897, and something of a Carnegie protégé. Acquiring ore properties, Schwab affirmed in early July, was "of the greatest importance."[72] Three days after Oliver's 27 July report, when Oliver had secured options on 56 percent of the Norrie's shares, Schwab reported his assessment of the Norrie and Tilden mines to Frick. For Schwab, the Norrie was a bargain too good to be passed over; fully three-quarters of the purchase cost could be recouped immediately through selling its "quick assets" (ore on hand, bills receivable, and so forth), which made the long-term cost of the ore exceptionally attractive. Schwab agreed with the favorable report of the company's mining superintendent, D. M. Clemson, that the property could reasonably produce a full 5 million tons of ore, half of which was already "in sight."[73] Adding the costs of mining and shipping rounded up to $2.33 per ton the ore's total cost at Cleveland or other lake ports. Yet Schwab's recommendation was contingent on whether Oliver could obtain a knock-down price for the remaining shares. The Tilden mine on the other hand was "objectionable" owing to excessive manganese that would muck up the firm's Bessemer converters. At the most, he favored leasing Tilden on a year-by-year basis. Schwab concluded: "We all agree that we should control our own ore supplies. The purchase and control of these mines would give us the kind of ore [low-phosphorus, Bessemer-grade] we want as cheaply as it can be produced in any locality known at this time."[74] Three weeks later, when Oliver had secured options on 75 percent of the Norrie's shares, and Frick's negative cable had landed in Schwab's lap, Schwab wrote to Carnegie that his recommendation for the Norrie purchase was "even stronger than before" and that the "objection on the account of the minority stockholders" was no longer relevant.[75]

Nine days before the Norrie options were to expire on 30 September, the cable from Carnegie in Scotland arrived at the Pittsburgh office: "Acquiesce." Carnegie had deferred to his managers, especially to Schwab. "We leave you to do what you consider best but views unchanged. Only an investment which can be however doubtless sold at profit." Oliver, unaware of the nod from Scotland, became desperate. With five days remaining on the options he cabled an appeal to Carnegie, who repeated his go-ahead: "Always approve unanimous action board after full expression views. Sure leasing true policy but if board decides this exception all right." After the board unanimously approved the purchase of the Norrie mines, Frick, who had changed his mind in the nick of time, cabled Carnegie their positive decision and concluded: "This removes the Carnegie Steel Company, Limited from the Bessemer Ore Market. Business broadening; outlook good."[76]

Only after this turning point was Carnegie fully convinced of the benefits of owning or controlling iron-ore mines, and then Oliver went to work. By the end of 1899 he and Carnegie Steel controlled through lease or ownership thirty-four working mines across the ore-producing region and sixteen exploratory sites on all the ranges except the Mesabi. When the Carnegie-Oliver properties were folded into U.S. Steel in 1901, they represented two-thirds of the known northwestern supply of Bessemer ores—approximately 500 million tons valued at more than $500 million.[77] Indeed, it was Oliver's strategic vision, realized through Schwab's support and Carnegie Steel's armor profits, that set the stage for the conclusive merger in the steel industry.

Creating the Corporation

*The situation is grave and interesting . . . A struggle is
inevitable and it is a question of the survival of the fittest.*
—ANDREW CARNEGIE

Just as the corporate consolidation of the Mesabi ore field encapsulated the nation's merger movement, the formation of U.S. Steel in March 1901 marked its apotheosis. Some 60 percent of the nation's steelmaking capacity was merged into one unit. Like General Electric and Westinghouse, but unlike Rockefeller's Standard Oil, U.S. Steel evaded Progressive-era trustbusters and survived to preside for a half-century over the industry's mature years. Like the vertical integration at Carnegie Steel, this definitive change in industrial structure involved an array of motivations.

Rational, functional, and technical motives to form U.S. Steel were clearly articulated by Charles Schwab. By many accounts a compelling public speaker, Schwab created a vision so powerful that in the end it put him out of a job. Schwab, like Carnegie be-

fore him, had meteorically risen through the ranks on the basis
of genuine talent, ceaseless drive, calculated flattery, and a bit of
luck. In 1897, after successful tours of duty as superintendent of the
Edgar Thomson and Homestead works, Schwab became company
president; in 1900, with the ouster of Henry Frick, he became the
company's most dynamic figure after Carnegie. Schwab thrived in
the steel industry of the twentieth century, first as essential courier
between Andrew Carnegie and J. P. Morgan, then as founding
president of U.S. Steel, and finally, after being ousted from the
steel combine, as president and chairman of Bethlehem Steel.[78]

Schwab unveiled his vision on 12 December 1900 at a dinner
given in his honor by two New York City financiers, whom he had
entertained in Pittsburgh. The University Club in Manhattan was
filled with notables that night. Among the eighty leaders of business and finance was J. Pierpont Morgan, seated at the right hand
of the evening's featured speaker. "Schwab," as Wall relates the
event, "was never in better form."

> In less than a half hour he laid out the blueprint for the de-
> velopment of a new organization within the steel industry.
> It was precisely the kind of plan that Morgan would heartily
> approve—a colossal organization of specialized plants, a super-
> trust of true verticality, made up of many parts. Each part
> would make, under the most rational and efficient methods
> possible, its particular contribution to the overall objectives of
> the organization, which were simply to produce the world's
> best steel and steel products—not at inflated monopolistic
> prices, but at prices so low as to bankrupt any existing com-
> pany or combination of companies producing steel at that time.
> No wasteful competition, no unnecessary duplication among
> individual plants, no faulty planning in plant location, no inade-
> quate transportation facilities.[79]

To Morgan, so it seemed, the vision was almost too good to be
true. In Schwab, Morgan had found someone with a perspective
congenial to his own and also one who could conduct the delicate negotiations with Andrew Carnegie. Ever since the *Corsair*
compact, Carnegie and Morgan had disliked and mistrusted one
another. Now, however, Carnegie's desire to sell off his steel holdings and launch a second career in philanthropy matched Morgan's
desire to hasten the steelman's retirement. With Schwab shuttling
back and forth the scant mile between the two men's mansions in
midtown Manhattan, agreement was soon reached on terms of sale.
Morgan was to pay $480 million for Carnegie Steel; for Carnegie's
personal share of $300 million Morgan quickly met his demand for

first-mortgage, 5 percent gold bonds, which precluded his influence over the steel combine. By February 1901 Carnegie Steel was
Morgan's. On 2 March, after consolidating it with his own Federal
Steel, a combine of steel, ore, and railroad companies centered on
Chicago, Morgan issued the formal circular announcing the formation of U.S. Steel, capitalized at $1.2 billion. Five weeks later, when
the American Bridge Company and Rockefeller's Lake Superior
Consolidated Mines were added to the combine, the capitalization
reached more than $1.4 billion.[80]

In a remarkable article for the *North American Review* that
May, Schwab arrayed the progressive forces of science, efficiency,
and national competitiveness behind the new steel combine. Supremacy in handling and transforming the raw materials of the
earth, he stated, is won by adopting "a scientific manner" that re-
duces costs to the lowest possible level. Since it was an "axiom" that
the larger the output, the smaller the cost of production, and that
this advantage increased "almost indefinitely" as the unit of production grew ever larger, "no great amount of harm" could be done
by anyone agitating against "the science of business consolidation."
Indeed, industrial combination was simply the "logical solution" to
the "wastefulness" of the individualistic, competitive business system. With consolidation, nothing is left to chance and all waste is
cut off, Schwab asserted. "Instead of being jerked here and there on
side tracks, and paying for the privilege, the material, from its raw
state to the finished product, is held under one control." The result
was a "superb industrial system" whose components ran "absolutely
without friction." Finally, Schwab refuted the "preposterous" idea
that the combination was a danger to the nation-state; rather, it
was a "distinct gain." Who doubts this, he said, should turn to the
foreign newspapers that were crying industrial alarm at the American nation. It was the industrial combination, not the individual
manufacturer, Schwab concluded, that had put the United States
in "industrial control of the world." "The politicians who attempt to
obstruct industrial development on the line along which it is moving are attempting to obstruct human progress," Schwab warned,
"and the result will be inevitable."[81]

Schwab's phantasm dissolves into mist, however, as soon as one
scrutinizes the financial and strategic motivations that lay behind
the steel combine. To begin, a grudge match was at play: Carnegie detested Morgan's libertine lifestyle, and Morgan abhorred the
steelman's buccaneering. But personality alone cannot explain the
timing and pace of events or square with the known documentary
evidence, nor can an analysis that deals only with the structural
properties of the industry.[82] The steel merger completed Morgan's

grand plan of stabilizing American industry and halting the waste-fulness of competitive capitalism; in the *North American Review* Schwab was chanting a Morgan mantra. The steel merger was triggered by the moves to integrate vertically that Carnegie Steel had undertaken after purchasing the ore lands. Once the company had committed itself to owning ore fields, a new dynamic took over, involving not only the fixed costs of massive investments but also a clash of two opposing strategies for organizing industrial affairs. On one side stood Morgan, who had labored diligently to forestall destructive competition among the railroads and steel companies under his control; on the other hand was Carnegie, the acknowledged master of destructive competition.

By 1900 the two men were on a collision course. By reorganizing the railroads and creating Federal Steel, Morgan had apparently achieved stability in the railroad and steel industries; then Carnegie initiated a calculated assault on this stability. Given the two men's rival investments and conflicting business philosophies, as well as the basic capitalist framework within which they worked, only a few resolutions were possible. The merger of mergers in steel was determined, in outline if not in detail, when the Carnegie company committed itself to the upper Midwest ore lands. Put in personal terms, no industry could have tolerated for long the presence of the expansive and systematic Morgan and the equally expansive but impulsive Carnegie. Once again, railroads and steel formed the matrix of conflict.

In 1899 several developments combined to destroy a delicate balance that had brought a measure of stability to the railroad and steel industries. In 1898, as noted earlier, a "community of interest" had been established among the Pennsylvania, the New York Central, and the Morgan-Hill roads. Furthermore, in 1896 and again in 1898, the Carnegie company and the Pennsylvania Railroad— Pittsburgh's largest single source of freight and largest single carrier of freight, respectively—had informally granted to the steel company generous rebates on the railroad's published freight rates, which Frick's coke firm had enjoyed since 1892; in return the steel company promised not to enter the railroad industry. Among the routes thus protected was the Pennsylvania's line running south from Pittsburgh to Frick's coke empire in Connellsville. From 1896 three-fourths of the Carnegie company's 15 million tons of freight flowed out of Pittsburgh on the Pennsylvania's tracks. This stable situation broke down in 1899 when the Pennsylvania's president (Frank Thomson of the *Corsair* compact) who had negotiated these stabilizing agreements died, and its new president, Alexander Cassatt, announced an end to all such agreements. Consequently, the railroad doubled its freight charges to the steel company. The truce that had kept Carnegie from building railroads, and hence upsetting the entire railroad industry, had broken down.[83]

While continuing to batter the Pennsylvania with demands for more favorable treatment, Carnegie began negotiations with the Gould syndicate, whose propensity for destructive competition in railroads matched his own in steel. By the end of 1900, he had signed an agreement with Jay Gould's eldest son to promise one-fourth of the Carnegie company's total tonnage as soon as Gould's projected new transcontinental line with its terminus at Baltimore reached Pittsburgh. One way or another, Carnegie's plans implied the destruction of the delicately structured "communities of interest" that had just been reestablished among the eastern trunk lines. To rationalizers like Morgan, the rumors that Carnegie and Gould were teaming up spelled trouble.[84]

Concurrently, Carnegie Steel was planning a new tube mill at Conneaut, on the shore of Lake Erie at the Ohio-Pennsylvania border. The new mill was Carnegie's offensive response to announcements by the Morgan and Moore steel syndicates that they would be cutting their orders of unfinished steel. The Moore brothers, William H. and James H., with the financial backing of John W. ("Bet a Million") Gates, had promoted mergers in matches, biscuits, tin plates, steel sheets, and steel hoops. Dealings with the Moores had never been to Carnegie's liking, especially after these speculators made an abortive secret attempt to purchase his own company in 1899. Instead, they had patched together thirteen steel transportation and ore concerns, centered in Ohio, and had formed National Steel, which became the nation's third-largest steel producer behind Federal Steel and Carnegie Steel. Faced with this warning, Carnegie directed Schwab to draw up plans for fabricating plants that would absorb the company's output of raw steel and thus sever its dependence on outside finishing mills. "A struggle is inevitable," Carnegie told him, "it is a question of the survival of the fittest." If built, these mills would be direct competitors of J. P. Morgan's Federal Steel and especially his National Tube company. "It was the desire to protect these companies from increased competition," writes his biographer, "that persuaded Morgan to seek a solution that would not only solve their own immediate problems but also help stabilize the entire steel industry, which the press repeatedly warned was on the verge of a 'steel war,' a great 'battle between the giants.' "[85]

Although its consequences were vast, the projected Conneaut mill was little more than a logical outgrowth of the Carnegie company's investment in ore lands. After 1897 the Carnegie company took determined steps to rationalize the transport network that joined its ore properties with its Pittsburgh mills. It completed this network by acquiring and revitalizing a defunct railroad that connected the port site of Conneaut Harbor with the rail terminus of Bessemer, Pennsylvania, at which point began Carnegie's Union Railroad with connections to its several Pittsburgh-area mills. Freight cars that had carried ore southbound to Bessemer had returned northbound empty; filling them with Connellsville coke for a new mill at Conneaut simply completed the system.[86] Already, the tight coupling of the transport network with steel mills gave the Pittsburgh–Lake Erie rail link a very high load factor, and its managers boasted of its desirable system properties.[87] As a bonus, this rail link directly threatened the Pennsylvania Railroad's stranglehold on Pittsburgh traffic, because the route crossed the lines of four of the Pennsylvania's east-west rivals. "We look upon our line to the northward as a mere connecting link with the trunk lines from which we were shut off and held under rates extorted by a combination," Carnegie wrote one such rival. "You will be large customers of ours, and we shall be large customers of yours."[88]

What was desirable, logical, and rational to Carnegie was of course madness to Morgan. These developments represented a serious threat to stability in Morgan's steel and railroad systems.[89] Carnegie's new tube mill at Conneaut threatened the financier's National Tube company, while his railroad plans threatened stability in that industry, too. "Carnegie," Morgan reportedly observed, "is going to demoralize railroads just as he has demoralized steel."[90] This was not the first time Morgan had intervened to halt a Carnegie venture in railroading, as related at the beginning of this chapter. Fifteen years after the *Corsair* compact, railroad directors knew that Carnegie was not bluffing in his threats to break the dominance of the established roads in western Pennsylvania, and they began calling on Morgan to forestall the impending calamity.[91]

This prologue—the Carnegie company's ore-property acquisitions and the transportation rationalization that brought Morgan and Carnegie into direct conflict—framed the negotiations in January 1901 that secured Morgan's dominance over the railroad and steel systems. It is not always appreciated that U.S. Steel was as much a raw-material-and-transportation combine as a steel-manufacturing one. It comprised the ore holdings of the Carnegie company, including the Oliver Mining Company; those of Federal Steel, including the Minnesota Iron Company; Rockefeller's Lake Superior Consolidated Iron Mines; large holdings of coal and limestone lands; a 112-vessel lake fleet; and more than a thousand miles of railroad.[92]

In steel, the early twentieth century witnessed a reversal of the relentless cost cutting and furious technical innovation that Carnegie had demanded of his managers. Perhaps only in the inveterately anti-union labor policy did the Carnegie practices persist; the steel combine used ham-handed tactics to break strikes in 1901, 1909, and 1919, which poisoned labor relations for decades.[93] Until the late 1930s U.S. Steel was really a loosely structured holding company; its central office remained small and had difficulty controlling the many subsidiaries, but it did manage to stamp out technical innovation in steel structures.[94] Carnegie Steel issued its last structural-steel handbook in the merger year of 1901, revised it slightly two years later, and then did not issue a successor until 1911. During those ten years, the company's designs for I-beams stagnated at 24 inches in height.[95] Moreover, despite Schwab's recommendation, U.S. Steel's finance committee turned down the most promising structural innovation for years, a wide-flanged H-beam devised by Henry Grey. The innovation was later acquired by Bethlehem Steel, which built a $4 million structural mill that opened in 1908. Carnegie Steel was eventually reduced to manipulating the Pittsburgh-plus pricing system (by writing out phony invoices that jiggered the freight-rate differential between its Pittsburgh and Chicago mills) to compete against Bethlehem's technically superior beams, the largest of which was a whopping 30 inches in height. Even with the same height and flange width, Bethlehem's beams were thinner and so featured less weight for equivalent strength compared to the standard Carnegie beams. In 1926, facing the severe erosion of its structural position, U.S. Steel made a forced technology choice and adopted a similar wide-flanged beam.[96]

Indeed, U.S. Steel was designed not to foster technical change but to promote stability in the industrial system.[97] At a lavish black-tie dinner celebrating the Morgan-Carnegie merger, the guests sat, appropriately enough, at a giant table in the shape of a *rail* (see the frontispiece to this chapter). The merger's acquisition of Tennessee Coal & Iron, the largest and most integrated southern producer, had little to do with the region's iron-ore reserves. Instead, it represented an attempt to bail out the brokerage house of Moore & Schley, which was overextended in the securities of Tennessee Coal & Iron in the panic year of 1907.[98] Even its steelmaking complex in Gary, Indiana, represented the evolutionary upscaling of the open-hearth process. The plant's coke ovens that collected valuable by-products and the use of blast-furnace slag to produce cement were notable but not novel. Although the plant first made steel in 1909, its fame really dates from the mid-1920s when the construc-

FIG. 4.6. Construction of U.S. Steel's Gary works (November 1908).
Courtesy Calumet Regional Archives, Indiana University Northwest.

tion of a tube works made it into a fully integrated facility.[99] With
60 percent of the nation's raw steel capacity, and far more in certain
sectors like wire, U.S. Steel set the pace and price for American
steelmakers during and after World War I; administrative centralization, technological innovation, and downward pressure on costs
and prices were of little concern.

The Carnegie-era cost-cutters committed to these policies were
forced out of the steel combine. In 1903 Schwab was ejected from
the presidency of U.S. Steel in the wake of trumped-up charges of
gambling at Monte Carlo. Later, in the years before World War I,
he helped reform and diversify Bethlehem Steel.[100] Schwab's successor as president, William E. Corey, along with other self-styled
"Carnegie boys," struggled within U.S. Steel to maintain the Carnegie practices. Eventually, they too ran afoul of the corporation's
policy of favoring industrywide stability over company-level profit
maximizing. Four years after their purge from the steel combine in
1911, several of the "Carnegie boys"—including Corey, William B.
Dickson, and Alva C. Dinkey—organized Midvale Steel & Ordnance, which combined three Pennsylvania plants: Midvale Steel
(Philadelphia), Worth Brothers (Coatesville), and Cambria Steel
(Johnstown).[101] All of these new concerns maintained a profitable
existence in the shadow of the corporation, as its directors intended;
this was the essence of the Morgan plan. But Schwab's vision of
an efficient steel combine selling at prices to keep out all comers
was never realized in the United States. Innovation, as in high-
speed steel for factories and alloy, sheet, and electric steels for
automobiles, came from beyond U.S. Steel.

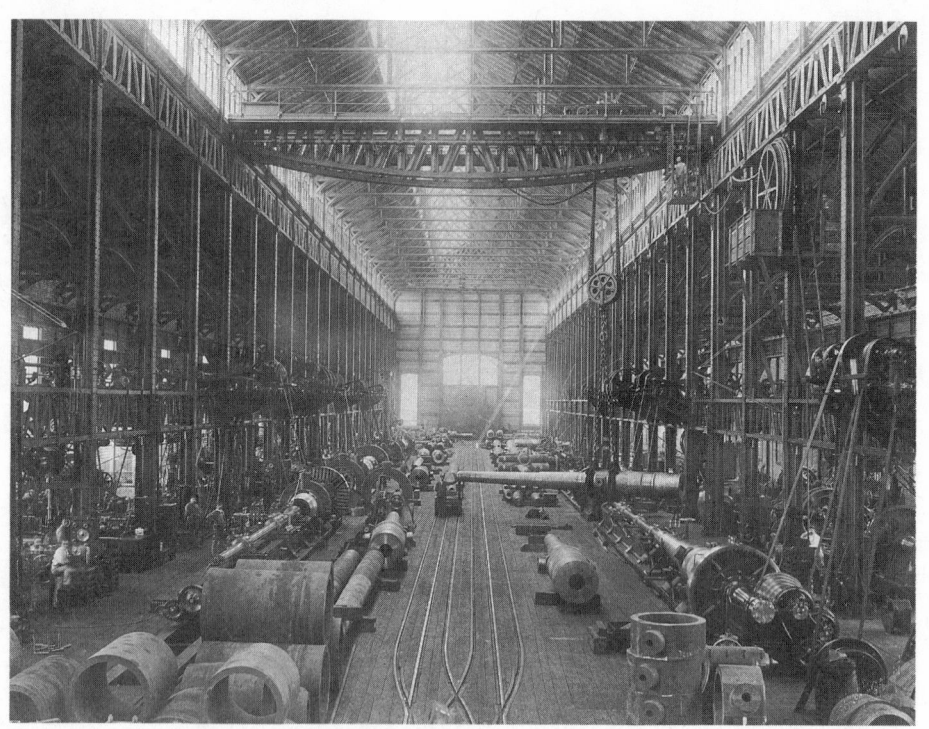

The birthplace of high-speed steel, c. 1900: Bethlehem Steel's Machine Shop No. 2, site of Frederick W. Taylor's metal-cutting experiments. *Courtesy Hagley Museum and Library.*

CHAPTER FIVE

The Reform of Factories

1895–1915

The crowd gathering in New York City on Tuesday evening, 4 December 1906, for the American Society of Mechanical Engineers presidential address was unusual in several respects. It was the largest group the society had ever attracted since its founding by Alexander Holley and Robert Thurston twenty-six years earlier. It included more scientists, and more engineers from abroad, than anyone could recall. Fortunately, New York Edison had offered its capacious auditorium to the society. ASME's old headquarters on West 31st Street was hopelessly small, and the grand United Engineering Societies building bankrolled by Andrew Carnegie was still under construction on West 39th Street. Advance copies of the address had created something of a sensation on both sides of the Atlantic. "On the Art of Cutting Metals" revealed work done in complete secrecy across the span of the society's existence. Its author, Frederick W. Taylor, was a captivating and compelling speaker.[1]

Two years earlier, on the strength of his technical talent and flair for publicity, Taylor had been chosen vice president (and president-elect) of ASME in a wave of reform promising greater efficiency in running the society's affairs. Thus far Taylor had installed several allies in key committee posts, and had created and personally funded a special assistant to the secretary for a detailed study of "possible simplification and standardization of the office routine." These changes were a pioneering effort to extend his "scientific management" techniques beyond the shop floor. On this evening, however, not office routine but the technology of metal working was Taylor's topic. Anticipating insight, perhaps even inspiration, 1,325 people gathered to hear his address.[2]

Taylor's delivery of the high points was sharp, clear, and confident. Rigorously, even dogmatically, he emphasized the scientific method in engineering research: resolving a complex problem into independent variables, then altering each variable—one at a time—to study its effect. Twelve such variables, he stated, completely described the cutting operations of any lathe, the fundamental machine tool of industry. Previous experiments to determine maximum cutting speeds, conducted by engineers, industrialists, and scientists in several cities and countries, had failed, according to Taylor, because the experimenters had altered two or more variables at once. With his method of altering a single variable, the specific effects of each variable could be identified and isolated. On and on the insights kept coming, piled up into an impressive edifice. It was, *Engineering News* reported, "the most remarkable presidential address ever delivered before an American engineering society."[3]

At the conclusion of his formal remarks, Taylor was thronged by the efficiency experts and management consultants rushing to the podium. In one step, according to Henry R. Towne, Taylor had lifted the art of forming and tempering metals "from the plane of empiricism and tradition to the high level of modern science." Later speakers jostled for attention by relating their experiences with the Taylor system of management. Then, as if recalling John Fritz himself, one member of the "old school of mechanics" expressed his fear that if carried to its logical conclusion, Taylor's line of reasoning would "ultimately eliminate" or at least "greatly emasculate" the real mechanic who possessed "finger wisdom" and brought to work each morning a salutary conscience and willingness to innovate. Taylor's metal-cutting research had the effect of concentrating the mechanic's craft knowledge on a special circular slide rule, from which nearly anyone—although in practice it was the shop superintendent—could read off the proper cutting parameters for any job at hand. Taylor's system, as the mechanic saw only too clearly, had "an inevitable tendency" to supplant the finger-wise, thinking mechanic with "unthinking automata."[4]

Following the mechanic's lone note of apprehension, the scientists had their say. H.C.H. Carpenter, a professor of metallurgy at Victoria University in Manchester, England, praised Taylor as an eminent authority, hailing his address as one of "the most remarkable practical scientific researches ever published." After clearing up some minor points, he launched into a fifteen-minute lecture on the metallurgical properties of high-speed steels. His evident aim was to engage Taylor's insistence that metallurgy had no insight into the special properties of "red hardness" possessed by these steels. Next, Henri Le Châtelier, the French physical chemist, told of his countrymen's pervasive disbelief on first hearing of Bethlehem's claims for high-speed steel. Nevertheless, he related, "we had to accept the evidence of our eyes" when the remarkable cutting tools went into action at the 1900 Paris Exposition. "We saw enormous chips of steel cut off from a forging by a tool at such high speed that its nose was heated by the friction to a dull red color." Like Carpenter, Le Châtelier praised Taylor's scientific method ("a model which every young engineer will have to study"), then embarked on his own spirited defense of metallurgical theory.[5]

In responding to his critics and supporters Taylor set up a discourse that at once maintained and reconstructed the communities of science and engineering, issues examined further at the end of this chapter. In concluding, Taylor observed that he and his associates had been "fairly overwhelmed" with the "commendation" that the advance copies of the address had received at home and abroad. He even expressed a "certain surprise and regret" that it had attracted so much more attention than their various papers on shop management, the "real vital subject" of interest. Taylor had formulated a comprehensive if controversial solution to the so-called labor problem in U.S. industry and believed that his tool-steel experiments were "merely one of the important elements" of the system. But in fact this technical element undoubtedly was his most wide-reaching and durable success.[6] This chapter focuses on high-speed steel and especially on the interplay between its science-inspired invention and its craft-oriented production. High-speed steel serves here as a window into the world of mechanical industry, for it not only nourished the scientific-management movement but also triggered a wave of factory reform that set the stage for the automobile-driven rational factory revolution considered in the final chapter.[7]

The Craft of Crucible Steels

The steel works of Sheffield [were] hermetically sealed . . .
the holders of the keys being apparently unconscious that
their most expert workmen were swarming over to the States.
—WILLIAM METCALF

By the turn of the century, the users and producers of steel included almost all the largest manufacturing concerns. In 1900 no fewer than twelve of the nation's fourteen largest factories (with between six and ten thousand wage earners each) were either the leading producers of steel—including Cambria, Carnegie's Homestead plant, Jones & Laughlin, Illinois Steel's South Chicago plant, National Tube at McKeesport, and Pennsylvania Steel—or the leading users in closely allied fields, including Baldwin Locomotive, General Electric's Schenectady works, Westinghouse Electric, Deering Harvester, and the shipbuilders William Cramp and Newport News.[8] For all these companies, and a host of smaller ones, the key to successful production was metal cutting, which fashioned great and small blocks of steel into useful articles. Throughout the nineteenth century and into the twentieth, the tools for metal cutting were made from crucible steel, an entirely different steel than the bulk products of the Bessemer converter and openhearth furnace. Crucible steelmakers, as John Ingham writes, "differed dramatically in workplace experience, social background, and temperament from the makers of Bessemer steel." While Bessemer charges weighed 30,000 pounds or more, the largest crucible charges weighed about 100 pounds, as much as a man could lift out of the furnace by hand. As the products of a high-grade, smallscale process, crucible steels were uniform, carefully monitored—and frighteningly expensive. The best grades ran about 60 cents a pound, or $1,300 a ton, at least three times the unit price of armor plate and forty times the price of rails. Into the 1920s, if not beyond, the high-speed tools that were literally and figuratively at the point of production were made from crucible steel.[9]

Metal working as the crucial link between the production and use of crucible steel began with its origins in eighteenth-century Sheffield, England, in the experiments of a clockmaker named Benjamin Huntsman. Huntsman found the available "blister" steel, made in a cementation furnace by baking strips of iron covered with charcoal for as long as a week, unsuitable for clock springs. He began experimenting in the early 1740s to melt pieces of blister steel in a coal-fired reverberatory furnace adapted from brass founding. Huntsman's key innovation was in devising clay pots, or crucibles, capable of withstanding the unprecedented heat needed to melt steel. About 1772 Huntsman established a works, where first he and then his descendants practiced steelmaking for more than a century. By the late eighteenth century the Sheffield district comprised nine companies making blister steel, eleven melting blister steel to cast ingots, and several separate water-powered mills hammering the cast ingots into usable rods.[10] Over time, Sheffield steelmakers developed an elaborate and arcane body of craft knowledge ("should a hard steel be required, for each hundredweight of metal the juice of four white onions is also added," disclosed one 1865 patent) to produce some of the finest and most expensive steels in the world.[11]

Early in the nineteenth century the Sheffield companies captured the growing American market for scissors, cutlery, and edge tools. The expansion of agriculture as the nation moved west opened a vast market, first in traps, knives, and axes, and then in agricultural machinery. Beginning in 1837 John Deere used strips of old Sheffield sawblades for his steel-bladed plows, which sliced through the sticky soil of the great prairies. He soon began importing the metal directly at the princely sum of $300 per ton. Even the steel manufactured by firms like Washburn and Moen into wire for pianos, scientific instruments, hoops for women's skirts, and umbrellas came from the English "metropolis of steel." Through extensive trade Sheffield and the United States became so closely linked that, as one observer put it, "the fortunes of the town rose and fell with the temperature of the American demands." At the peak of the trade about 1860, a third or more of Sheffield's total output, nearly 22,000 tons, was exported to the United States. For Americans the imports cost $3.25 million, about as much as contemporaneous steel rail imports. Would-be American crucible steelmakers in turn depended on skilled labor imported from Sheffield.[12]

Skilled English steelworkers drawn by high wages began arriving in Pittsburgh in the 1860s. When Anderson, Cook & Company erected the Pittsburgh Steel Works in 1865, the company used Sheffield methods for converting iron into blister steel, melting the blister steel in crucibles, casting the ingots, and hammering the ingots (under a Sheffield-made steam hammer). The open-hearth pioneer Samuel T. Wellman related that he first came to Pittsburgh as the assistant of a "big black-whiskered Englishman" who started and operated Anderson, Cook & Company's works. Then at Singer, Nimick & Company's "Sheffield Works," Wellman built two crucible-steel melting furnaces "of the same type." Wellman later installed similar furnaces at the Chrome Steel Works in Brooklyn, New York. In Pittsburgh, the Black Diamond and Crescent works also followed Sheffield lines.[13] Few people recall that crucible, and not Bessemer or open-hearth, was the steel city's first steel.

These efforts significantly boosted domestic output of crucible steel. Between 1867 and 1877 U.S. imports dropped by more than half, to 2,946 tons, while production more than doubled, reaching 36,098 tons. The Pittsburgh region now emerged as America's center for crucible steel. In 1877 fourteen crucible-steel works in Allegheny County, Pennsylvania, produced 69 percent of the U.S. total. Still, available evidence suggests that Sheffield firms maintained their dominance for the highest grades of steel, including tool steel. As demand for their steel mounted steadily, and as rising tariffs hampered exports, Sheffield steelmakers themselves founded concerns on American soil. Sheffield steelmakers that attempted to erect, buy, or manage crucible-steel plants in the United States included William Butcher (Philadelphia, 1867), Sanderson Brothers (Syracuse, 1876), Thomas Firth & Sons (Firth-Sterling in Pittsburgh, 1896), and William Jessop & Sons (Washington, Pennsylvania, 1901).[14]

In setting up early crucible-steel works Americans had adopted the well-tried English methods of coke-fired furnaces, clay crucibles, one daylight shift, and hammering the cast-steel ingot. Beginning in the 1870s, however, American crucible steelmakers increasingly employed new practices, which can be traced in part to the abundance around Pittsburgh of natural gas and the absence anywhere in the United States of high-quality clay for crucibles. As the American practice matured in the 1880s, gas-fired furnaces heated crucibles made from domestic plumbago (graphite); two shifts worked continuously from Monday morning to Saturday morning; and the cast ingots were rolled rather than hammered into usable bars of steel. The shallower gas-fired furnace allowed American workers to heft crucibles weighing 90–110 pounds, while Sheffielders averaged 70-pound crucibles with the deeper coke furnace. In the 1880s almost all Americans, and many Sheffielders as well, abandoned the time-honored preliminary step of cementation for the so-called direct process in which sources of iron and carbon were melted together in the crucible. These American modifications—including hasty pouring procedures labeled by one Sheffielder as the "American slop method"—boosted output if not always quality.[15]

By exporting steel or setting up American works, Sheffield companies enhanced their already prominent role in the making of high-grade crucible steels. Even when a domestic company produced crucible steel, it often came from a crucible either owned or managed by a Sheffield steelman or his American pupil. When the Pittsburgh crucible-steel industry established itself and began providing significant competition, the Sheffield companies rebounded in the 1880s with a string of innovations. These included Robert A. Hadfield's steel containing 10–15 percent manganese (an exceedingly tough material used for safes, rock crushers, and railroad switches) as well as his steel containing up to 5 percent silicon (later used in electrical transformers). Production of Robert F. Mushet's tungsten tool steel also expanded in the 1880s, as steelmakers in Sheffield and Philadelphia, including Edgar Allen, Sanderson Brothers, Midvale Steel, and others, joined the pioneering Sheffielder Samuel Osborn in its manufacture.[16] The dominance in the U.S. market that Sheffield companies enjoyed throughout the nineteenth century did not last far into the twentieth. The shift was a result of the development first of high-speed-tool steels and then of the electric furnace.

Overall crucible-steel production in the United States climbed from 85,000 to 131,000 tons between 1892 and 1907. Although reliable statistics for total production of tool steel are unavailable, an estimate of annual consumption of high-speed-tool steel does exist for the year 1907 (fig. 5.6). Of the total consumption of just under 4,000 tons, domestic sources accounted for 2,300 tons. Sheffield companies accounted for imports of 1,600 tons (40.6 percent) and thus figured prominently, but not as strongly as in the market for carbon and self-hardening tool steels (where their share of the U.S. market exceeded 50 percent).[17]

Especially in the United States, the crucible-steel industry did not withstand competition from electric furnaces. Crucibles held their dominant position in the manufacture of alloy steel only until about 1914. After that date, the huge demand of the automobile industry for high-quality alloy steels sealed the fate of crucible-steel manufacture in the United States (see chapter 6). Compounding these liabilities, crucible-steel manufacturers were unlikely to employ scientists. Albert Sauveur, the Harvard metallographer, observed in 1900 that "with the exception of the crucible steel makers . . . there are now few important steel makers or steel consumers which do not subject that metal to a microscopical examination." Two years later the Crucible Steel Company of America, a recent amalgamation of thirteen companies, was the sole crucible concern to join eleven other steel companies in installing metallographic outfits in their laboratories.[18] The president of Columbia Tool Steel stated in 1906 that "nearly all of the crucible-steel makers follow the old methods." Only a select "few . . . are conforming their practice to scientific principles with as much care and research as has been expended on the Bessemer and open-hearth processes."[19]

Experimenting with Tool Steels

*A few years ago at the Bethlehem Steel Works some person,
whether he was a blunderer or genius history does not say,
revolutionized the whole machine business.* —WILLIAM
METCALF

Frederick W. Taylor has been portrayed by a biographer as an
"unlikely revolutionary," an energetic reformer of technology and
factory organization yet a diehard reactionary in relations with
workers. Taylor's split temperament is traced to his childhood in
a wealthy and cultured Unitarian family in Philadelphia. Tutored
to follow his father to an Ivy League law school, Taylor instead
trained in the "shop culture" of machine shops, recalling Alexander
Holley and William Sellers. In finding a middle way between the
dirt-and-grime of industry and the respect of a gentlemanly profes-
sion, Holley and Sellers were among the moving spirits behind the
founding of the American Society of Mechanical Engineers in 1880.
Taylor joined ASME at his earliest opportunity. His contacts with
Sellers and other members of Philadelphia's business and social
elite helped propel his career. While an apprentice machinist and
patternmaker at the Enterprise Hydraulic Works, he enjoyed membership in the prestigious Young America Cricket Club. His flair
for acting found expression in amateur productions; his compelling performance of female roles (including the redoubtable "Miss
Lillian Gray"), along with his obsessive personality, have inevitably
captivated a later generation of psychohistorians. In 1881 Taylor
won the first U.S. Lawn Tennis Association doubles championship
with his boyhood friend Clarence M. Clark, who happened also
to be his future brother-in-law and financial adviser and the son of
Edward W. Clark, patriarch of the wealthiest family of German-
town.[20] Taylor later hymned the formula of luck, pluck, and virtue,
but his career testifies to the advantages of place and privilege.

Finishing his apprenticeship in 1878, Taylor began a fruitful decade with Midvale Steel. Six years earlier Edward Clark and Sellers
had taken over and rejuvenated the ailing William Butcher Works,
heir to a Sheffield name but not to its talent. The turnaround at Midvale owed much to Clark's preeminence in banking and Sellers's
in machine technology. The company's improved fortunes owed
something also to the managerial capabilities and scientific insights
of two Yale classmates, superintendent Charles A. Brinley, who
had trained in metallurgical chemistry, and Russell W. Davenport,
who served as general manager and chief chemist. But the military

orders that began in 1875 really transformed Midvale from an also-ran into Philadelphia's largest steel plant and America's pioneer military contractor. By the time Taylor began work as a subforeman in the machine shop, his first managerial position, the firm had 400 employees (a fivefold increase in as many years).[21]

Taylor flourished in Midvale's machine shop, rising quickly to shop superintendent in 1881, then to master mechanic, and by about 1884 to chief engineer. By the end of the decade he had shown himself to be a creative if not brilliant technician, having invented or improved several of Midvale's machine tools. Some-time before 1885 he hired additional clerks to monitor and control the flow of jobs through the machine shop, introduced an incentive wage, and pioneered stopwatch time study; all these steps later found expression in his system of scientific management. He also imposed the use of standard tools—each one ground to uniformity, numbered, and then deployed by the foreman—and reinstated the special tool room. Also at Midvale, according to his later accounts, in 1880 or 1881 he began investigating the "art of cutting metals." [22]

As Taylor told the story, his metal-working experiments at Mid-vale centered on locomotive axles and tires, but his work on heavy rifled guns was far more important. Midvale was the nation's pio-neer manufacturer of ordnance steel in the 1880s, and beginning with his 1881 promotion to machine-shop superintendent, Taylor was at the center of the company's efforts to manufacture the huge steel tubes and hoops that made up modern cannon. In 1883 he re-designed the machine shop to accommodate work on the large gun forgings and also discovered that a stream of water directed onto the lathe cutting tool increased its cutting speed by up to 40 percent. The next year he conducted an early example of stopwatch time study, again in relation to machining ordnance steel. "Mr. Taylor seemed to think that there was a future in selling ordnance to the government and was very desirous of getting such notations," re-called machinist William Fannon, who worked on this task for three years. "My work was under a stop watch which was in constant possession of a man known in his official capacity as an observer." Here and later Taylor's efforts to boost cutting speeds focused on the roughing cuts that prepared forgings for the final machining to contract specification. The results were notable, and Commander Chauncey F. Goodrich sought without success to recruit him as manager for the Washington Navy Yard's new Naval Gun Factory. Other of Taylor's mechanical innovations that can be traced to the military work at Midvale include a 75-ton steam-driven forging hammer, as well as special tools for machining the oversize ordnance forgings. "Taylor's accomplishments were achieved through Midvale's innovative efforts to manufacture large cannon forgings," concludes the most detailed study of this episode. "Taylor's metal-cutting experiments were primarily made to improve the production of these cannons."[23]

By the end of the 1880s, however, Midvale was losing its position as the premier ordnance manufacturer, and Taylor had lost his insider status. The senior Clark sold his Midvale holdings in 1886 to Charles J. Harrah, Sr., who had made his career in Brazil promoting railroad, trolley, and shipping concerns before returning to his native city of Philadelphia. As Harrah consolidated his position in the company during 1887–88, Taylor lost his closest supporters, including William Sellers, the junior Clark, and Davenport, who became the head of Bethlehem's armor-plate department. At the death of his father in 1890, Charles J. Harrah, Jr., was elevated from manager to president. In the short term the Harrah reorganization scuttled the company's ability to compete in the ordnance field.

Although Harrah and he were friendly enough, Taylor's prospects at Midvale were obviously dimming. In 1889 Taylor received a tempting offer from Simonds Rolling Machine of Fitchburg, Massachusetts, a pioneer ball-bearing maker that had purchased rights to Taylor's forging patent. But his real break from Midvale came through William C. Whitney, who as secretary of the navy during 1885–89 had been impressed by his ordnance achievements. In May 1890 Whitney persuaded Taylor to sign a three-year contract as general manager of the Manufacturing Investment Company, a concern formed to produce wood pulp from lumber-mill waste using a patented sulfite process. Also associated with this venture was Chauncey Goodrich, who later helped promote Taylor's management system within the navy. Apart from frustrating technical difficulties, Taylor's experience confirmed his poor impression of "financiers" who ran such companies. He returned to Simonds as engineer, then struggled to launch his career as an engineering and management consultant.[24]

In his unsettled years between Midvale and Bethlehem, Taylor's most important technical work was done at the William Cramp & Sons Ship & Engine Building Company in north Philadelphia. He worked there for approximately six months during the winter of 1894–95 with additional material support from William Sellers. In a report, Taylor alerted Cramp's superintendent to a breakthrough in machine-shop practice. For generations, hardening tools made of regular carbon steel had been a highly skilled and esoteric craft.[25] By comparison, tools made from Robert Mushet's tungsten steel could be hardened simply by letting them cool in air. In the early 1890s such self-hardening or air-hardening tools were being widely adopted by American machine shops; among the suppliers of the new steel was Midvale. Machine shops typically used the new tool steels to cut especially hard forgings and castings at normal speeds. Taylor, however, investigated their ability to cut regular mild steel at much higher speeds. Here for the first time he equipped an experimental lathe with an electric motor "so as to obtain any cutting speed."[26]

Taylor's experiments in Philadelphia yielded two significant results as well as one anomalous result that was his first halting step toward high-speed-tool steel. First, compared with tools made of carbon steel, the self-hardening tools could cut hard steel forgings at speeds 45 percent faster and regular steel at speeds almost 90 percent faster. Clearly, their greatest value lay in speeding up the pace of production on regular steel, not simply cutting hard steel. Second, Taylor found that directing a heavy stream of water onto a self-hardening tool's cutting edge boosted its cutting speed by 33 percent. Finally, early in 1895 Taylor also experimented at the shop of William Sellers to compare the Mushet and Midvale self-hardening steels, but these tests were inconclusive. If overheated the Mushet steel "crumbled badly" when struck even lightly at the anvil and so could not be formed into a useful tool. Although the Midvale steel cut faster and did not crumble after high heat, the high heat "apparently permanently injured" it. Taylor thus confirmed traditional shop practice about the proper heat for treating tool steels: anything hotter than "a bright cherry red caused them to permanently fall down in their cutting speeds." These results left Taylor "in doubt" which steel to recommend as a standard.[27]

Taylor reported his achievements to Edwin Cramp on 12 March 1895. He had installed special furnaces and a steam hammer in the smith shop for dressing and tempering machine-shop tools. In addition, he had centralized all tool dressing, introduced automatic tool-grinding machines, and partly completed the tool rooms.[28] Taylor's mounting expenses worried Cramp's managers, however, and they soon fired him. From the vantage point of South Bethlehem three years later, Taylor remarked, "Judging from what Mr. Cramp has written, I should have thought that my work was worse than useless."[29]

With this string of promising if inconclusive experiments, Taylor welcomed Bethlehem's call in 1897 as an opportunity for a full-scale trial of his technical and managerial system. For nothing less than the company's core business was at stake. As related in chapter 3, Bethlehem had triumphed over Midvale as the country's premier manufacturer of gun forgings and armor plate, and had preserved its secure and profitable position by arranging cooperation with the Carnegie company, its only domestic rival. But persistent charges of extortionate armor prices, collusive contracting, and rigged inspections reached a climax in 1897 when Congress threatened to slash armor prices nearly in half and to build a government-owned armor plant. Carnegie and Bethlehem jointly refused to bid at the congressional demand of $300 per ton and soon compromised with the navy at $400 per ton. Still, the message was clear: the days of open-ended armor prices were over. In an effort to cut production costs, Bethlehem turned to Taylor and systematic management. At the time, Taylor enjoyed a growing reputation—his essay on "A Piece Rate System" (1895) had been discussed in the engineering community—but had not yet achieved national fame. Russell Davenport, Taylor's former colleague from Midvale, suggested his name to Bethlehem's president, Robert P. Linderman. In November 1897 Davenport contacted Taylor "in connection with the proposed establishment of [a] piece work system in our machine shop."[30] With piece rates as an ultimate goal, Taylor sprang into action.

As with Midvale and his consulting work, Taylor's tactics to systematize Bethlehem combined careful study and traditional "driving" methods.[31] Early in 1898 Taylor wrote Linderman that to introduce piece work successfully it would be necessary to take "entirely out of the [workers'] control many details connected with the running of the machines and management." He also urged restarting the metal-cutting experiments begun at Midvale and Cramp. "A careful study of each type of machine should be made so as to ascertain its driving and feeding power . . . and a table should be made for each machine which indicates the best cutting speed, feed, etc. for doing work as well as the time required to do it." To solidify his position, he attacked General Superintendent Owen F. Leibert and proposed a new "superintendent of manufacture" to discharge his duties more effectively. Despite Leibert's opposition to this brash transfer of power, Taylor installed Davenport in the new post that summer.[32]

With his survey of the works complete, Taylor focused his efforts on the linchpin to Bethlehem's success—Machine Shop No. 2. The company had established this huge shop (1,250 by 116 feet) when it entered the armor-plate and gun-forging business, but a decade of piecemeal expansion had resulted in a production system prone to backlog. A serious problem was the flow of open-hearth castings between the machine shop and the forging shop; the heavy ingot lathes seemed always to be behind. To rationalize this shop Taylor proposed and enacted a series of changes. Backed by Davenport, he first pressured Harry Leibert, superintendent of Machine Shop No. 2 and brother of the general superintendent, to resign, replacing him with his pliant assistant E. P. Earle. Taylor quickly moved to install other components of his system, including appointing "functional" foremen, introducing his metal-cutting innovations and belting practices, using soda water to cool the cutting tools, installing a planning department, and employing his costing system. Edward S. Knisely soon began collecting time-study data. Taylor resumed his investigations of metal cutting. In fact he seems to have done everything except introduce piece work.[33]

Beginning in the fall of 1898, Taylor's tool-steel experiments at Bethlehem illustrate a systematic, empirical, and quantitative style of invention. "During the original development of modern high speed tools," Taylor later said, he and Maunsel White "accurately heated chromium-tungsten tools of standard shapes, one set after another, to different temperatures, and then determined their standard cutting speeds . . . This accurate practical heating and running of the tools (without any theory on our part as to the chemical or molecular causes which produced the extraordinary phenomena) led to the discovery that marked improvements in the cutting speed . . . were obtained by heating the tools up close to the melting point."[34] Because Taylor's method was an early and paradigmatic case of "parameter variation" in engineering research, it is instructive to examine the tool-steel experiments in some detail.[35]

"To standardize the cutting tools," Davenport reported, "was one of the first requisites in connection with improving our Shop practice and the introduction of piece work under Mr. Taylor's system." Taylor had lathe tools made up and resumed his quest to find the "one best" tool steel (figs. 5.1, 5.2). Setting up a lathe to cut at speeds ranging from 3 to 300 feet per minute, Taylor revisited his effort at the Cramp shipyard to identify a standard air-hardening tool steel, this time to be used exclusively in the Bethlehem shop. Begun in early October 1898, the tests compared the self-hardening steels of Midvale, Mushet, Firth Sterling, Atha, and Sanderson. Taylor conducted a shopwide exhibit on 22 October to ratify his choice of Midvale's tool steel, but during this trial it exhibited "a very unaccountable irregularity." Modest speeds of 28–30 feet per minute ruined the Midvale tools in 3 minutes or less (one survived only 90 seconds). Taylor concluded that the Midvale tools had been somehow injured during their preparation, by too much or perhaps too little heat, and on 27 October he repeated the trial with three Midvale tools that Maunsel White, Bethlehem's testing engineer and metallurgist, had redressed at low heat. These tools fared even worse.[36]

After this "rather humiliating" episode, Taylor quickly secured Davenport's approval to identify a heat treatment that would produce superior lathe tools. In an attempt "to restore the grain to a good condition," White and his assistant heated four of the Midvale tools to incandescent heats ranging from a "blood red" to a "full cherry" (estimated as 1,000°F), then allowed each of the tools to cool in air. On the morning of 31 October, when these tools were first tested in the lathe, none survived the modest speed of 28 feet per minute for more than 6 minutes.[37]

FIG. 5.1. Blacksmith shaping round-nose cutting tool, c. 1900.
From F. W. Taylor, "On the Art of Cutting Metals," ASME Trans. 28 (1907): packet 1.

That afternoon, the same four tools were heated even more: to "bright cherry," "dull salmon," "salmon," and "bright salmon," or "yellow" (estimated as 1,400°F). Again they were allowed to cool in air. The two tools that reached the heats of salmon and yellow became so hard that they could not be "touched" or nicked with a file. At the lathe that afternoon the tool treated at light cherry heat was ruined in 30 seconds, apparently confirming received wisdom that treating self-hardening tools above a cherry heat was disastrous. But received wisdom was shattered by the tools treated at the three highest heats ("dull salmon" through "yellow"), which survived full 20-minute lathe tests at speeds of 25–29 feet per minute. The yellow tool even threw off chips that had turned blue from the ferocious heat of cutting. In lathe tests across the next five days another tool heated to "salmon" proved capable of cutting for 20 minutes at speeds up to 42 feet per minute; for 15 minutes the tool withstood even 50 feet per minute. A "yellow" tool briefly withstood speeds up to 45 feet per minute. In a second phase of experiments, lasting an additional four days, the men found that annealing the tools, by

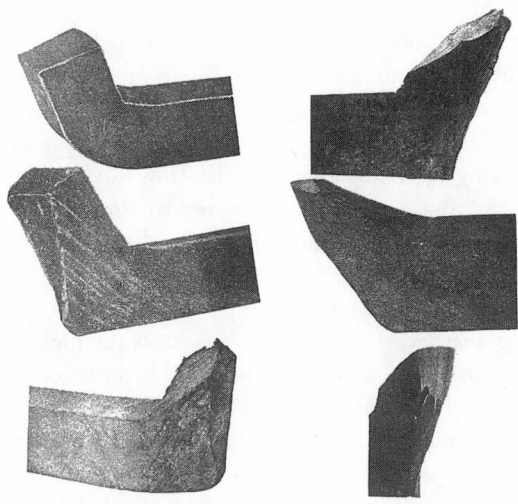

FIG. 5.2. Forging a round-nose cutting tool, c. 1900. *Left, top to bottom:* nose turned up on the anvil; heel drawn down to give support close under cutting edge; tool cut roughly to proper lip angle; *Right, top to bottom:* front corners cut off so as to approximate to proper curve; diagonal view looking partly down upon the tool after grinding; end view of ground tool showing nose properly set over to one side.
From Taylor, "On the Art of Cutting Metals," packet 2.

placing them in a hot container of lime for 12 or more hours to cool slowly, increased the tools' hardness even further yet maintained their performance at the lathe.[38] These two steps—cooling from a high heat, then annealing at a medium heat—were the core of the Taylor-White process.[39]

Taylor and White had made an important discovery, and they knew it. About the middle of November they forbade all other employees to treat tool steel by the new method. "From that date," White remarked, "the treatment of tools has been exclusively under the care of Mr. Taylor and myself." Their assistant Joseph Welden noted simply, "Further work on this report discontinued—as treatments were kept by Mr. White and Mr. Taylor, the results of tests being turned over to them."[40] The tools treated during those five days from 31 October to 4 November 1898 had produced such striking results that when later pressed by patent lawyers, White, among others, pointed to them as the first "high speed" steels.[41] Conceptually, this was accurate. Nevertheless, contemporary documents reveal a slow, steady process of trials and improvements; not until about a year later did Taylor and White's tools achieve really impressive cutting speeds.

Beginning in late November, Taylor brought representatives of the principal manufacturers of self-hardening steel—including Midvale, Sanderson, Carpenter, Atha, Crescent, and possibly others—to Bethlehem for demonstrations of their preferred methods for making lathe and planer tools. How Taylor convinced the manufacturers to divulge their preferred treatments is not entirely clear. In the event, Taylor and White gained a comprehensive view of treating tool steels, and used this "best practice" as a standard against which to measure their own innovation. Among other things, they learned that "bright cherry" was the highest heat anyone recommended for self-hardening tools and that other manufacturers all aimed in their treatments to produce a fine grain in the steel and to avoid the overheating that produced a coarse grain. In lathe tests that spring, they found to their delight that their special treatment for Midvale steel might also increase the cutting speed of similar steels by up to 300 percent. For instance, Atha tools treated by the manufacturer's representative passed 20-minute lathe tests at about 20 feet per minute, while the same tools treated by Taylor and White's secret, two-step treatment achieved speeds of nearly 60 feet per minute. Equally important in Taylor's mind (if not equally clear from the raw data) was the realization that their treatment made the performance of the tools much more precise and predictable.[42] Such reliability had the practical effect of ensuring that any given tool could perform at a very high level, which lent legitimacy to the broader cause of setting objective standards for shop work (fig. 5.3).

Two aspects of Taylor and White's experimental method had important consequences far beyond the confines of the Bethlehem machine shop. First, they imposed rigorous, controlled criteria on the metal-cutting experiments. By altering only two of the many variables that influenced the lathe-cutting trials, they gained quantitative data that could be used to persuade potential customers or fellow engineers, whether or not they had seen the tools in action. Arguments about whose tools and which treatments were superior could now be framed in quantitative terms. The Bethlehem experiments produced a mountain of quantitative data alongside the mountain of lathe shavings (which eventually came to more than 200 tons). Second, Taylor and White switched from judging heats to measuring temperatures—that is, they moved from craft skills (judging heat) to scientific measurement. These quantitative elements allowed Taylor and White, and Bethlehem, to place their findings within the larger framework of scientific measurement and universal standards.

Taylor's assistants collected data by testing the tools in the experimental lathe. The lathe feed (the distance the cutting tool moved across the piece per spindle revolution) was set at $\frac{1}{16}$ inch and its depth of cut at $\frac{3}{16}$ inch. They even corrected for variations in the

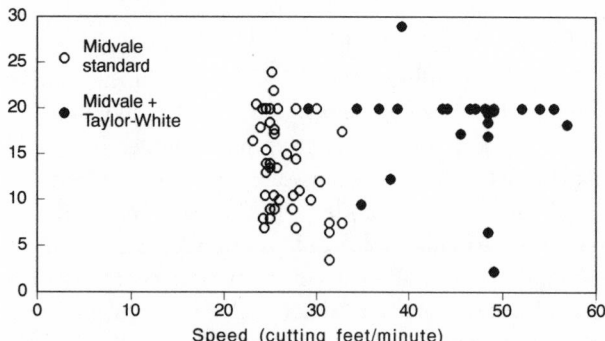

FIG. 5.3. Lathe tests of Midvale tools (with Taylor-White treatment).
Data from M. W. Welsh, "Report on Heat Treating and Lathe Testing Tool Steel (Midvale)," 22 November 1898, Lot 241—E. S. Knisely, BS.

FIG. 5.4. Taylor-White tools after lathe tests. *Left to right:* steel mounted on lip surface; ruined; fair; good; tool guttered.
From Taylor, "On the Art of Cutting Metals," packet 3.

hardness of the steel being cut. Only the speed of the cut, besides the cutting tool itself, remained as variables. Finally, Taylor defined the efficiency of a tool as the maximum cutting speed it would withstand for 20 minutes without being "ruined" (fig. 5.4). The 20-minute figure, as Taylor later justified it, was intermediate between shorter durations that yielded unreliable data and longer durations that consumed too much of the test forging and thus hampered useful comparisons.[43]

Taylor and White also expanded their quantitative regimen to the heat at which the tools were treated. Craft-oriented blacksmiths had long realized the importance of heat-treating steel. Yet for blacksmiths heat was not a quantity to be measured analytically (by temperature) but a quality to be judged with a practiced eye (by color). They had long observed that when heated, a piece of iron or steel began to glow, first dimly red, then various hues of red, orange, yellow, and, near its melting point, white. By judging the color of a particular heat—for example, "bright cherry"—blacksmiths could accurately reproduce it.[44] For craft metallurgists and blacksmiths, a particular color did not *represent* an analytic quantitative temperature. Rather a color *was* the heat at which a blacksmith performed a particular operation, such as bending, forging, welding, or heat treating.

The analytical pyrometers that provided a quantitative alternative to the blacksmiths' qualitative judgment were temperamental instruments that had yet to become straightforward "black boxes." In 1895 Maunsel White wrote to Henry Howe, the advocate of carbon standards for steel who was then lecturing on metallurgy at MIT, to inquire about pyrometers. Howe, in reply, cautioned that without a good deal of common sense, care, and intelligence, "you can get along better with the eye alone. In short, except in careful hands I think [a pyrometer] is more likely to mislead than to enlighten." Howe noted that the passing of a team of horses was enough to disrupt his instrument, so that the tremendous vibration from Bethlehem's steam hammer would render a pyrometer nearly useless. For overcoming these difficulties, Howe recommended a photographic recorder recently invented by the English metallurgist W. C. Roberts-Austen. White also asked Henry Souther of the Pope Manufacturing Company about his experience with the Le Châtelier pyrometer. White collected various booklets and brochures describing pyrometers built by French instrument makers, including a catalogue of pyrometers from Queen & Company of Philadelphia. Precise or not, the Le Châtelier pyrometer that White purchased was put to good use with Taylor's arrival and the beginning of the tool-steel experiments.[45]

Taylor and his helpers used this sensitive instrument to measure the temperatures of the heating furnace and the lead bath used for annealing. Blacksmith John Nowak, his two helpers, Taylor, and Welden judged the heat ("light cherry") at which Michael Welsh of Midvale Steel had dressed his tools. For each of six heats the men recorded the pyrometer reading and calculated the temperature. On 13 December 1898 the *heat* "light cherry" became the *temperature* 845°C, or 1,553°F.[46]

Taylor and White went on to correlate the incandescent color of hot steel with its quantitative temperature. Assembling their observations, Taylor and White reported on "colors of heated steel corresponding to different degrees of temperatures" before a meeting of ASME in December 1899. "There is . . . nothing more indefinite in the industrial treatment of steel," they began, "than the so-called color temperatures [used daily] by thousands of steel workers." Although they allowed that colors varied due to individual interpretation, the time of day, and even the weather, they nevertheless assigned temperatures to heats ranging from "dark blood red" (990°F) to "white" (2,200°F).[47] By 1901 their work had yielded a patent on a new pyrometer to be used in connection with the Taylor-White process of treating tool steel.[48] However strained, these efforts must be seen in light of Taylor and White's other attempts to systematize shop practice, invent standards, and impose uniformity.

Their effort to correlate color with temperature resonated with

Henry Howe. Beginning in 1897 Howe built up a world-renowned research and teaching laboratory at Columbia University. Comparing Taylor and White's results with his own, he found considerable agreement, in all cases within 55°C (98°F). But when Howe compared their data to that of an eminent French pyrometer expert, he found striking differences: the differences *averaged* some 167°C (300°F). Howe pointed to "gross errors of color-naming" on the Frenchman's part and bitterly complained that this "extremely misleading" data still contaminated textbooks.[49] For Howe, Taylor and White's data was an opportunity to standardize the so-called color temperatures on a uniform scale. Now Howe's metallurgical laboratory at Columbia University became linked to Taylor and White's experiments at Bethlehem.

The efforts of Taylor and White, Howe, and other pyrometer advocates were multiplied when Samuel Wesley Stratton, director of the recently established National Bureau of Standards (NBS, founded 1901), decided that the new federal agency would promote uniform temperature measurement. A physicist from Illinois Industrial University and the University of Chicago, Stratton in 1899 joined the federal government's fledgling effort in weights and measures as a vigorous advocate for scientific standards. The founding legislation that Stratton wrote for the NBS two years later authorized $250,000 for a new laboratory to support research in chemistry and physics needed to develop standards. The NBS addressed the metals industry by setting up a special laboratory to test pyrometers and to investigate "such pyrometric problems as are of technical and scientific interest." After preliminary work, the NBS advertised itself as prepared "to standardize almost any type of pyrometer, and to place in the hands of American engineers certified and reliable standards of temperature." As one NBS scientist noted, standardizing pyrometers fulfilled the agency's mandate "to bring about uniformity in the scales of temperature that are used by scientists, engineers, and technical men, that each may profit to the fullest extent by the experience of the other."[50] In short, accurate pyrometry promised a powerful method to link scattered local efforts into a coordinated national technical effort.

Anyone trying to dispute the efficacy of high-speed-tool steels now faced a formidable array. To challenge the tool steels was no longer a simple matter of dislodging a social network, disrupting the machine shop, or even bankrupting a steel company. The effort now required doing battle with ASME, which validated the research; Henry Howe and his laboratory at Columbia University, which verified the results; and NBS, which promoted analytical measurement to standardize U.S. industry. It is unclear that the quantitative pyrometer measurements were, in narrowly technical terms, superior to the qualitative judgments they replaced. For years Taylor and White experienced difficulties in calibrating their instrument. But the local associations behind the craft-oriented blacksmiths were no match for the coordinated technical effort of the professional groups that invented, tested, measured, verified, and validated the high-speed-tool steels.

Reforming the Factory

High-speed tool steel [was] the direct cause of the now widespread movement toward the reorganization of industrial methods. —JAMES M. DODGE

Once perfected, the new steels cut at impressive and unprecedented rates. Under the experimental conditions, tools made of the best brand of self-hardening steel could withstand cutting speeds ranging from 20 to 30 feet per minute. When treated by the new process, these same tools could be run at 60 feet per minute, doubling or even tripling their cutting efficiency. "It is highly interesting," Davenport remarked, to note that "the edge of the tool becomes red hot, and the chip or turnings, is so highly heated as to take on a dark blue color on its surface."[51]

Change was evident at Bethlehem as early as the spring of 1899, and within a year the new tool steel decisively altered the company's shop practice. The new cutting power erased the long-standing production bottleneck at the No. 3 ingot lathe. From December 1898 to June 1899, monthly production at the lathe nearly tripled, while the hours of lathe time fell by almost one-third; the output per hour increased dramatically (fig. 5.5). By January 1900 similarly impressive results were evident across the machine shop. Compared with the conditions that had existed before Taylor's arrival in 1898, the new steel increased the cutting speed of round-nose roughing tools by 183 percent, their depth of cut by 40 percent, and the pounds of metal they removed per hour by 340 percent. Individual pieces sped through the shop: to bore and face a 12-inch trunnion hoop took 12 hours instead of 75; crank pins were turned in 3 hours instead of 6–8; and 6-inch navy tubes were turned in 16 hours instead of 33–36.[52]

"This wonderful increase in the efficiency of cutting tools," as Davenport reported to President Linderman on 25 January 1900, "is of very great and far-reaching importance." Formerly, in all well-designed machine tools the cutting tool had been the weakest component, the cause of backlog in production. Now, with the sharply increased cutting speeds, the new tool steels required "corresponding modifications in the design of machine tools and the apparatus driving them." Furthermore, because each machine tool was so costly ($10,000–$30,000) and because each required a large amount of shop room and the services of expensive cranes, the new steels demanded factory-level modifications to achieve their impressive savings. Altogether, concluded Davenport, "the great advantage accruing by an increase of product . . . is self evident."[53]

Reaping the benefits required making an investment. In developing the new tool steel, Taylor and White performed some 16,000 experiments over eight or nine months at an estimated cost of $50,000–$125,000.[54] Although the tool steels' striking performance was now a matter of record, the Bethlehem managers knew that the technical capabilities of the innovation would not be sufficient to persuade others to adopt it in the form they hoped. Unauthorized duplication was a danger recognized as early as November 1898, when Taylor and White imposed secrecy on the project. To retain proprietary control, the Bethlehem managers constructed a legal framework through patenting and licensing agreements. In recognition of Bethlehem's support, Taylor granted the company a "perpetual shop right" to the new tool steel, although as early as April or May 1899 he successfully negotiated with Linderman to retain the right to patent the innovation. About this same time Midvale Steel, having heard of the unusually good results Taylor obtained with its steel, informally asked him for information on

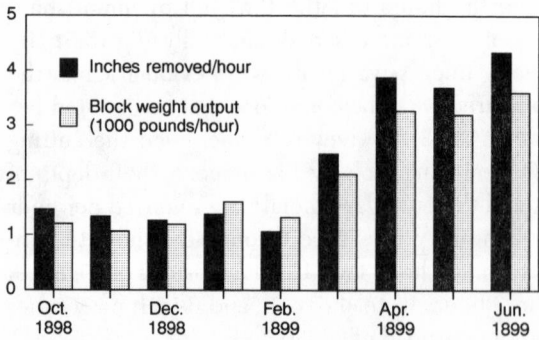

FIG. 5.5. Taylor-White ingot lathe output (October 1898–June 1899).
From E. S. Knisely to E. P. Earle, 6 July 1899, Lot 241—E. S. Knisely—Machine Shop, BS.

the process, which he refused to disclose. In October 1899 he applied for a patent; two closely related patents were granted sixteen months later.[55]

Even as Bethlehem publicized the tools, it closely guarded the secret heat-treating process used to make them. Taylor himself maintained secrecy by deflecting further inquiries from the Franklin Institute and Midvale Steel. In January 1900 he hinted that the asking price for his and White's patent might be $50,000, but he declined to make this a formal offer. Davenport wrote Linderman that Taylor's "principal claims are fundamental" and that the patent could be successfully defended. Although he recommended purchasing Taylor's patents, Davenport admitted there might be "difficulty . . . in ascertaining whether the patent was being infringed by other people, owing to the fact that the treatment of the tools can easily be done in a small space guarded from public view." On the other hand, to treat the steel samples successfully one needed to control the temperature accurately, and Davenport concluded that "it would scarcely pay concerns of importance and capital to spend much time or money in working up the details of the process" if it could be purchased at a reasonable price. Davenport hoped to delay the issuance of Taylor's patent for three years or more by taking "the proper steps" as well as keeping the process secret. By June 1900 Taylor had sold Bethlehem his patent, still tangled up in the patent office at Bethlehem's behest, retaining a half-interest in the European rights.[56]

The attention the new tool steel brought to the Bethlehem shops further complicated control. By early 1900 four companies, including Westinghouse Manufacturing, had approached Bethlehem about procuring the new tools. Six more had directly approached Taylor, and he had been negotiating with William Sellers & Company, which soon began using tools made from Midvale steel that Taylor personally had treated. Taylor and Sellers agreed that the machine-tool company would pay $6,000 for the information necessary to conduct the heat treating itself. In June 1900 Taylor noted that "we have a daily string of visitors to our large machine shop to see it in actual operation."[57]

In July at its works Bethlehem held an exhibition whose results *American Machinist* described as "startling." *Iron Age* detailed the tools' cutting power. After a heavy cut lasting 15 minutes on an unusually hard piece of cast iron, the Taylor-White tool showed no signs of wear, while a Mushet tool lasted a mere 22 seconds. While cutting a steel shaft at the unusually high speed of 150 feet per minute the Taylor-White tool glowed red yet was uninjured after 4 minutes, while a Mushet tool broke down completely in just 6 seconds. "A most important result, which must follow the more or less universal introduction of the Taylor-White process," forecast the *Iron Trade Review*, "will be the practical exclusion of the foreign brands of self-hardening steel which are now supplying over one-half of the demand and will place the market entirely in the hands of the American steel maker."[58] In fact, within seven years, the efforts of American makers of crucible steel trimmed the Sheffield companies' share of the American market for high-speed-tool steels to just 40.6 percent (fig. 5.6).

That fall, the new tools impressed crowds at the Paris Exposition. There Henri Le Châtelier saw the evidence that converted him to the Taylor system. Among others, the leading German engineer Franz Reuleaux viewed the compelling "dull red heat" of the cut chip. "The most important fact," he reported, "was that the tool hardly appeared to suffer in any manner; when afterwards removed its cutting edge proper appeared sharp and uninjured."[59]

High-speed-tool steel exerted great influence at Bethlehem and beyond even after Taylor was fired in April 1901. His firing was

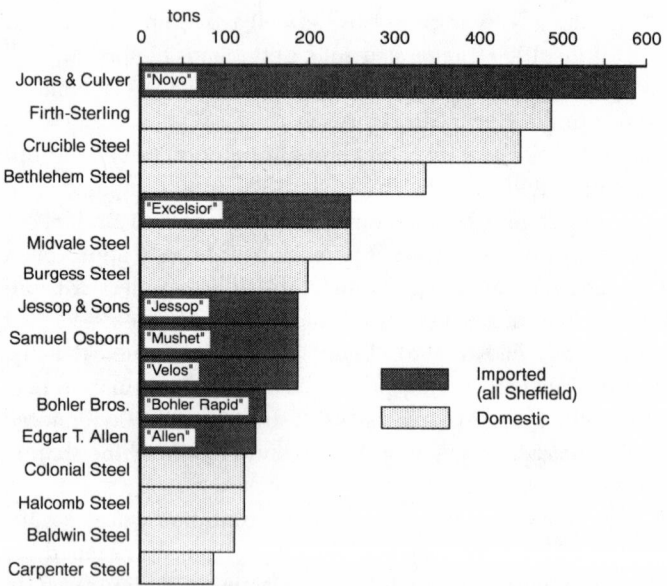

FIG. 5.6. Domestic production and imports of high-speed steel, 1907.
Data from: H. S. Snyder, memo, 6 August 1907, 17:2, AJ; and K. C. Barraclough, Steelmaking before Bessemer (London: Metals Society, 1984), 2:140–41.

the result of "continuous strife" and the uncertain economic return attending his managerial system; it happened just two months after Taylor and White's two patents were finally issued. Bethlehem was again in the throes of change. After buying the controlling interest of Bethlehem Steel in June 1901, Charles M. Schwab ordered Taylor's labor system thrown out. Although Taylor claimed that falling output compelled certain department heads to covertly carry the slide rule, time-study men, and planners on the payroll as mechanics, a Bethlehem mechanic stated that Taylor and his assistants had been entirely replaced by "practical" men by the autumn of 1901. "The [Taylor] system obtains in part, but it is gasping—evidently the dying swing of previously acquired momentum." [60]

Since Taylor and White had assigned their patents to Bethlehem, its licensing effort proceeded apace even in their absence. By the end of 1903 no fewer than sixteen companies, as well as the government of Cape of Good Hope, had paid (typically) $3,000 for a shop license. [61] Sales of foreign patent rights netted much larger sums. Bethlehem collected $100,000 each from Vickers Sons and Maxim, Ltd., for rights in Great Britain, and from Isador Loewe & Gebruder, Bohler & Company for rights in Germany, Austria, and Hungary. Rights for the Scandinavian countries and Canada fetched $11,250 and $15,000, respectively. Taylor received half these handsome sums, thereby securing his fortune. [62]

For those $3,000 payments, shop licensees learned the "secret" treatment at the core of the new process. The patents outlined the general principles: the chemical composition of chromium and tungsten or molybdenum, the "high heat" treatment near the steel's melting point, and the "low heat" treatment of annealing. The licensing description further provided a modular set of instruction to ensure the reliability of results. The Taylor-White process did more than increase the maximum speed of cutting tools. The results of the process were reproducible because of the analytical temperature measurements, standardized heat treatments, and temperature-control apparatus (figs. 5.7, 5.8). In this respect the Taylor-White process was to tool steels what the Krupp process was to hardened Harvey armor. The Taylor-White licensing description first listed fifteen "elementary operations," including variations on the high- and low-heat treatments. The shop superintendent chose one of nine numbered treatment sets, depending on the desired characteristics of the tool, and directed the tool treaters to execute the numbered treatment and then stamp the tool with that number. [63]

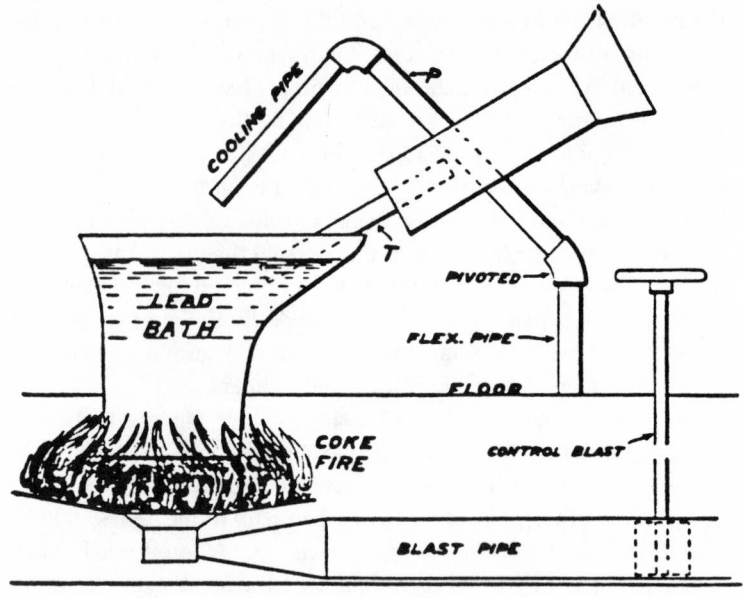

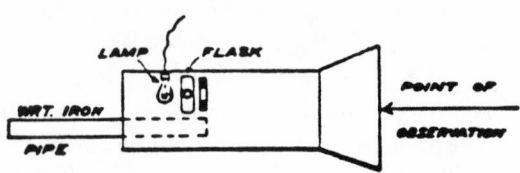

FIG. 5.7. Taylor-White heat-treating process. Taylor and White achieved
uniformity with high-speed steels by devising special apparatus for the
heat treatments. After being heated up to a yellow heat, the tools were
plunged into the bath of molten lead shown here. The temperature of the
bath could be raised by the air blast for the coke fire or reduced with the
cooling pipe P, which had cold water circulating in it. The bath temperature was determined using the metal pipe T whose lower end projected
into the lead bath and whose upper end was surrounded by a blackened
wooden box. The incandescent heat of the bath was visually compared
with flasks of colored liquid lighted by an incandescent lamp. See figure
5.8.

*From Charles Day et al., "The Taylor-White Process of Treating Tool Steel,"
Metallographist 6 (1903): 134.*

FIG. 5.8. Taylor-White heat-treating laboratory, c. 1900. The two molten lead baths (on either side of the central chimney) were used for heat-treating high-speed tools. The bent cooling pipes and the wooden temperature-viewing boxes are clearly visible; compare with P and T of figure 5.7.

Courtesy Hagley Museum and Library; Acc. 80.300, box 2, lot 338, 22463A.

Tool treaters now faced a set of decontextualized instructions. Treatment No. 4, for instance, directed them to execute the elementary operations P H V M O L K:

Preheat the tool to any heat above incandescence;

Heat the tool to the point at which it *dents or crumbles* readily when tapped with a light steel rod;

Quench the tool all the way over in *lead bath* held at *1450°F.* for from three to ten minutes, according to the size of the tool, until tool and *bath* are at exactly the same temperature. Take great care that *bath temperature* does not rise while the tool is cooling; slight fall in *bath temperature* will not injure the tool;

Bury whole tool as soon as possible all the way over in *lime;*

Preheat the tool to above 450°F.;
Reheat tool after it has cooled from previous treatment in *lead bath held at 1160°F.* for about ten minutes;
Lay tool down to cool *in air.*

Only the shop superintendent was in a position to know that this treatment produced especially tough tools capable of resisting unusual strain or shock.

Such compartmentalization served to maintain secrecy about the process; the italicized words in the instructions, as well as all other crucial steps and temperatures, were handwritten or typed into the instructions sent to individual licensees. The form of these instructions clearly reinforced the "collar line." Blue-collar tool treaters were not told which numbered treatment produced which set of tool characteristics; if they left for another job they possessed only the shop-floor rumors they could scavenge on this essential matter. These instructions were therefore a severe blow to tool treaters, highly skilled workers who had once possessed unique craft knowledge. The elementary operations also served to reinforce the Taylorist split between the white-collar workers who planned and made decisions and the blue-collar workers who carried out their orders.[64] In both these respects, Taylor and White's work foreshadowed the automatic, mass-production heat-treating technology pioneered by the automobile industry in the 1910s.

Machine builders and users alike were eager to sign up for Bethlehem licenses, for by 1902 a revolution in machine design was under way, impelled by high-speed steel. Bullard Machine Tool of Bridgeport, Connecticut, the traditional center of machine building, offered a 26-inch rapid-reduction lathe designed especially for the use of high-speed cutting steel. "This is a lathe of radical departure from existing design," the company stated, "made necessary" by the inability of traditional lathe designs to handle work "as rapidly as it can and should be done." Lodge & Shipley Machine Tool of Cincinnati, the up-and-coming center of machine building, pitched its 48-inch lathe to "the up-to-date manufacturer." The "magnificent efficiency" of the new cutting steels, the company stated, "calls for" lathes of increased power, rigidity, and wearing qualities. The following year, 1903, the company described high-speed lathes that had been designed to "meet the conditions and requirements" brought about by high-speed-tool steels. Soon it became a commonplace to discuss the necessity of redesigning all machine tools—lathes, planers, milling machines, and more—to take full advantage of high-speed steels.[65] While some writers felt that increases in weight and stiffness would be sufficient, many others argued that changes in the very form of machine tools were required.

In a broad view, not only individual machines but also the design of the factory itself was at issue. One Cincinnati machine builder stated that high-speed steels were responsible for "a great onward and upward movement in the industry of machine making." The dramatic increases in output obtained with the new tool steels required greater attention to the routing of work through shops, as well as to the design of stockrooms and storage facilities. Attention also had to be given to the fast-moving ribbon of metal thrown off the lathe: at a speed of 3 feet a second or more, "the chip comes writhing and twisting, almost red hot, in a continuous length, shooting here and there, everywhere but the chip box." One company had to hire three men just to cart away the 4 tons of metal chips produced each day, "so that the men won't get lost." The new machines required two or three times as much power as the old, pressing overhead-belt drives beyond their capabilities and eventually hastening the changeover to electric-motor drives. Moreover, since increased output was no longer limited by the tools but rather by "the endurance of the workman," the new tool steels spurred the development of labor-saving, self-acting mechanisms, such as automatic magazine feeds. Finally, high-speed steels, rapid-reduction lathes, and motor drives were a package that could not be "intelligently considered separately, nor should a plant be designed and equipped without knowledge of the type of management to be employed."[66]

Taylor had begun his tool-steel experiments to develop managerial knowledge about how quickly work could be done. Appropriately, then, the possibilities of high-speed steel revived and extended the managerial impulse behind factory reform. "It is indeed a pleasure to see the new type of machine tool operating with high-speed steel, and treating the work it has to turn out in such a businesslike way," wrote one shop superintendent, "imparting a life and energy to the whole establishment [and] exerting an influence on everybody therein to get 'a hustle' on that is positively exhilarating in its effects." James M. Dodge, a past president of ASME, traced the factory-reorganization movement directly to high-speed-tool steel. "We cannot attach too much importance to this new and grand era of reform, which means billions of dollars added to the wealth of the world," added Oberlin Smith, "and which started with the incomparable work of Mr. Taylor and Mr. White."[67]

Indeed, high-speed steel affected the traditional balance of power and authority in the shop. "The new conditions imposed by" high-speed steel "rendered obsolete the long-treasured experience and accumulated data of the tool-maker." It was now "imperative" to approach the design of lathes and other machine tools "upon a basis of experimentally ascertained fact." The traditional, empirical design for machine tools must give way to a process that began with kinematics and worked through the mechanism of cutting to arrive at a "rational" design. Blacksmiths, who had previously been in charge of forging and treating shop tools, were the losers. Instead of the blacksmith, "some responsible person"—a superintendent, foreman, or speed boss—should be in charge of introducing high-speed steel in a shop.[68]

Given the level of interest in high-speed steels and the force of their impact, high-speed steels were quickly reinvented by U.S. and Sheffield crucible steelmakers eager to enter this promising market. When their efforts interfered with Bethlehem's attempt to commercialize its patented product, the company mooted the ambitious plan of forming a tool-steel cartel. In August 1907 Vice President H. S. Snyder outlined how such a cartel would aid the tool-steel manufacturers with "economies" in producing and selling the steel as well as put "the price on a considerable [sic] higher level than at present." Snyder observed that "a combination of the largest tool steel manufacturers . . . would be certain to exercise a very great influence upon consumers of steel manufactured by concerns who were not licensed." Although a cartel could not be formed explicitly for the purpose of setting prices, Snyder felt that a uniform price scale could "no doubt be established" through monthly meetings. The chief problem "essential to the success of any combination" would be excluding imported high-speed-tool steel. Imports from Sheffield arrived, duty paid, for 45 cents per pound, undercutting the domestic price of 60 cents per pound. To exclude imports, Snyder suggested either signing up the importers or "proceeding against their customers, the consumer." To coordinate these activities Snyder suggested capitalizing a new entity, the Bethlehem Tool Steel Company or the United Tool Steel Company, at $100,000. While retaining full ownership of the Taylor-White patents, Bethlehem Steel would give an exclusive license to the new company and receive in return 51 percent of its paid capital stock. The principal producers of high-speed-tool steel joining in the arrangement would subscribe the rest of the capital stock. The financial return to Bethlehem would be $80,000 per year in royalties.[69]

Snyder also proposed that the cartel pay legal costs. This was no idle matter, for in the previous year the steel company had initiated a patent-infringement suit against the Niles-Bement-Pond Company to force this leading machine-tool manufacturer to honor its patents. It was a test case whose outcome would determine the commercial future of high-speed steel. Based in large measure on the expert testimony of a Sheffield professor of metallurgy, Niles-Bement-Pond constructed a case that destroyed Bethlehem's claims of infringement. What is more, when Bethlehem appealed the decision in 1909 the court nullified the Taylor-White patents, and Bethlehem lost control over the innovation. In the legal realm, if not in the machine shop, scientific expertise was highly prized.

The European metallurgists who first investigated the structure of the new tool steel provided scientific accounts that could have strengthened Taylor and White's claim to novelty. Introducing reports of two leading French metallurgists, *The Engineer* of London commented in 1904 that although high-speed steels "have sprung into popularity with surprising suddenness . . . little has been heard as to the causes which give them their astonishing properties." Floris Osmond and Henri Le Châtelier directly addressed these "causes." Osmond adapted his general "carbon" theory of steel by hypothesizing that the presence of chromium, tungsten, or other substances delayed the solidification of carbide (which rendered steel hard) during cooling as well as the solution of this compound during heating. Thus, tools made of high-speed steel remained hard at high temperatures because they retained significant amounts of the low-temperature carbide. From the rival "allotropic" perspective, Le Châtelier explained that high-speed-tool steels remained hard, even at the red heat that softened typical tool steels, because the additional elements elevated the "critical point" of the high-speed steel. According to Le Châtelier, even the dull red heat of high-speed cutting did not raise these steels to the critical point, at which molecular rearrangement occurred. Here then were two explanations for the performance of the high-speed-steel tools, based on the new structure-oriented metallurgy. These essays appeared in the same American journal that reprinted Taylor and White's essay on correlating temperature with incandescent color.[70] Two years later, in 1906, Carpenter and Le Châtelier both gave metallurgical explanations of high-speed steel—in Taylor's presence at the ASME meeting in New York. In the event, Taylor and White's failure to incorporate these or any other scientific accounts of why high-speed-tool steel remained hard at red heat, which was certainly a new metallurgical property, crippled Bethlehem's legal case.

When testimony began in 1906 the case looked promising for Bethlehem. "The lawyers seem to think that I left our case in good shape," White wrote Taylor after giving some ninety pages of testimony. One of the defendant's witnesses had even admitted that "he did not begin his experiments until after he had heard of the results obtained by our tools!" Taylor was extremely reluctant to testify, however. As he wrote to White, "My inclinations are strongly to decline on any terms to testify in the Taylor-White suits, as this work would . . . be a severe nervous strain for me . . . I do not propose to bother myself and possibly injure my health in worrying over the affairs of the Bethlehem Steel Co., unless someone beside the Bethlehem Steel Co. is going to receive somewhere near an adequate return." Taylor's reluctance did not help Bethlehem's case. The petulant and sometimes incoherent testimony that he eventually gave probably did damage. Even worse, from Bethlehem's perspective, Niles-Bement-Pond enrolled Bethlehem's many domestic competitors and formed a transatlantic alliance with the Sheffield steelmasters, who rightly feared that Bethlehem's cartel would close up the American market. Most devastating was the activity of the Sheffield Steel Makers Ltd., a "more or less secret tribunal" represented in the United States by Colonel Herbert Hughes. The Sheffield Steel Makers produced witnesses such as John Oliver Arnold, a Sheffield consultant and professor of metallurgy, who denied Taylor's priority or novelty in high-speed steel. By March 1908 testimony had been completed, and parties on both sides of the Atlantic anxiously awaited the verdict.[71]

Remarkably, the judge presiding over the suit denied nearly every claim that the Patent Office had allowed. His decision on 29 January 1909 stated that

> the prior art was radically different from what the patent would
> lead us to believe. There was no such thing known to the art
> as a breaking-down point, on the contrary, it is established
> that given a new steel, particularly if the composition were
> unknown, it was customary to experiment with and test and try
> it and thereby ascertain the best method of treating it; that in
> pursuing this course, the method of the patent was substantially
> followed, and temperatures as high as any mentioned therein,

or higher, were, not in isolated cases, but in ordinary practice resorted to.

The judge concluded that "if in the race, the patentees have surpassed others, it has not been through novelty or procedure, but by means of special facilities, apparatus and methods not embraced in the patents."[72] Chief among these methods, as we have seen, was the obsessive control of temperature. The judge moreover nullified Taylor and White's two principal patents on which the suit hinged. Whatever the merits of the final decision, the suit broke Bethlehem's legal monopoly on high-speed-tool steel. Here was a conclusive pronouncement that not the artifact but the analytical apparatus—the standardized lathe tests, the pyrometers measuring temperature, the elaborate controls—was at the core of the Taylor-White process.

The Art of Cutting Metals

The discoveries of Messrs. Taylor & White . . . have simply proved again that scientific methods lead to much better results than guess-work.—CHARLES DAY

The issue of whether metal cutting was an art or a science was left unresolved by Taylor's appropriating both labels in his 1906 presidential speech to ASME. Indeed, this essential tension—whether engineering was a practical art requiring judgment or a science requiring calculation—was a fundamental point of debate in professional engineering at the time. In his writings, if not always in his work, Taylor boldly asserted that science should become the impartial authority that intervened in the raging battle between labor and capital. As Taylor himself phrased it, the core of "scientific management" was not the stopwatches and routing sheets or any specific details or devices but "a complete mental revolution" effected through the engineer's practice of fairness and objectivity in arbitrating the competing demands of workers and employers.[73]

How an engineer was to achieve this vaunted objectivity was an object lesson in Taylor's 1906 ASME presidential address. He pointed out that "the most important lessons taught by these experiments, particularly to the younger men are: That several men when heartily cooperating, even if of everyday caliber, can accomplish what would be next to impossible for any one man even of exceptional ability; that expensive experiments can be successfully carried on by men without money, and the most difficult mathematical problems can be solved by very ordinary mathematicians, providing only that they are willing to pay the price in time, patience and hard work; that the old adage again is made good that 'all things come to him who waits,' if only he works hard enough in the meantime." In spite of such pat homilies about luck, pluck, and virtue, Taylor crafted the lengthy published version of his address—which ran to 319 pages in the ASME *Transactions* of 1907—as a model of professional engineering practice. Henri Le Châtelier suggested that Taylor's essay was "a model which every young engineer will have to study." What lessons did it offer?[74]

One is tempted to label the new tool steels "scientific," given Taylor's method and reputation. A more insightful approach, however, recognizes that the experiments at once represented a scientific style of engineering research and an attempt to delimit scientific metallurgy from mechanical engineering. Generally, Taylor's rhetoric and presentation were intended to construct separate professional communities of science and engineering. Taylor presented his results in engineering tables rather than scientific equations (the fancy mathematics of his assistant Carl Barth produced only an instrument, the metal-cutting slide rule). More striking is Taylor's artful interpretation of the new structure-oriented metallurgy known as metallography, which first gained application in U.S. industry with the contemporaneous rail crisis (see chapter 4) and featured in European discussions of high-speed-tool steel. Taylor was reasonably well informed of these advances in understanding the crystalline microstructure of a metal and its resulting physical properties; what he did was conceive and interpret them in a way that minimized their impact on engineering practice.

Fully acknowledging the metallographic work on high-speed steels by Floris Osmond, Henri Le Châtelier, and John E. Stead, Taylor in his essay complained that their use of "special technical expressions make it difficult for even an educated engineer to clearly understand their articles without very considerable study."[75] Yet he discussed at some length H.C.H. Carpenter's work on high-speed steel at the National Physical Laboratory in England. Taylor even grasped metallography's key insight that the temperature history of the metal, not merely its chemical composition, determined its physical properties. In discussing this insight he used the terms *hardening carbon* and *softening carbon* (in preference to the accepted scientific terms *martensite* and *pearlite*) to represent the two principal microconstituents of steel.[76] And he accepted that molecular rearrangement occurred at the metal's "critical point" (the temperature at which the cooling or heating was temporarily arrested). Finally, he noted that high-speed steel contained austenite, an extremely hard formation of steel known to exist at extremely high temperature but rarely seen in industrial applications of steel. Taylor had clearly grasped the principal insights of metallography.

But Taylor blunted the impact of the new science in three revealing ways. First, he maintained that high-speed steels had two separate, independent properties of hardness, one at low temperatures, the other at high temperatures. Metallography, he argued, did not account for or explain the "red hardness" that typified high-speed-tool steels. He also dismissed the heating and cooling curves that were the metallographer's primary analytical tool. These curves, he claimed, "throw no light upon the important question as to the exact heat at which the tempering or softening of the high speed tool begins, or as to what range of temperatures this softening process extends." Finally, he maintained that there was "no traceable relation between the highest cutting speeds and any particular one of the microscopic structures."[77] Taylor had little incentive to take up the metallographer's microscope and discard the forty thousand experimental lathe tests his career rested on. In this light, his heterodox interpretation looks less like an ill-conceived misunderstanding of the new science than a carefully conceived strategy to blunt its influence.[78]

A second general insight that can be gained from this discourse surrounds the potent issue of secrecy, reconstructing again the divide between the arts and science. For a quarter-century Taylor the practical artisan achieved the remarkable feat of "keeping almost all of these laws secret." Such a frankly commercial, not to say mercenary attitude to knowledge raised eyebrows among the scientists—but not the engineers. "The inventor [Taylor] informs us with legitimate pride that during all these years none of the results of these experiments had become known to outsiders," stated Le Châtelier. "We cannot too greatly admire such an example of probity in the arts." In contrast to Taylor's secrecy, at the Illinois engineering experiment station the tool-steel experiments had been rapidly conducted and openly published with the aim of benefiting the state's manufacturers. For his part, Taylor bristled but did not budge when one engineer associated with the Manchester trials "insinuates that we are holding back some much needed data."[79]

Such secrecy resulted from the imperatives of engineering consulting. Taylor and his associates did not fashion their consulting careers by selling outright their hard-won information on metal cutting. Rather, they advanced their project through ten manufacturing establishments that paid their salaries, set up special machines, and provided workmen, in exchange for the valuable data on cutting metals. "All of the companies who were given this information, and all of the men who worked upon the experiments, were bound by promises to the writer not to give any of this information away nor to allow it to be published." Presenting this paper to the engineering fraternity, Taylor emphasized, "will cut off our former means of financing the experiments."[80] Seen in this light, taking steps to install Taylor's shop-management scheme was the (only) means of acquiring knowledge of the invaluable metal-cutting experiments, including the latest on high-speed-tool steel. This hypothesis helps explain why so many companies were only incompletely "Taylorized." It also positions high-speed-tool steel as the motive force behind the shop-management and scientific-management movements, where contemporaries like James Dodge and Oberlin Smith placed it. Taylor himself stated his "surprise and regret" that the metal-cutting experiments had eclipsed his writings on the "real vital subject" of shop management.

Finally, the metaphors Taylor employed in describing his management system revealed his activist stance for engineering. "Our whole system of management," he stated, entails "the substitution of an elaborate engine to drive and control the shop in place of the old fashioned foot power." Of comparable importance to the fireman who tended the steam engine or the draftsman who prepared accurate drawings was the new "human managing machine" necessary for maintaining standard shop details, using standard shop methods, and giving each workman a definite task to be finished in a given time. "The efficiency of our engine of management . . . compared with the old style single foreman is like a shop engine as compared with foot power."[81]

In the years after his death in 1915, Taylor's potent image of a "human managing machine" spread irresistibly from its original realm of mechanical engineering, first into business education and then to entirely new realms of hospital management, home economics, nursing, secondary education, government reform, and public administration. The international appeal of Taylorism was no less impressive.[82] As with many founding figures, Taylor did not see the full implications of his work. He remained committed to the byzantine system of belt drives and opposed to driving machine tools with electric motors, recognizing that he was considered reactionary for taking that stand. He was artfully conservative in interpreting metallurgical science. These developments in machine tools, factory design, and metallurgy culminated in the rational-factory movement in the automobile industry. In responding to a new and insistent user, the automobile makers, the U.S. steel industry was prodded into its fully mature form.

Confronting the Rocky Mountains by automobile, 1910: departure
of brothers Floyd and Bill Clymer from Denver for Spokane in a
20-horsepower Studebaker.

From Floyd Clymer, Henry's Wonderful Model T, 1908–27 *(New York: McGraw-Hill, 1955), 25.*

CHAPTER SIX

The Imperative of Automobiles

1905–1925

Floyd Clymer began a lifelong career in the automobile industry by selling cars out of his father's medical office in a small town north of Denver, Colorado. After three years of steady salesmanship, during the summer of 1910, he set out on a cross-country publicity stunt for Studebaker. Floyd was then a precocious fourteen-year-old; his aide-de-camp and younger brother Bill was just eleven. After hatching a scheme to drive up and over the Rocky Mountains, concurrent with their family's move to the Pacific Northwest, the two boys sought backing from the manager of Studebaker's local branch. Their scheme was well timed. Just then Studebaker was chasing Ford and Buick for market share, its 30-horsepower Flanders model priced at $1,250 and selling 17,000 units a year. To capture the attention of prospective buyers, Studebaker was backing several publicity stunts; among its recent successes were a marathon from Montreal to Mexico City, speed runs at Indianapolis, hill-climbing trials, and entries in the prestigious Glidden Tour. The Denver manager instantly saw that a trip over the Rocky Mountains by such supposedly inexperienced automobilists would generate interest along their route and, with any luck, favorable newspaper copy across the country. For their stunt, he gave the Clymer brothers a brand-new, 20-horsepower Flanders, priced at $1,050 and soon to be selling nearly 10,000 units a year. They departed without a mechanic or spare parts, with only, as Clymer later put it, "a load of trust in a comparatively new make of car."[1] The brothers permitted themselves a collapsible canvas bag for filling the radiator and two 100-foot-long strips of canvas (sewn together by their mother) for getting out of sandpits. With a send-off from the governor of Colorado and a letter of introduction from the mayor of Denver to the mayor of Spokane, Washington, the two young automobilists set off.

All went well at first. "Our engine sounded invincible," wrote Clymer, "even when confronted with the dry terrors of the desert and the worst imaginable roads." At small towns along the way the two boys refueled, checked the tires, and dreamed of the cheering Spokane residents whom they imagined lining the roads in welcome. In the flat country west of Cheyenne, Wyoming, the headlights of their car illuminated jackrabbits. The sheepherders and cowboys they met up with offered them hospitality.

With the Far West stretching out around them, it was a fine and inspiring adventure—until the Studebaker's transmission developed trouble climbing into the hills west of Laramie. First it ground, then it howled, finally it locked up tight. "We were two tired and disgusted kids but we couldn't just sit there and rust," Clymer explained. "We found the railroad tracks leading west and walked to the nearest flagging station, where we sent a frantic telegram to Studebaker in South Bend, Indiana." The factory dispatched a mechanic who installed a new transmission at the railroad station in Walcott, near the Continental Divide, gave the car a tune-up, and sent the brothers on their way. A hundred miles farther on, when the second transmission gave way, a Flanders mechanic from Denver came out and fixed the car. But when the third transmission broke a gear, Studebaker decided that the cost of keeping the two boys on the road outweighed whatever publicity value remained. The brothers shipped the ill-fated car back to Denver and took the train onward to the Clymer family's new home in eastern Washington. Once again Floyd Clymer sold Studebakers, until he realized that "if I was ever going to get anywhere in the car business, I had better join up with a fast-moving company . . . The Model T Ford, I felt, would be a rich vein."[2] Ford was advertising a new alloy-steel planetary transmission, as if calculated to appeal to the transmission-plagued Clymer brothers, and Floyd joined those whose ranks helped sell some 15 million of the "Tin Lizzies." Meanwhile Studebaker, in what amounted to perhaps the first-ever factory recall, spent nearly $1 million sending out mechanics across the country to fix the defective transmission of the Flanders 20, a remarkable technical effort that helped boost total sales of this model by 1914 to just over 30,000.

With the achievement of these unprecedented production figures, the leading automobile manufacturers—Studebaker, Maxwell, Willys-Overland, the General Motors companies, and Ford—became the steel industry's leading consumers. Unlike the railroads and the navy, which had exerted force on the steel industry principally through one or two products, such as rails, armor, and gun forgings, the automobile makers required a multitude of products, services, and support. An automobile handbook of the mid-1920s required five hundred pages just to describe the components, from exhaust valves to steel wheels. A comprehensive account of steel in automobiles would be even longer and more ponderous. Instead, this chapter explores five key interactions between the producers of steel and its new leading user, the automobile industry.[3]

The first of these interactions was the establishment of standards for steel. Setting standards for steel proved to be one of the automobile engineers' most durable achievements. The Society of Automobile Engineers (SAE) debated and discussed specifications for steel throughout the 1910s, echoing earlier discussions by the railroad engineers with one signal difference: the automobile engineers' standards persist today. The SAE faced stiff competition in this effort, since both the American Society for Testing Materials (ASTM) and the American Society for Steel Treating (ASST) were attempting to set standards. The triumph of the SAE standards, then, signifies the society's dominance in the automobile industry as much as the automobile industry's dominance in steel.

The second interaction was the use of alloy steels—steels containing small amounts of desirable elements beyond carbon and manganese. These steels became a defining product of the motor age, and their use connected the steel industry, for the first time, directly to individual consumers. Automobile manufacturers did not exactly discover alloy steels, of course. As we have seen in previous chapters, nickel had toughened battleship armor since the 1890s, while chromium and tungsten kept high-speed tool steels hard at incandescent heats. But automobile manufacturers brought alloy steels to public awareness. The Ford Motor Company's Model T (1908–27) introduced mass production and also the direct-consumer use of vanadium steel. Ford retained vanadium steel into the 1920s, while other automotive metallurgists experimented with alloys of chromium, nickel, silicon, tungsten, and molybdenum. When added in amounts from 1 percent to 15 percent, these elements imparted new physical properties to steel and lent a high-tech image to the product, an association promoted by automobile sales departments.

Proper heat treatment—the third point of contact—was necessary to secure the full strength from alloy steels. "Until the advent of motor-car production in commercial quantities," remarked one industry observer, "heat-treatment of steel in commercial work was little practised excepting in the treatment of tools."[4] Companies that merely added an expensive alloying element to a batch of steel secured only its high cost. As if recalling the obsessive process control of the Krupp armor shop, and more recently the Taylor-White process, automobile metallurgists worked out exact procedures for the heat treatment of the forgings, stampings, and castings that composed an automobile. By the 1910s automobile metallurgists were discussing the effects of different alloys on the "critical point" of the resulting steel—the temperature at which molecular rearrangement occurred. A decade later, some metallurgists argued that proper heat treating alone could make normal carbon or "low alloy" steels the equal of steels containing high proportions of alloys.[5] For producing large numbers of carefully heat-treated parts, whether high- or low-alloy, automobile metallurgists combined the science of steel treating with the technology of mass production to devise a new technology of heat treatment.

Achieving the continuous production of steel sheets was the fourth interaction. This removed the sole remaining labor-intensive step in producing automobile components. When car production began to soar in the 1910s (fig. 6.7), a great deal of skilled hand labor was still needed to roll the steel sheets that became fenders, hoods, and radiators. Continued growth in the following decades, as well as the shift to all-steel enclosed bodies, was made possible by a continuous process that yielded long sheets of steel just 1/16-inch thick. The sheet-steel process was the climax of nearly twenty years of patient engineering experimentation at the American Rolling Mill Company, a medium-sized specialty producer. American Rolling Mill had a record of developing latent market niches through extensive interaction with consumers and effective use of scientific research.

The fifth point of contact was electric steelmaking, which rose dramatically as a result of the distinctive demands of the American automobile industry. Wartime demand for armor-piercing projectiles and other alloy products initially boosted the use of electric furnaces, but viewed across the 1915–25 period such demand appears as a temporary stimulus rather than as a fundamental determinant.[6] The observations of contemporaries as well as production figures reveal the underlying pattern: steelmakers adopted electric furnaces to serve the demand of the automotive industry for large volumes of high-quality alloy steel. Electric furnaces yielded an exceptionally "clean" steel, free from the gas bubbles that weakened other steels. They also could achieve high temperatures while maintaining a chemically reducing atmosphere, conserving the expensive alloying elements that other steelmaking processes wasted by oxidizing. The imperative of automobiles, as these five developments illustrate, brought the steel industry to its modern form.

Standardizing Steel

The matter of co-operation between the makers and the users is the important thing. —JOHN A. MATHEWS

The paradigmatic standardization of automobiles had its origins in the mechanical branch of the Association of Licensed Automobile Manufacturers (ALAM). ALAM distinguished itself from the several other trade associations that attempted to organize the early automobile industry by controlling the patent of George B. Selden, which claimed the fundamental concept of a gasoline motor car. ALAM's organizers, Henry B. Joy of Packard and Frederick L. Smith of Olds Motor Works, hoped to use the Selden patent to achieve a degree of price and volume stability by restricting licenses, much as the Bessemer Association had done. Founded in March 1903 with nineteen members, mostly the makers of the most expensive lines of automobiles, ALAM by 1907 had grown to comprise thirty-two automobile manufacturers responsible for 80 percent of the industry's total output. Yet excluded from ALAM in that year were 186 other automobile manufacturers—including Henry Ford, who disputed the notion that the Selden patent had anything to do with his ever-multiplying number of car models. After a long legal battle (1904–11) that reportedly cost him nearly a million dollars, Ford won a court decision that ended the association's effective control over the industry.[7]

Efforts to standardize automobile components began in the midst of this legal dispute. In 1905, from the headquarters of ALAM's mechanical branch in Hartford, Connecticut, Coker F. Clarkson, a patent attorney and editor, initiated a résumé of automotive literature. Tests for the association were conducted in a special metallurgical laboratory established by Henry Souther, a graduate of MIT who had headed the engineering laboratory of the Pope Manufacturing Company, the prominent bicycle manufacturers of Hartford. Until 1909 Clarkson and Souther's efforts centered on standardizing sparkplugs, wheels, screw-threads, and (not least) steel specifications; in that year ALAM, smarting from a reverse in its patent battle with Ford, abruptly discontinued the mechanical branch and transferred its equipment and records to the recently established Society of Automobile Engineers (founded 1905). Thus it was SAE, a professional society, and not ALAM, a trade organization, that oversaw the setting of industrywide standards.[8]

The SAE standardization effort was initiated by Howard E. Coffin. Coffin was the technical expert behind the Hudson Motor Car Company, a forerunner of American Motors. Formal recognition for his medium-priced, technically excellent automobile designs came with his election as the society's president in 1910, and he successfully recruited Clarkson and Souther to continue work under the auspices of SAE. Coffin argued that the absence of standards caused needless expenses and most production problems. The issue of standards especially concerned the many smaller manufacturers. Without standards these companies risked commercial failure if a supplier went out of business; conversely, suppliers were at risk if a manufacturer suspended business, a frequent enough occurrence. By 1908 a larger number of automobile manufacturers had failed (262 since 1899) than were presently active (253). From the first, the political-economic nature of standardization was evident in the smaller companies' vigorous support of the SAE effort; only later, when the momentum behind standardization became irresistible, did the larger automobile (and steel) firms join the effort. The smaller companies hoped standardization might offset the economies of scale already enjoyed by the larger manufacturers, such as Ford, General Motors, and Studebaker. Beyond economic benefits to manufacturers, standardization gave purpose and direction to SAE; the society's membership tripled from 1910 to 1911.[9]

The bicycle engineer Souther completed the task Coffin had begun. From the first set of standards in 1910, Souther chaired the SAE standards committee. This committee oversaw the effort to establish standards for oil, rubber, forgings, and steel. Using test results from the ALAM laboratory, Souther assembled the first set of specifications discussed by SAE. As printed in the society's *Transactions* of 1910, Souther's specifications introduced many of the topics over which later committees agonized. He gave specifications for eleven grades of steel, valve metals, steel and iron castings, several types of bearing metals, aluminum alloys, and a lubricating oil.

Souther struck a note of moderation in introducing his specifications. For one thing, he avoided "impractical" specifications. While working in a steel works he had been charged with scrutinizing the specifications on orders that came into the office. "Some of the specifications received were so impractical that the orders were refused," he said. If the chemical specifications were met the steel would fail the physical specifications, or vice versa. Souther elected to provide chemical rather than physical specifications "because none of these steels is likely to be used in the shape or condition in which it is received. All are likely to be forged or heat-treated, or both." (Although they later drew criticism, the chemical specifications remained.) Souther emphasized the importance of good handling for any of these grades of steel. Good open-hearth steel if properly treated "may equal" the best crucible steel, while the best crucible steel if improperly melted, rolled, or hammered could "easily be worse than" open-hearth steel. Moreover, he recommended the tensile test and the alternating stress or endurance test as particularly appropriate because automobiles were subjected to "violent alteration of stress under regular use."[10]

Moreover, Souther was a well-known advocate of pyrometers, as Maunsel White at Bethlehem Steel had appreciated in the tool-steel experiments. Especially for alloy steels having a narrow range of safe heat-treating temperatures, the pyrometer was "an absolute necessity," Souther affirmed. Nonetheless, he believed that something of a fad had overtaken the automobile-steel industry. "Only a few years ago it was impossible to persuade a manufacturer of automobiles to pay anything like 10 cents or 15 cents per pound for alloy steels. It was equally impossible to get pyrometers installed," Souther observed. "Now the pendulum has swung in the other direction and fancy prices for steel are paid without rhyme or reason, and pyrometers are installed where there is no necessity for them and in such a way that they are practically useless and perhaps misleading." Souther faulted manufacturers who bought alloy steels but did not heat-treat them in any way. Untreated alloy steels were only slightly better than good carbon steels, he believed, although they might cost three or four times as much. Souther concluded that "moderation is to be used in all things, and particularly in the selection of material and the heat-treatment of them for automobile purposes."[11] Souther's cautious, consensual, and conciliatory attitude provided a model for the standardization effort.

Members of SAE vigorously discussed the specifications for iron and steel over the next five years. In fact, nothing else that SAE undertook in its early years attracted more attention or demanded such effort. Consequently, by 1920, when many other automobile-component specifications were still undecided, the iron and steel specifications determined the daily practice of scores of design engineers, purchasing departments, and steelmakers, to say nothing of the driving experiences of automobilists themselves. What motivated the SAE engineers to articulate specifications? Why did the specifications take the form they did? And what accounted for their success? Attempting to answer these questions is more useful than recapitulating the SAE's protracted discussion of the standards.

On several occasions SAE president Coffin articulated the professional rationale for the society's effort. "There is no use of any argument by the chair as to the necessity of the conduct of this standards work," he stated. There was "unanimous opinion" that this effort "must be undertaken and will be undertaken, and that the Society of Automobile Engineers, as the one engineering body in the field of the motor car industry, is the organization which should undertake it." Coffin outlined the society's plan to appoint a legion of committees and subcommittees to devise and refine standards and to compile the results as a textbook of automobile engineering. "We have no such engineering work at present in the American industry," he observed. "The German engineers are ahead of us in the matter, having a more or less comprehensive textbook of automobile engineering." [12]

"Gentlemen, your Standards Committee has found that it has a tremendous task ahead of it," continued Souther, suggesting how the standardization effort would challenge, stimulate, and define the society. The great number of industries contributing to automobile manufacture required many divisions and committees, and there would be enough work for all interested engineers. Although Souther initially chaired each of the dozen committees, the work soon proved too taxing and he retained only the chairmanship of the iron and steel committee. Six months into the effort, some of the committees already had preliminary reports to be discussed at the society's annual meeting. "We want it clearly understood that what we place before you we do not consider infallible," he emphasized, "we want the heartiest and most open criticism." [13]

A sample of the discussion at the 1911 meeting helps explain why the specifications took the form they did. A key point was how the standards would mediate between users and producers. When one member questioned the committee's choice of chemical over physical standards, Souther replied that as a former steelmaker, he might prefer physical specifications, because there would be "several ways of getting around them." But from the consumer's point of view, ordering steel by a certain analysis meant that a simple chemical test could always reveal whether this steel had actually been received. Moreover, since most steels were forged, brazed, annealed, or case-hardened before use, "the physical condition of the steel as received by the consumer cuts very little, indeed." W. P. Barba of Midvale Steel echoed Souther's point, emphasizing a proper division of responsibility between users and producers. Otherwise, if chemical and physical specifications were linked, Barba maintained that the steel manufacturer would have the absurd necessity of "full access at all times and partial control of the heat treatment which is and should be entirely in the hands of the man working up the steel." Representing a company that produced a cold-finished screw steel, Thomas Towne raised the issue of how chemical standards could ensure this humble material's machining properties. Steel that was too low in phosphorus or sulphur would never cut freely in an automatic or hand-screw lathe. Not all the participants were so accommodating. William H. Tuthill, a manufacturer of springs, objected to the "mandatory" character of standards. "I say it is none of your business, Mr. Coffin, if I make my springs of pot metal. What is it to you if they carry a car and never break?" Tuthill, as Coffin explained, presented "rather the extreme . . . phase of the matter."[14]

The specifications under discussion in 1911 set the form for those adopted later. Each of the twenty-three iron and steel specifications listed the permissible chemical composition. Specification No. 14, for instance, for a chrome-vanadium alloy steel, specified the following ingredients: carbon (0.25–0.35 percent), manganese (0.40–0.70 percent), silicon (0.10–0.30 percent), phosphorus (not over 0.04 percent), sulfur (not over 0.04 percent), chromium (0.80–1.10 percent), and vanadium (not less than 0.10 percent). The specifications also gave general guidance for using the steels. Two high-carbon steels were "primarily for springs," a nickel steel and a chrome-vanadium steel were "primarily for case hardening," while a chrome-nickel steel was "primarily for frames." The appended notes and instructions gave greater detail. For a medium-carbon steel, for example, the notes stated that basic open-hearth furnaces were the ordinary way of producing this steel; that the steel could be case-hardened, forged, and machined; that it was a soft and ductile material; and that it could be drawn into tubes and rolled into forms, such as frames. Finally, the notes suggested proper heat treatments.[15]

Large steelmakers approached the SAE effort in 1911. Until then, three of the four steel firms in the society's iron and steel division had been crucible- or electric-steel manufacturers. Manufacturers of ordinary commercial steel gained a stronger voice in January 1911, when Pennsylvania Steel and Union Drawn Steel joined the standards committee. Five months later, shortly after joining SAE, nine large steel concerns convened a special meeting in Pittsburgh to discuss the proposed standards. Believing the specifications "unsatisfactory in a number of particulars," they dispatched a representative to the society's June meeting in Dayton, Ohio. The representative outlined the steelmakers' many suggestions: from removing a diagram of a test specimen to altering levels of carbon, nickel, manganese, chromium, and other elements in the specifications. These suggestions were "coincident with the commercial practice to-day." Understandably, Coffin welcomed the steelmakers, who, he noted, had previously shown "some reticence" toward standardization. Now, their active participation helped build momentum behind the standards.[16]

Partly as a result of the large steelmakers' suggestions, the SAE standards took their modern appearance in 1912. They were given descriptive numbers, which could be written onto shop drawings and blueprints. Now, the entire class of simple carbon steels bore the prefix "10-" and a suffix indicating the percentage of carbon. The label "10-40" thus indicated a carbon steel with 0.40 percent carbon, while "61-25" indicated a steel containing 0.90 percent chromium, 0.18 percent vanadium, and 0.25 percent carbon.[17] The SAE shorthand especially simplified the ordering and specifying of alloy steels. By classifying both carbon and alloy steels, the society standardized a significant portion of the entire steel market (table 6.1).

Responding to the "sharp demand from the membership," Souther expanded the notes and instructions accompanying the specifications. For each steel the SAE specifications now listed two physical characteristics—elastic limit and toughness, expressed as percentages of elongation and of reduction in cross-sectional area—

TABLE 6.1. Society of Automobile Engineers steel specification, 1912

Steel Type	Specification Number
Carbon steels	10-
Carbon steel, screw stock	11-
Carbon steel, castings	12-
Nickel steels (3.50% nickel)	23-
Nickel-chromium steels, low nickel	31-
Nickel-chromium steels, medium nickel	32-
Nickel-chromium steels, high nickel	33-
Nickel-chromium-vanadium steels, low nickel	41-
Nickel-chromium-vanadium steels, medium nickel	42-
Chromium steels (1.00% chromium)	51-
Chromium steels (1.20% chromium)	52-
Chromium-vanadium steels	61-
Silico-manganese steel	92-
Valve metal (96% nickel)	296-
Valve metal (30% nickel)	230-

Source: Third Report of Iron and Steel Division, SAE Trans. 7, pt. 1 (1912): 32.

to help design engineers select an appropriate steel. This addition appeared to combine the chemical and physical specifications, but Souther explained a division of labor: "The purchasing department does not need and must not have the Notes and Instructions . . . They are none of the purchasing department's business, if I may say so. If the purchasing department will tell the steel manufacturer what it wants in a chemical analysis, the steel manufacturer will deliver it if it knows its business. Then it is up to the manufacturing department to make a proper use of the steel; that is where the Notes and Instructions are intended to come in. They should be placed in the hands of the manufacturing department, and not in the hands of the purchasing department."[18] Souther evidently failed to appreciate the point made a generation earlier by the Carnegie structural-steel handbook: that consumers wanted and needed more and not less information about steel.

By 1912 the specifications had begun to win over steelmakers and engineers alike. "We have no fault whatever to find with the specifying which you have done," stated George F. Fuller, a manufacturer of forged crankshafts. "In fact, we endorse it heartily." "I think I have used the reports of the Iron and Steel Committee for information and guidance in selecting the proper materials in designing, more than any other reports of the Standards Committee," stated the engineer B. B. Bachman, who recommended adding a list of physical properties to aid engineers in choosing between the carbon and alloy steels. A crucible steelmaker, John A. Mathews, captured the spirit of the proceedings when he emphasized the importance of cooperation between the makers and users of steel.[19]

The SAE specifications were soon cited as authoritative statements of the physical properties of different steels. "While a great many of the published figures referring to the chemical analysis and heat treatment of steel, were supposedly recommended practice only, they have assumed in our relations with the trade and engineers the form of specifications," stated Fuller, the crankshaft manufacturer. He related that engineers had not only specified physical properties and demanded chemical analyses but also had itemized the heat treatment to which the steel was to be subjected.[20]

In 1915, when the standardization effort was essentially complete, K. W. Zimmerschied, a General Motors metallurgist, replaced Souther as chair of the steel standards committee. In Souther's own words, the achievement of industrywide standards was the result of the "most exhaustive discussion and careful consideration by both consumer and producer." After World War I, the rival American Society for Testing Materials brought its steel specifications "into . . . exact conformity with those of the Society of Automotive Engineers."[21]

By the mid-1920s the SAE standards broadly influenced steel-making and steel purchasing. The specifications circulated as Section D of the five-hundred-page SAE *Handbook*—the automobile textbook that President Coffin had called for at the beginning of the standardization effort. In 1924 the society's *Journal* published a survey of 122 automobile manufacturers representing 98 percent of the nation's production. Only 8 of the respondents did not use the physical-property data. Fully 114 manufacturers (93 percent) stated that they used the physical-property data; 95 (78 percent) stated that they used the charts for the "guidance of the designing and heat-treating departments"; and 67 (55 percent) used the charts for purchasing to physical specifications. The SAE specifications had profoundly influenced the purchasing of forgings, stampings, and parts machined from bar stock: a sound majority of manufacturers purchased these products according to chemical specifications or combined the chemical and physical specifications. Evidently, few had listened to Souther's pleas to keep the notes and instructions out of the purchasing departments. "The work of the Iron and Steel Division has been most valuable in standardizing automotive steels," observed one survey respondent. "We think it would be a very grave mistake to discontinue the physical-property charts," commented another, "not so much on account of the larger companies that maintain their own metallurgical departments, but on account of the smaller companies that depend almost entirely on information obtained from the Society of Automotive Engineers and similar organizations and representatives of the steel companies for their guidance in the selection and treatment of steels."[22]

A variety of motives and activities thus characterized the SAE standardization effort. The standards-setting process showed SAE to be a professional engineering body committed to the open discussion of complex technical issues. The process also gave the fledgling organization a "tremendous task," in Souther's words, that defined its purpose and guided its activities. The success of the specifications owed much to the society's ability to embrace controversy during several years of intense debate and yet to achieve closure supporting the specifications. The success also signaled the automobile industry's new role as a leading steel consumer. With such a market at risk, the large steelmakers found it imperative to join the standards-setting effort. Indeed, a compromise among steelmakers, automobile purchasing departments, and design engineers determined the form of the standards. As originally designed, the physical specifications were to guide only the engineers as they chose an appropriate steel, while the chemical specifications were to guide the purchasing of steel. In practice, the chemical and physical standards did much more. Together they defined steel— by determining how engineers chose the material, how purchasing departments ordered it, and how manufacturers made it.

The Innovation of Vanadium Steel

> *Willis told Mr. Ford that for this new steel and heat treating process we would have to have a university graduate metallurgist. Ford said, "No, make one out of Wandersee!"*
> —JOHN WANDERSEE

The automobile industry posed a supreme challenge to steel metallurgists. Bumpy roads and the engine itself subjected automobiles to incessant shocks and vibrations that could cause a tiny imperfection in any part to crack from fatigue. Inexperienced drivers tortured clutches, gear boxes, and rear axles. Exhaust valves inside the engine reached temperatures typically used for heat-treating steel. The consumer character of automobiles further complicated innovation. Strength, as a metallurgist measured it by tensile or compression tests, represented only one criterion for evaluating a new alloy steel. The machine shop often found otherwise promising steels impossible to drill or shape; the forging shop might experience excessive rejects; the executive office might judge the distant source of the alloying element as insecure or the element as too costly. If anything went awry, the sales department suffered. Yet the allure of alloy steels proved irresistible. Throughout the 1910s, while the Ford Motor Company adopted vanadium steel, other automobile metallurgists experimented with various alloy steels before developing a "universal steel" with chromium and molybdenum. The efforts of these automobile metallurgists to shift their focus from production-oriented concerns to innovation in product and process marked the rise of industrial research in the steel industry's most important consumer.

The Ford Motor Company learned of vanadium steel through J. Kent Smith, a British consulting metallurgical engineer active in Pittsburgh. Kent Smith had discovered the properties of vanadium steel about 1900 while investigating an especially tough and durable piece of Swedish steel for Messrs. Willans & Robinson of Queensferry, Scotland. While the steel appeared unexceptional under a microscope, it turned out to contain a small amount of vanadium.

At this point Kent Smith initiated, as he put it, "an exhaustive study of the application of vanadium to the steel industry." Beginning in 1901 he experimented with crucible carbon steels and open-hearth vanadium steels. Three years later he presented his results to the Institution of Mechanical Engineers of England, establishing himself as an expert on vanadium steel along with John Oliver Arnold, the Sheffield metallurgist known in connection with high-speed tool steel, and Léon Guillet, a French metallurgist. In tiny amounts, less than 0.2 percent, vanadium imparted to steel the capability of withstanding dynamic stresses—a crucial advantage for automobiles. To measure these dynamic properties, Kent Smith devised an alternating impact machine that subjected a test sample to a grueling trial of impact-and-rotation cycles. A high-carbon steel withstood 100 such cycles and an excellent nickel steel survived 270, while a vanadium steel whose static properties equaled those of the nickel steel lasted 570 cycles. For Kent Smith, the union of chromium and vanadium produced "the finest steel the world has ever seen for moving-machinery parts." By 1905 French and English automobile manufacturers "were already considerable users" of this vanadium "anti-fatigue steel."[23]

Kent Smith brought vanadium to the attention of Ford's shop-trained lieutenant, C. Harold Wills, who retained him as a consultant. Wills learned that vanadium steel could be machined readily and that it was as strong as nickel steel and less expensive. In July 1906 the Ford Motor Company arranged to have the United Steel Company (later United Alloy Steel) of Canton, Ohio, make an experimental 25-ton open-hearth heat of vanadium steel (fig. 6.1). Rolled into bars and billets, then forged into shape, this new steel was first used in prototypes of the Model T. Ford's man at the United Steel plant, a resident user as it were, was John Wandersee, the man "hired as a sweeper and converted into a metallurgist." Wandersee related that he began studying metallurgy and heat-treating in the fall of 1906, after the first heat of vanadium steel. Until that time, automobile companies had known little about heat-treating, which was mostly a concern of tool treaters and licensees of the Taylor-White patents (see chapter 5). "I don't exactly know how I happened to be picked out as a man who would be a metallurgist. I understand that Wills told Mr. Ford that for this new steel and heat treating process we would have to have a university graduate metallurgist. Ford said, 'No, make one out of Wandersee!'" Preparing to set up an analytical laboratory, Wandersee studied the methods, chemicals, and equipment of the United Alloy Steel laboratory for three months in the fall of 1907.[24]

FIG. 6.1. Birth of chrome-vanadium steel, 1906. In July 1906 United
Alloy Steel of Canton, Ohio, made the first batch of vanadium steel for
Ford Motor Company. J. Kent Smith, the metallurgist who brought
vanadium to Ford, is third from left; Henry Ford is third from right.
*From the Collections of Henry Ford Museum and Greenfield Village; Acc. 1660,
negative 0-10456.*

Metallurgical work for the Model T took place early in 1908 at the
Ford plant on Piquette Avenue in Detroit. Built in 1904 as the Ford
Motor Company's first factory and expanded in 1907, the three-
story structure served mainly as an assembly site for the Model N,
a precursor of the Model T. Wandersee's analytical laboratory was
housed in a small building next to the main factory, separate from
the metallurgical and heat-treating "experimental station" located
above the parts storage area. There Wandersee, Fred Allison, and
August Degener assessed the effects of different heat treatments
using a standard Olson universal testing machine. For about six
months, the men assessed which heat treatments were appropriate
for the Model T's different parts.

Wandersee's experiments informed the design of the heat-
treating plant built at the Piquette plant by Degener and Charles
Hartner. Edward Haines took charge when heat-treating moved to
the Highland Park factory in the summer of 1911. "I wouldn't say
that it was vanadium steel only that was an improvement," Wander-

FORD

Four Cylinder Touring Car
$850.00

W HEN you can couple the Ford guarantee — the guarantee of the best known automobile manufacturer in the world, whose imprint is already on more good cars than any other concern has made or promised with the Lowest price ever announced for a Touring Car, it's a mighty safe buy.

1908

MODEL T TOURING CAR. $850.00
High Priced Quality in a Low Priced Car

Here is the first and only chance ever offered to secure a touring car at a reasonable price, a price any man can afford to pay. It is a big, roomy, powerful car of handsome appearance and finish, at a price lower than you are asked to pay for any 4 cylinder runabout excepting the "FAMOUS FORD." This car sounds the death knell of high prices and big profits.

When this now famous FORD runabout was announced, other manufacturers, despairing of being able to compete, knocked—"A good car could not be built at the price, much less sold." Every knock was a boost. With twenty thousand cars Ford proved the car was right. Ford's financial standing to-day proves the price was right. And the knocking that will be done on this new car will be silenced in the same fashion.

Henry Ford promised three years ago to build a high grade touring car and sell it at a heretofore unheard-of price and now, just as surely as every claim made for the small car was made good, just so surely has Ford made good this promise.

The Model T is that car, the car that was promised, a four cylinder, twenty horse-power touring car — a roomy, commodious, comfortable family car, that looks good and is as good as it looks.

This is no imitation car. Henry Ford has never found the need to copy, and the fact that he has never designed a failure is your security that in the Model T touring car Ford continues to be two years ahead of any car manufactured to-day.

We have no high-sounding names with which to charm sales. It's the same old name, "plain as any name could be," it's just "FORD" but it has the advantage of being already on twenty thousand cars (real automobiles) that have "Delivered the Goods."

You know that Henry Ford can build a better car for less money than any other manufacturer on the face of the earth. You know it because he has always done it, and that is your guarantee of his ability and your security in dealing with him.

Our organization is made up of that same force of men that you have learned to know as being connected with the manufacture and sale of an honest product. It's the same old Company that has been doing business right along; the same old successful organisation and plant.

Not having anything new but the car; and, because it is unnecessary for us to spend money to solicit confidences in our Company (that being already secured by past performances), we can devote all our time and money to taking care of the orders for the car that people have actually been waiting for—a family car at an honest price

The Model T is a four cylinder, twenty horse-power car of graceful design, powerful in appearance and reality and of sturdy construction. It is not a new car in the sense that it has been conceived, designed and built all in a few weeks. Mr. Ford started on the car three years ago. Two years were spent in designing, experimenting and research. A year ago the first cars were shown and for twelve months have been in constant service.

These experimental cars have been run under every conceivable condition. All last winter they were tried on the snow and slush covered country roads—all summer they have worked on the sand and mud roads, in good and bad weather.

It is a new car, however, in the sense that it is not an old chassis with a new body or an old engine in a new chassis, but it is a new car built throughout to meet the requirements of a family car. The engine is new (twenty horse-power), cylinders, pistons, valves, etc., are get-at-able. The chassis is new, a hundred inch wheel base and thirty inch wheels. The transmission is of new design (planetary, of course) and altogether silent. $150,000.00 worth of new machinery and tools were added before we could start to build.

Vanadium steel, the strongest, toughest and most enduring steel ever manufactured, is used throughout the entire car. The only reason every automobile manufacturer is not using Vanadium steel is either that they cannot afford to or do not know how. To use it in Fords required two years of experimenting and an expenditure of two hundred thousand dollars, plus the increased price of steel.

FORD—THE CAR THAT LASTS LONGEST

Ford Motor Company

254 Piquette Ave., DETROIT, U. S. A.   "Standard Manufacturers, A. M. C. M. A."

NEW YORK CITY   BOSTON   PHILADELPHIA   BUFFALO   CLEVELAND   CHICAGO
ST. LOUIS   KANSAS CITY   DENVER   SEATTLE   PARIS   LONDON
CANADIAN TRADE SUPPLIED BY FORD MOTOR CO., LTD., WALKERVILLE BRANCH AT TORONTO

FIG. 6.2. Vanadium steel in Ford Model T advertisement, 1908. "Vanadium steel, the strongest, toughest and most enduring steel ever manufactured, is used throughout the entire car."

From Floyd Clymer, Henry's Wonderful Model T, 1908–27 *(New York: McGraw-Hill, 1955), 153.*

see recalled. "With the proper kind of heat treat you could improve any kind of steel. Any steel that had enough alloy or carbon content to permit hardening could be tempered to the requirements." The advantages of vanadium steel were low cost, even though the forgings were done outside the Ford Motor Company, and ease of machining. Wandersee estimated that chrome-vanadium steel, "our chief alloy," composed fully 50 percent of the Model T—including gears, crankshafts, connecting rods, springs, and drive shafts. Manganese-carbon steel was used for secondary parts, such as fender irons.[25]

Henry Ford himself launched the sales campaign touting vanadium steel as a tough and durable material, in an essay published in *Harper's Weekly* in 1907. A year later, one of the earliest advertisements for the Model T trumpeted that "Vanadium steel, the strongest, toughest and most enduring steel ever manufactured, is used throughout the entire car. The only reason every automobile manufacturer is not using Vanadium steel is either that they cannot afford to or do not know how. To use it in Fords required two years of experimenting and an expenditure of two hundred thousand dollars, plus the increased price of steel" (fig. 6.2). A year later, Ford advertisements made a more specific claim that surely pleased Floyd Clymer, plagued by those broken Studebaker transmissions: "Transmission—New design Ford spur planetary, bathed in oil—all gears from heat-treated Vanadium steel, silent and easy in action." "Ford made much to-do about his use of vanadium steel," Clymer later wrote. "For years his advertising and dealer displays played up that feature, often showing a vanadium-steel axle twisted like a pretzel but unbroken to prove the toughness of the material forged into the T."[26] The American Vanadium Company aided Ford's publicity effort by distributing a half-dozen booklets authored by Kent Smith and John A. Mathews, the crucible steelmaker.[27]

Successful concepts rooted themselves deeply in the Ford organization. Just as the notion of mass production proved overly rigid in the 1920s, when General Motors exploited flexible mass production and innovative marketing to surpass Ford in sales, vanadium steel proved difficult to abandon even when its cost went through the roof during the war (vanadium was imported from Peru) and after axles made of the material started breaking in the 1920s. Ford banned the alloy, Wandersee said, only after its economics became unbearable. "When vanadium went up to $7.50 a pound after the war, which was quite an item, we did everything we could to get around it, to equal or better the results for less money. That was a gradual change. You don't take the whole thing and change it. You just have to try one thing after the other and see which works out best. I think we had eliminated vanadium by the time the last Model T came off the line in '28. It was all chrome carbon and chrome manganese."[28] Vanadium, he remembered, "was probably harder to get rid of than to bring in." In 1922 Henry Ford affirmed that "vanadium steel . . . is our principal steel. With it we can get the greatest strength with the least weight." In the light of Wandersee's recollection, Ford's statement should be considered not as an expression of technological progressiveness but, like the company's inflexible mass production generally, as an indication of a technological cul-de-sac.[29]

Ford's difficulty in finding a replacement for vanadium steel was not from lack of trying. "All during this period from 1908, when vanadium had been perfected in the heat treat process to be used in the Model T car, we were experimenting constantly on metals," Wandersee recalled. Apparently, the idea to use chrome-manganese steel resulted from experiments indicating that chromium, not vanadium, imparted hardness. Vanadium acted merely as "a scavenger which refined the grain," and so could be omitted if a heat treatment to refine the grain were adopted.

> On a lot of parts, like axle shafts, we got much better service and life on the manganese carbon steel properly heat treated than you would get out of a fancy alloy steel. For instance, when we were using chrome vanadium steel for the rear axle of the Model T car, after we had about 10,000,000 cars on the road [c. 1924], you could drive from here downtown any day you wanted to and see two or three rear wheels laying on the pavement.
>
> We had to stop that. We couldn't redesign the Model T car but we had to stop that. We cut out the chrome vanadium. It wasn't so much cutting out the chrome vanadium as it was going to a different heat treat. We devised by proper analysis manganese carbon steel and then heat treated it to give us the desired results. We just dried up everything. We dried up our rear axle trouble and our crankshaft trouble. We had a lot of crankshaft breakage. We substituted manganese carbon there too, properly heat treated.[30]

In the 1920s, while Ford tried to rid itself of vanadium steel, other automobile companies began investigating a spectrum of new alloy steels. A new spurt of activity was taking place in the laboratories of automobile metallurgists.

The Innovation of Airplane Alloys

The aircraft experience of the last two years should have taught engineers how to revolutionize car construction and performance.—O. E. HUNT

As Wandersee's experience with the Model T suggests, the immediate problems of design and production were the first focus of automotive metallurgists. Helping stanch the flow of broken axles and crankshafts and transmissions was a useful activity, of course, but about 1920 a pronounced shift to innovative activities took place. The formalization of research activities at General Motors in 1920 is well known to historians.[31] An earlier and broader influence on the community of automobile metallurgists, however, was the effort to build lightweight airplanes during World War I. The shift can be seen in the new name of the SAE, adopted in 1917: "automotive" rather than "automobile" engineers.

Before the war automobile metallurgists remained tightly tied to production. In a paper presented at an SAE meeting, Ralph H. Sherry, metallurgist for General Motors, outlined a production-oriented role for the automotive metallurgist. Acknowledging a close relationship to the engineering, experimental, purchasing, and factory-management departments, Sherry claimed that the metallurgist's "own peculiar field" was the specifying, inspecting, and testing of materials plus the "scientific control" of their treatment. He said nothing concerning the innovation of new materials.[32]

Sherry propounded the theme of "the connection of the laboratory with production." He first indicated that metallurgists should help select designs that facilitated the machining, hardening, and finishing of each part. Such subtle imperfections as nicks and sharp corners often occasioned fatigue failures; some types of heavy rear axles had even been known to break on account of tool marks. These problems disappeared when the design did away with sharp corners and when tool marks were ground off the part in question. "Some of these considerations may appear ridiculous, but they are all drawn from life," Sherry noted. "They may serve to show that metallurgical considerations should enter into design." He next discussed the selection of materials, emphasizing steel and alloy steel. He noted the "well understood" point that alloy steels furnished greater strength and thus permitted lighter construction or greater hardness. Suitability and cost were the most important issues. "It is a waste of material to use a high-priced alloy steel when a cheaper steel would serve," he stated, "and it is equally a waste to try to economize on parts for which the best material is required." He further claimed that the metallurgist, as an expert, was uniquely fitted to inspect the material once it had arrived; only careful testing could ensure that questionable materials from the war-crowded steel mills were not used. For these tests, the metallurgist presided over an array of physical tests (transverse, tensile, torsion, compression, alternating stress, shock) as well as more subtle tools like the microscope and the Brinell and scleroscope hardness tests. Finally, Sherry discussed the control of heat-treating processes, including case-hardening (cementation), the heat treating of high-speed steel ("an art in itself"), and the proper use of pyrometers.[33]

Sherry's paper was praised for showing "the tremendous technique involved in the detail of manufacturing" and prompted a wide-ranging discussion of production concerns. SAE members were chiefly concerned with such matters as how different grades of steel threw off characteristic sparks from grinding wheels, how heat treatments distorted parts, why transmission gears were sometimes pitted, and how workers misused pyrometers. Only at the end did Sherry briefly field three questions on new materials (on tungsten steel for valve heads, on nickel steel, and on the absorption of carbon by alloy steels). Clearly, before World War I the problems of production were the main concern of Sherry, General Motors, and the SAE members.[34]

The experience of World War I helped shift automobile metallurgists from production to innovation. Using its expertise in mass production, the automobile industry geared up to produce the materials of war. The manufacture of airplanes most directly influenced metallurgists. As the head of the Army Signal Corps Division of Engine Design, Charles B. King oversaw the work of Jesse G. Vincent, of Packard Motor Car, and E. J. Hall, of the Hall-Scott Motor Company, to build the Liberty engines for airplanes. Companies such as Packard, Cadillac, Buick, Marmon, Ford, and Lincoln manufactured some 24,000 of these lightweight engines.[35]

The challenge of the Liberty engines sharpened automotive engineers' thinking about materials. As Harold F. Wood, metallurgical engineer for Central Steel (Massillon, Ohio), wrote: "It was only through the careful adherence to these fundamental principles that it was possible to produce 20,000 Liberty Engines, . . . the most highly stressed mechanism ever produced, without the failure of a single engine from defective material or heat treatment." Describing the wartime effort to standardize aircraft materials, Charles M. Manly, chairman of the aeronautical division of the SAE standards committee, observed that "much valuable information" on using light alloys had been gained, "and we hope to use it in connection with researches which are in progress in this country." To some, the airplane-engine program seemed almost too successful. "There seems to be an impression in some quarters that the aircraft experience of the last two years should have taught engineers how to revolutionize car construction and performance," observed O. E. Hunt, chief engineer at Packard Motor Car, in 1919.[36]

Hunt outlined how airplanes and automobiles differed in so many ways that developments in one industry rarely influenced the other. In a car the chassis was the "backbone" of the whole structure, but it was a "parasite" for an airborne structure. Moreover, airplanes were "almost entirely devoid" of the power transmission systems so critical in automobiles. Further, the design criteria for airplane and automobile engines differed dramatically.[37] The designers of wartime airplane engines exploited unusual light materials, such as aluminum, reduced all parts to the smallest thickness, ignored labor costs, and used all-steel instead of cast-iron cylinders. Many of the performance-boosting "stunts" used in airplane engines—including overhead camshafts, very high compression ratios, abnormally large valves, and open exhaust pipes—threatened to "sacrifice results that the public insist upon as necessary to car performance." Finally, the designs for airplanes, in which first cost was a low priority, were ill suited to automobiles, in which first cost was of prime importance. Hunt flatly denied "that airplane experience will suddenly and immediately revolutionize cars."[38]

On the other hand, Hunt observed that the aircraft experience had an indirect but powerful influence on automobile engineers and metallurgists. "I believe . . . that the most important contribution which the airplane has made to the automobile is the stimulus to the thought of the industry as a whole that has resulted from the study of its design and manufacturing problems," Hunt stated.

Our war experience has set engineers to dreaming of ways to produce a light-weight result at moderate expense. It has taught an increased respect for good metallurgical practice as a necessary factor in successful design and manufacture. It has given the average manufacturer, who engaged in airplane engine building, a new conception of the value of high-grade workmanship and of the small margin of cost that, in many cases, separates it from the ordinary variety. I should say that in setting up new ideals of design and workmanship for the industry to strive toward, this study has given us an inspiration of far greater value than any design details could possibly yield.[39]

The examples discussed in the following pages amply confirm Hunt's conclusion that a "new conception" of innovation had overtaken the automobile industry.

The Innovation of Chromium-Molybdenum Steel

Chrome-molybdenum steel of a single type in two or three carbon grades develops under suitable heat treatment a sufficient variation in physical properties to meet all of the demands of motor car construction.—ARTHUR H. HUNTER

Developing a "universal" steel became an overriding concern of automobile metallurgists in the early 1920s. A broad cross-section of the automotive community made published contributions to this effort, including Climax Molybdenum Co. of New York; Parish Manufacturing Corp. of Reading; Crucible Steel Company of America, Pittsburgh; Carbon Steel Co. of Pittsburgh; Pierce Arrow Motor Car Co., Buffalo; United Alloy Steel Co. of Canton, Ohio; Molybdenum Corporation of America, Pittsburgh; Studebaker Corp., Detroit; Atlas Crucible Steel Co., Dunkirk, New York; and Maxwell Motor Corp. of Newcastle, Indiana.[40] The absence of Ford reflects that company's obsession with vanadium. Indeed, C. Harold Wills, *after* leaving Ford Motor Company in 1919, conducted some of the earliest experiments with molybdenum and nickel-molybdenum steels, with the assistance of his metallurgist Henry T. Chandler, and it was the Wills-Sainte Claire vehicle that in 1921 first used molybdenum steel.[41] The absence of General Motors, however, remains puzzling.

Moreover, this diverse community sought new means to compare its results. Previously, metallurgists had often used the hardness test developed by the Swedish engineer J. A. Brinell as an index of steel strength. A more holistic measure of strength became available in 1920, when John D. Cutter of Climax Molybdenum announced his "merit index" equation at an ASST meeting.[42] Extending the definition of efficiency to include "man, machine and material," Cutter combined into a single equation four standard measurements of physical strength. He began with the logical idea that a given material "had merit in proportion to the ability of a unit mass thereof to withstand work upon it." Therefore, when a test piece was broken in the ultimate-limit test, this merit was calculated by dividing the work done in breaking the test piece by the mass on which the work is done. A steel's overall "efficiency" equaled its merit index divided by the actual expenditure for alloying elements per pound of steel.[43] Although identifying fair prices for the alloys in question was not always easy, the equation itself allowed metallurgists to make quantitative comparisons between different alloy steels.

The quantitative technical-economic reasoning behind the merit index made it clear that chromium-molybdenum steels were especially well suited to the American context. These steels, according to Arthur H. Hunter, president of Atlas Crucible Steel Co., were especially favored by automobile manufacturers for several reasons. These included their relative ease of manufacture; superior physical properties; capability for forging, heat-treating, and machining; and, surprisingly enough, low total cost. In 1920 a pound of molybdenum cost five times as much as a pound of nickel, yet chromium-molybdenum steels containing a tiny amount of these alloying elements were cheaper than equivalent nickel steels. In 100 pounds of a chromium-molybdenum steel with 1.0 percent chromium and 0.4 percent molybdenum, the alloying elements cost $1.13, while those in an equal amount of 3.5 percent nickel steel cost $1.40.[44]

Chrome-molybdenum had economic benefits even beyond the cost calculation, since it was a "universal steel." "Chrome-molybdenum steel of a single type in two or three carbon grades develops under suitable heat treatment a sufficient variation in physical properties to meet all of the demands of motor car construction," Hunter observed. "This fact has resulted in the adoption of molybdenum steel for the manufacture of all vital parts by some of the leaders in the automotive industry." Such a standardization on molybdenum steels was especially attractive because if a manufacturer needed fewer kinds of steel, he would also have lower costs in production, purchasing, and inspection.[45]

A crucial consideration was the domestic availability of molybdenum. Large deposits of molybdenum ore existed in Colorado and New Mexico; in 1921 the United States reportedly possessed four-fifths of the world's molybdenum supply. Chromium too was available from domestic sources. Rival alloying elements, as Navy Secretary Tracy had appreciated in 1890, were imports from Canada (nickel) or Peru (vanadium). "The war showed us this dependence upon foreign supply," Hunter concluded, "and entirely aside from technical considerations, this fact alone is of sufficient importance to warrant the domestic industry in lending its support and cooperation to the development of this unique natural resource."[46] As an innovation of the automobile metallurgists, chromium-molybdenum steel was especially well adapted to American automobiles.

The Innovation of Heat-Treating Technology

*Mechanical contrivances in heat treatment are becoming
universally used now, which corrects a great many evils of
defective heat treatment by regulating necessary time, mass
and temperature.*—HARRY E. HEMSTREET

"Possibly the greatest example of the application of scientific methods to the industry," asserted George B. Waterhouse in 1912, "is . . . the heat treatment of steel." Waterhouse, a Ph.D. engineer for the Lackawanna Steel Company of Buffalo, went on to say that however inspiring was the research of "pure physicists," automotive manufacturers experienced many difficulties in scaling up heat-treating procedures from laboratory bench to production line. A shortage of skilled heat-treaters was chief among these problems, according to one automobile manufacturer. "The demand for heat treatment of steel has been developed so rapidly that the supply of men, with the practical experience and the fundamental knowledge of the underlying theories necessary for this work, has not been available." There was no general agreement on the desirability of human labor. "The human element may very materially modify results," wrote one General Electric engineer, so that "every opportunity offered to minimize the effect of human manipulation" should be taken.[47] To address these problems automobile metallurgists devised a new technology of heat treatment by coupling the science of heat-treating to the technology of mass production.

Automobile metallurgists devised at least three solutions, applied at different levels. For small numbers of varied parts, a production-oriented laboratory preserved maximum flexibility while ensuring adequate output. For medium production volumes, metallurgists devised batch production systems using larger, often electric furnaces. For the largest production volumes, they adopted the model of a flow-through assembly line. Using conveyor belts, drop slides, and continuously circulating oil for quenching, this assembly-line treatment, appropriately, produced automobile parts for the assembly line.

Each of these systems featured an elaborate control technology. To harden steel involved heating it to just above its "critical point," the temperature at which metallurgists understood that a desirable molecular rearrangement occurred; it was then quenched in cool oil. If the piece did not reach its "critical" temperature before quenching, it would be soft; if the "critical" temperature was far exceeded, the quench produced a coarse grain in the steel, yielding a hard and brittle part. The strains produced by hardening

were relieved by tempering (or "drawing") the part at a temperature below the "critical point" and cooling it slowly. Traditionally,
a highly skilled hardener had used his eye to judge the heat at
which these steps occurred. Pyrometers that quantitatively measured temperature began a shift from craft skills toward scientific
control (see chapter 5). The heat-treating technology devised for
the automobile industry completed this shift, so that in place of the
"old hardener" stood an elaborate amalgam of scientific knowledge
and production technology.

The heat-treating plant of the Automobile Machine Company of
Cleveland typified small-scale, production-oriented heat-treating
laboratories. At this plant thirty workers carburized, hardened,
tempered, and inspected each day approximately 300 rear axle
units, transmissions, and steering mechanisms for the Chandler

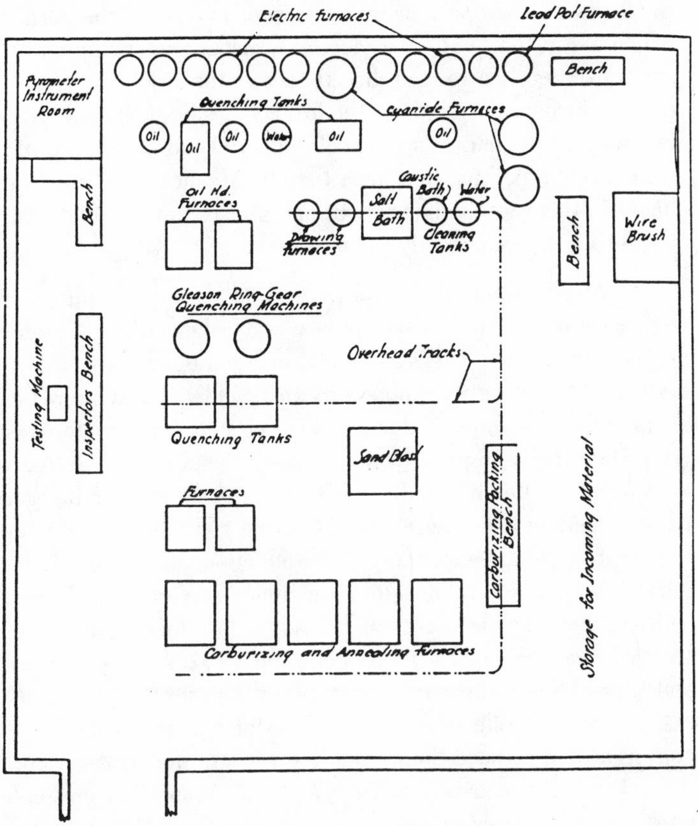

FIG. 6.3. Production-oriented heat-treating laboratory.
*From A. H. Frauenthal and C. S. Morgan, "Heat Treatment of Automotive Parts
and Description of Equipment Used," ASST Trans. 8 (1925): 854–55.*

Motor Car Company. A two-story building 80 feet square housed twelve electrical and fourteen oil-fired furnaces, seven water-cooled oil quenching tanks fed by a 3,000-gallon storage tank, six hardness testing machines, and a shot-blasting machine (fig. 6.3). An "instrument man" used a bank of pyrometers to monitor the electrical furnaces and a light signal system to control the process. He observed the pyrometer charts in the instrument room, waiting for the "hump" in their reading (a measured slowdown in the temperature rise that signified the parts were passing through their "critical" range). He permitted the temperature to rise 70°F beyond the "hump" before giving the signal to quench.

> When the instrument man observes that the parts have been properly heated he rings a bell, then pushes a button which lights the red light over the potentiometer and likewise over the furnace. The operator then proceeds to quench the parts which are in that particular electric furnace. Quenching tanks are conveniently arranged directly across the aisle from the electric furnaces. As soon as the furnace has been unloaded, the operator again charges it with cold material and pushes a button which lights the blue light over the potentiometer and the furnace. In this way the instrument operator is then notified that the furnace is again ready to be turned on.[48]

This system exercised control by vesting all responsibility with the "instrument man" who determined when the "operator" quenched each part. This system also avoided the inaccuracies inherent in calibrating pyrometers to a particular temperature. Instead it relied on the slowdown in heating that a pyrometer measured when the steel parts passed through their "critical" range.

A variety of "batch" production systems using electric furnaces characterized a medium-scale heat-treating plant. To persuade engineers of the relative low cost and production advantages of electricity, C. L. Ipsen, a designing engineer at General Electric's industrial heating division (Schenectady), described several large electric furnaces that were in use in the early 1920s. These furnaces treated products as diverse as gun tubes, steel wire, large iron castings, and automobile parts. Ipsen described a large ring-shaped rotating hearth furnace then being used at an (unnamed) automobile factory for heat-treating each hour 2,700 pounds of cam shafts, connecting rods, front and rear axles, steering knuckles, and other parts. These electric heating furnaces facilitated integrated factory planning. "It is not necessary with the electric furnace to restrict it to a furnace room apart from the manufacturing space on account

FIG. 6.4. Conveyor-type heat-treating furnace (small parts).
*From J. M. Watson, "The Heat Treatment of Automobile Parts," ASST Trans. 6
(1924): 718.*

of its noise, obnoxious fumes and intense waste heat," Ipsen noted.
The electric furnace "can be so located as best to fulfill its mis-
sion as a link in the manufacturing chain."[49] These larger furnaces
also used pyrometers and light signals to control the temperature.
Yet even rotary electric furnaces, widely used as a labor-saving ex-
pedient, did not entirely do away with workmen. As one furnace

engineer stated, "They still need keenness of supervision in order to produce the results."[50]

Beginning about 1920, the mounting demand for uniformly treated parts prompted several parts manufacturers to devise assembly-line heat-treatment plants. "Mechanical contrivances in heat treatment are becoming universally used now, which corrects a great many evils of defective heat treatment by regulating necessary time, mass and temperature," affirmed one manufacturer of springs and axles.[51] "The very much increased production of cars . . . calls for mass production. Such being the case it has become necessary to design heat-treating departments that will give quality results on a quantity basis," noted J. M. Watson, metallurgical engineer of the Hupp Motor Car Corporation (Detroit), which built a large assembly-line heat-treating plant in 1923.[52] The company had been dissatisfied with its stationary furnaces, which required excessive space as well as a "large corps of men" working around the clock. The new plant captured the logic and spirit of mass production. In one unit of Hupp's new heat-treating plant in Jackson, Michigan, about seventy miles west of Detroit, sliding pans carried small parts, such as steering knuckles, equalizer pivots, and drive shafts, through a normalizing furnace in a medium-temperature treatment that relieved strains. A second unit was designed for hardening small parts. These were placed into a continuous pan conveyor composed of a special heat-resistant alloy and fed through the furnace. At the discharge end the parts dropped immediately into a bath of quenching oil. A second pan conveyor picked the parts out of the oil bath and deposited them on the conveyor that fed the tempering furnace (fig. 6.4). After passing through the 18-foot-long tempering furnace in 20–25 minutes, the parts were emptied into a container and cooled slowly in air.

The plant contained a third unit for hardening front axle forgings. In this unit a moving hearth pushed the axle forgings through the furnace 6 inches at a stroke.[53] Upon reaching the discharge end, they dropped onto a pair of counterbalanced "fingers" that guided them into the quenching oil to prevent warpage (fig 6.5). Again, a separate conveyor picked the axle forgings out of the oil and delivered them to the tempering furnace (fig. 6.6). And again, after the axle forgings passed through the furnace, they dropped into containers and cooled in air.

The hardening units at the Hupp plant used two control systems. One monitored the temperature of the furnace by means of pyrometers coupled to lights that signaled low, correct, and high temperatures (workers manually adjusted the furnace). Recording

FIG. 6.5. Moving hearth hardening furnace (axle forgings).
From Watson, "The Heat Treatment of Automobile Parts," 720.

instruments in the foreman's office kept a permanent record of all
operations. A separate control system monitored the continuous cir-
culation of 4,000 gallons of quenching oil. Using an array of valves,
returns, and bypasses, workers adjusted the flow of cool water and
hot oil to maintain the oil temperature at or below 110°F. Cool oil
returning from the spray-pond cooling system entered the quench
tank where the hot parts were placed into it.

FIG. 6.6. Conveyor-type tempering furnace (axle forgings).
From Watson, "The Heat Treatment of Automobile Parts," 722.

The entire system of automatic feeding and discharging had definite limitations as well as striking advantages. "Small production
does not warrant the automatic features," commented F. J. Ryan, a
furnace engineer who had assisted the Hupp staff. "A furnace of an
automatic nature is inherently a quantity production instrument,
and if you have not a quantity production it is really expensive and
useless to attempt to use an automatic furnace." Yet for high-volume

production, Watson found at the Hupp company that "we not only get fewer rejections, but we get much more uniform results all the way through the entire batch of material that goes through."[54]

The Innovation of Continuous-Sheet Steel

*The biggest job was to get up the nerve to recommend
the enormous expenditure required to test the idea.*
—JOHN B. TYTUS

The manufacture of continuous-sheet steel gave the automobile industry an assured supply of steel for body parts. It also illustrated a basic pattern of innovation in the postmerger steel industry: firms outside the orbit of U.S. Steel were most successful at developing new techniques and processes. Rolling a continuous sheet of steel had been a frustrating quest ever since 1857, when Henry Bessemer patented a method of rolling "liquid steel" into sheets. Perfection of such a process eluded Bessemer and many others who followed. In 1904 U.S. Steel abandoned its effort to develop a continuous mill using perfectly cylindrical rolls. Continuous rolling was finally achieved through a remarkable development effort conducted by a medium-sized specialty sheet producer—the American Rolling Mill Company, known universally as Armco.

By the time it entered the automobile market about 1910, Armco had established a solid record of identifying promising but undeveloped uses for steel and collaborating with customers to develop special products. The result of a consolidation of two Cincinnati sheet-metal companies, Armco installed at its Greenfield site in Middletown, Ohio, an open-hearth furnace, a bar mill and sheet mill, a galvanizing plant, and a factory department that fabricated sheet metal for building materials. With the start of operations in March 1901, the Middletown plant consolidated four previously distinct branches of manufacture. The company later expanded production by acquiring a sheet mill in neighboring Zanesville and in 1911 built a new factory in Middletown known as the East Side Works. There Armco centralized its previously scattered innovative activities in a special three-story building devoted to research, development, and process control. The East Side Works embodied the company's aims for growth by departing from the traditional "small mill practice" of beginning the rolling process with open-hearth ingots 8 inches square and 4 feet long. Instead, it adopted the "big mill practice" of using ingots 24–30 inches square and 6 feet long.[55]

Compared to the really large producers, however, Armco was

quite small. Before its East Side Works came into operation, the company required nine months to produce what the Carnegie mills churned out in twenty-four hours (fig. 7.2). Making a virtue of necessity, Armco looked for markets that the larger producers had missed or ignored. At the turn of the century, one such undeveloped market was the manufacture of sheets for the windings of electrical motors, generators, and transformers. Large producers disdained this market because of its finicky requirements and small orders. As a result electrical manufacturers were forced to import European sheets at high cost or use common commodity sheets with poor magnetic properties. In a cooperative research effort conducted with Westinghouse during 1902–3, Armco metallurgists developed a special silicon steel with extremely desirable electrical and magnetic properties. Another niche product developed by Armco was a $99^{87}/_{100}$ percent pure iron product that resisted rust, the bane of steel before the development of stainless steel. Its equally impressive welding and enameling properties made it perfect for the bevy of new consumer products, including stoves, refrigerators, washing machines, and other household appliances. To build recognition among consumers, "Armco Ingot Iron" was the first primary steel product marketed through direct-consumer advertising, beginning with a two-page spread in the *Saturday Evening Post* of 8 August 1914. Advertisements for finished consumer products sported blue-and-gold triangles proclaiming "Made of Rust-Resisting Armco Ingot Iron," an early use of so-called collateral advertising.[56]

When it began making sheets for automobile bodies in 1911, Armco learned of requirements that other steel manufacturers were ignoring, including stringent stamping and drawing properties. The first sheets, for the Case and Kissel automobiles, had to be cold-rolled on Saturdays when the hot mills were shut down and the workers could devote special care to them. To gain information on the automobile manufacturers' problems and requirements, Armco dispatched a representative to make the rounds in Toledo, Flint, Lansing, Pontiac, and Detroit; soon the company established a district office in the motor city with a mandate to listen attentively to the "unreasonable demands" of the car manufacturers and to relay them to the research and development laboratory at Middletown for action. The Chalmers and Cadillac motor car companies were among those that lodged complaints against the available grades of sheet steel, and Armco's research department went to work. A special Silver Finish sheet steel was the first of several such customer-driven products for the automobile manufacturers; it had exceptional drawing qualities that withstood the punishment of the new machines pounding out the curves of fenders and other body parts. By 1913 this steel was being used in half of all Ford bodies.[57] From this time onward the demand for sheets and plates closely followed automobile production (fig. 6.7).

Among the mill superintendents responsible for producing the new steels developed by Armco's researchers was John B. Tytus, who successfully developed a continuous-sheet mill after nearly two decades of effort. His mental model for the continuous flow of material was his grandfather's paper mill, which used Fourdrinier paper-making machines to transform wood-pulp into a continuous sheet of paper. His only engineering training (his degree from Yale was in English literature) appears to have been with a Dayton, Ohio, bridge builder for a year or two following a stint in the family paper mill. In 1904 Tytus returned to his native Middletown and landed an entry-level job at Armco. He stated later that after watching sheets being handled twenty-two different times, he reasoned that a business with so much lost motion had plenty of future for a young man with an aptitude for and interest in engineering.[58] As figure 6.8 indicates, the basic steps in rolling thin sheets of metal required a great deal of heavy physical labor. Workers fed the sheets in and out of the rolls, "doubled" the sheets over into a stack, fed the stack back for more rolling, and eventually peeled thin finished sheets off the stack. Working regular night and day shifts, Tytus

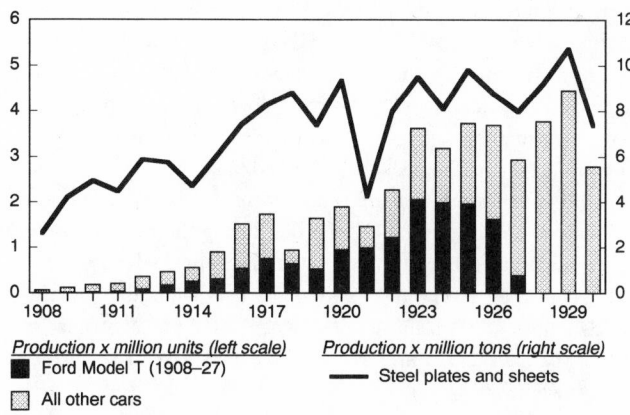

Production x million units (left scale)   Production x million tons (right scale)
■ Ford Model T (1908–27)   —— Steel plates and sheets
▨ All other cars

FIG. 6.7. Automobile and sheet steel production, 1908–1930.
Data from AISI, Annual Statistical Report *(1915–30); Clymer,* Henry's Wonderful Model T, *134.*

FIG. 6.8. Hand rolling of steel sheets, c. 1920. Roughing down on soft mill (*top, left*): "When the bars have been heated to the proper temperature they are brought out of the furnace with tongs and pulled over a cast iron floor to the 'roughing mill,' where they are given a number of 'passes' to prepare them for the finishing rolls." Doubling (*top, right*): "In rolling light gauge sheets, the bars are roughed down and matched, then placed in sheet furnaces and again heated to the proper rolling temperature and given one or two passes through the rolls, after which they are 'opened' and 'doubled.' Heavy sheets are finished either singly or in 'pairs,' lighter ones are 'doubled' and finished in 'packs' of four, six or eight sheets to the pack." Sheet heating (*opposite, top*): "The 'matcher,' 'doubler' and 'sheet heater' work together in putting a 'pack' that has just been doubled into the sheet furnace The man with long tongs is the 'sheet heater' and much of the success in rolling depends upon his knowledge and skill." Shearing and opening (*opposite, bottom*): "After the packs of sheets have been rolled, they are distributed on the floor between the rolls and shears to allow them to cool sufficiently to handle with gloves and hand pads for shearing. The shearman sets gauges on his shears to conform with the size of sheets ordered and the packs are sheared on all four sides. After being trimmed it is necessary to open the packs of sheets as the surfaces of the sheets have a tendency to stick together during the rolling process. The opener bends up a corner of the pack with his tongs, which generally causes the sheets at that point to spring apart sufficient to allow the insertion of his tongs and then the sheets are separated one at a time."
Courtesy Hagley Museum and Library; from Armco in Picture and Fact *(1921), pp. 62 (A), 64 (B), 66 (C), 67 (D); 92.220.1–4, po. 92-176.*

gained personal experience with each of these strenuous tasks. Within eighteen months, he became assistant to Superintendent Charlie Hook and soon confided his idea of making steel as paper was made, in long strips.

Tytus's continuous process was the product of careful engineering experimentation. He gained experience with mill design and operation as superintendent of Armco's new mill at Zanesville from 1906, as Hook's operations chief of the new East Side Works at Middletown from 1909, and as assistant general superintendent under Hook over all the Armco mills from 1913. Tytus studied the exact effects of rolls on sheets by measuring their cross-section with a micrometer. During the war Armco was transformed into a huge forging shop to make projectiles, idling the sheet-mill machinery and thus allowing Tytus to experiment with it at night. In March 1916 he described continuous rolling in the *Armco Bulletin,* and in 1919 he presented blueprints to President George M. Verity. A postwar flurry of orders was met by four standard sheet mills, however, which Tytus helped install at Zanesville during 1919–20.

On January 1, 1921, Armco purchased the Ashland Iron & Mining Company in Ashland, Kentucky, and decided to build a continuous-sheet mill there for approximately $10 million. Tytus recalled that the biggest challenge was daring to recommend the "enormous expenditure" necessary to test his idea. "The best informed men in the industry had by now conceded that what we were trying to do was possible if one had the nerve and money."[59] Early in 1922 he began drawing up plans, and soon six engineers were doing layout and design work; by October construction at Ashland had begun. In September 1923 the engineering effort was transferred from Middletown to Ashland, and about a hundred skilled workers followed. By 31 December 1923 the mill was ready for testing; in the month of February 1924 the mill produced 9,000 tons, half the output necessary to recoup the huge investment. Within three years the mill was producing a profitable 40,000 tons a month.

Tytus perfected and patented a "controlled pass" imparted through fourteen sets of rolls, as figure 6.9 suggests. A red-hot, 5-inch-thick slab emerged from the first set of rolls with a convex section, and became progressively thinner and less convex at each succeeding set of rolls, until it was flat and as thin as 1/16 inch. Large electric motors up to 5,000 horsepower powered the rolling trains. In spring 1927 Armco brought industry leaders to Ashland for demonstrating and licensing the process; the same year Tytus became

FIG. 6.9. Continuous rolling of steel sheets, c. 1927. *First Continuous
Hot Strip Mill Built at Ashland, Kentucky, 1923,* painting from *Steelways*
magazine.
Courtesy Hagley Museum and Library; Acc. 80.300, po. 92-159.

vice president of Armco, first for operations and later for technical development. To secure its position Armco soon purchased the
rival patents held by the Columbia Steel Company, whose mill design achieved an even greater degree of continuous operation using
four-high rolling trains. That design has been called "the prototype
of the modern continuous wide hot-strip mill."[60] Again in concert
with automobile manufacture, the production of hot rolled steel
strip expanded. Between 1927 and 1940, Armco licensees built
twenty-six continuous mills at a total cost of $500 million. Across
the same period, the price of top-quality 16-gauge steel sheets fell
from $135 to $60 a ton.

Electrifying Steel

*Electric steel will find its market in . . . automobile and
aeroplane parts.* —JOHN A. MATHEWS

The rapid expansion of electric steel production after 1915 reflected
the large demand of the automobile industry for a high-grade alloy
steel that was low in carbon. "Each process has its particular field,"
commented John A. Mathews, president of the Halcomb Steel

Company of Syracuse, New York, a crucible-steel company that in 1906 had operated the first electric furnace in the United States. Although some of the tonnage previously filled by crucible steel might be "diverted" to electric steel, Mathews stated in 1916, "it is more likely that electric steel will find its market in the most exacting requirements of steel for structural and tensile purposes . . . namely, for automobile and aeroplane parts." Crucible steelmakers could not fill the large demand of the automobile industry, nor (because of their carbon-containing crucibles) could they produce the high-grade, low-carbon alloy steels much in demand for automobiles (table 6.2). As Mathews put it, "The electric furnace was opportunely invented to meet a new demand rather than to replace an old process." Six years later, in 1922, Mathews aptly noted that the electric furnace yielded a "clean steel." "The electric furnace is opportune," he continued, "to meet the new and exacting requirements for ordnance, automobiles and aeroplanes, and other devices in which alternating stresses are very severe."[61]

The furnaces that Americans installed were invented by Europeans, continuing the durable pattern of technology transfer begun with the Bessemer converter. The names of the furnaces available in 1911—Stassano, Héroult, Girod, Kjellin, Röchling-Rodenhauser, and Hiorth—suggest their European origins. These furnaces were of two general types. Arc furnaces generated heat by maintaining an electric arc between an electrode and the steel bath, between an

TABLE 6.2. Annual alloy steel production by process, 1916–25
(in thousands of gross tons)

Process	1916	1917	1918	1919	1920	1921	1922	1923	1924	1925
Open-hearth	1,155	1,353	1,378	1,197	1,284	668	1,436	1,764	1,757	2,049
(%)	84.7	82.3	77.1	80.8	77.3	82.5	85.8	83.8	86.7	84.2
Bessemer	90	111	65	80	102	77	102	130	74	82
(%)	6.63	6.75	3.62	5.41	6.12	9.46	6.11	6.19	3.64	3.39
Electric	71	131	291	182	246	63	125	195	189	294
(%)	5.22	7.94	16.3	12.3	14.8	7.81	7.49	9.26	9.31	12.1
Crucible	46	50	54	23	39	1.8	9.5	17	7.4	7.2
(%)	3.41	3.05	3.04	1.53	1.78	0.23	0.57	0.79	0.36	0.30
Total alloy	1,363	1,644	1,788	1,481	1,660	810	1,673	2,106	2,026	2,433
(%)	3.19	3.65	4.02	4.27	3.94	4.09	4.70	4.69	5.34	5.36

Source: American Iron and Steel Institute, Annual Statistical Report (New York: AISI, 1916 and 1925).

Note: Percentages are of total alloy steel, except for total alloy. "Total alloy (%)" is the alloy proportion of that year's entire (alloy plus carbon) steel production. (The source did not define the criterion used to identify "alloy" from carbon steel.)

electrode and the hearth beneath the steel, or between two or more electrodes above the steel bath. In 1916 Héroult arc furnaces were used in about 70 percent of American electric steel production. Induction furnaces, on the other hand, resembled a giant electrical transformer. For the Kjellin induction furnace, for example, the cylindrical primary winding was surrounded by a ring-shaped hearth of refractory material containing molten metal, which acted as a single-turn secondary winding. A modest current in the primary winding induced an exceptionally large current in the hearth, where resistance "losses" heated the steel.[62]

Whether arc or induction, electrical furnaces possessed striking technical advantages. Most of all, electrical furnaces could achieve high temperatures while maintaining a reducing atmosphere. All other furnaces required an oxidizing environment; otherwise the fuel would simply fail to burn. (Crucibles with lids could also maintain a reducing atmosphere.) "Oxygen," as William R. Walker of U.S. Steel explained, "in combination with carbon (carbon monoxide), silicon, iron, aluminum and manganese, and also the combination of silicates with these oxides, are very deleterious to steel."[63] The reducing environment had two signal advantages: it did away with the need for elements, such as aluminum, that removed excess oxygen from steel but had no other beneficial effect; and unlike an oxidizing environment, it retained virtually all the alloying elements introduced to the steel, permitting the economical use of expensive elements, such as molybdenum. The shop floor was affected too. Previously most furnace operators had waited to add alloying elements until the steel had left the furnace and was in the casting ladle. With an electric furnace a metallurgist could add the alloying elements to the steel while it was in the furnace. This practice ensured that the alloying element dispersed itself throughout the steel. Electric steels were also remarkably free from occluded gases. Observers who broke open electric-steel ingots invariably marveled at how uniform the inside appeared.

The determination of electric light and power companies to raise their load factor also spurred the development of electric furnaces. Mathews, for one, ascribed an "epidemic" of furnace installations "to the activity of power companies in looking for desirable business." He tried to combat the widespread but mistaken notion that electric furnaces were automatic machines guaranteed to produce crucible-quality steel merely by "throwing on a switch."[64] In fact, electric furnaces required great skill to erect and maintain the electrical apparatus as well as to attend to the metallurgical manipulations. "The present success is due not only to the original

TABLE 6.3. Electric steel furnace capacity by state, 1916–25

State	1916 Capacity (gross tons)	Plants (N)	1920 Capacity (gross tons)	Plants (N)	1925 Capacity (gross tons)	Plants (N)
Pennsylvania	135,120	15	268,790	40	274,390	44
Illinois	81,270	3	227,500	12	212,300	11
Ohio	72,700	5	302,840	24	163,423	20
New York	39,150	6	147,950	20	152,082	13
Michigan	31,700	6	58,800	10	84,840	8
Wisconsin	13,400	5	68,735	12	49,875	13
California[a]	4,600	3	53,450	13	38,940	16
New Jersey	7,500	1	31,000	4	37,133	4
Great Plains[b]	2,770	2	23,780	7	20,708	10
New England[c]	—	—	31,900	10	19,337	9
Utah[d]	5,250	1	4,550	5	16,749	12
Louisiana, Texas	—	—	3,000	3	13,050	7
South[e]	35,000	1	62,000	2	8,636	5
Virginia	12,000	1	12,900	4	7,781	4
Minnesota			10,100	3	7,100	3
Oregon	1,000	1	—	—	6,050	3
Colorado	—	—	—	—	5,250	2
Maryland[f]	10,300	2	34,550	4	4,500	2
Indiana	7,200	2	18,800	6	3,400	2
West Virginia[g]	4,800	1	11,800	2	3,000	2
Washington	1,500	2	17,200	11	—	—
Total	465,260	57	1,389,645	192	1,128,544	190

Source: American Iron and Steel Institute, Annual Statistical Report (New York: AISI, 1916 and 1925).

Notes: Capacity given in gross tons for 31 December of year. Of total capacity, castings averaged 23.1% in 1916, 37.0% in 1920, and 37.4% in 1925.

[a] Includes Oregon for 1920; includes Alaska for 1920 and 1925.

[b] Comprises Iowa and Missouri in 1916 and 1920; adds Arkansas and Nebraska for 1920; adds Kansas and Oklahoma for 1925.

[c] Massachusetts, New Hampshire, Connecticut, and Rhode Island.

[d] Includes Oklahoma and Colorado for 1920; includes Washington for 1925.

[e] Comprises Alabama in 1916; adds Georgia in 1920; adds Kentucky, Tennessee, and Florida in 1925.

[f] Includes District of Columbia; includes Delaware for 1925.

[g] Includes South Carolina for 1920.

inventors," Mathews noted in 1922, "but also to the active co-operation of the great manufacturers of electrical equipment, furnace designers, makers of refractories and electrodes and a group of earnest metallurgists in many plants who have studied every detail of operation."[65] Table 6.3 summarizes the geographical pattern of electric steel production from 1916 to 1925. It illustrates the transient nature of wartime production (note the sharp contraction after 1920 in several states including West Virginia, the site of a government ordnance plant) and the durable nature of automotive demand.

Steelmakers in the early 1910s assumed that electric furnaces would be used only for refining already melted steel, but after 1920 electric furnaces were increasingly used to melt steel. While the cost of electricity dropped, the cost of other fuel sources, such as natural gas and oil, climbed sharply. In the early 1920s a computation of the relative cost of various energy sources might even have favored electricity.[66] As the steel industry matured in the 1920s, electricity joined these well-established fuels. In fact, electric furnaces were the last major production technology that American steelmakers adopted in advance of external competitive pressures.

The high tide of industry, 1950: U.S. Steel's Gary works (*foreground*) and Gary, Indiana (*background*). *Courtesy Calumet Regional Archives, Indiana University Northwest.*

CHAPTER SEVEN

The Dynamics of Change

W hat is remarkable about the steel industry following 1925 is not how it expanded, though production roughly tripled from then to the peak in 1969. What is remarkable is how *little* the patterns of using and producing steel changed. After the rise of automobiles, the principal consuming sectors remained remarkably stable. Already evident by the mid-1920s was a substantial decline in railroads and an offsetting rise in containers (principally tin cans) and consumer durable products like household appliances (fig. 7.1). For the most part, the major consumers of steel continued to rely on the established steel-producing companies. Very few major consumers—Ford and International Harvester among them—built their own, integrated, primary steel-producing plants. By and large, steel had become a mature commodity; demand was generated by a relatively small number of large companies placing large orders.[1] No significant changes occurred in this pattern of production and use, in spite of the disastrous decline during the Depression (steel production in 1932 was one-fourth that in 1929), passage in 1935 of the Wagner Act mandating union recognition and collective bargaining, and even the great but temporary wartime boom in shipbuilding. Steel consumption by the late 1940s once again strikingly resembled that of the mid-1920s.[2]

Antitrusters, for all their bluster, had little impact on the structure of the steel industry. Even for U.S. Steel the threat of antitrust litigation was more significant than its direct results. As a result of public outcry over U.S. Steel's 1907 agreement to lease James J. Hill's Great Northern ore properties, the sole remaining large supply of Great Lakes iron ore, U.S. Steel canceled its controlling leases. Consequently, after 1915 three independent ore

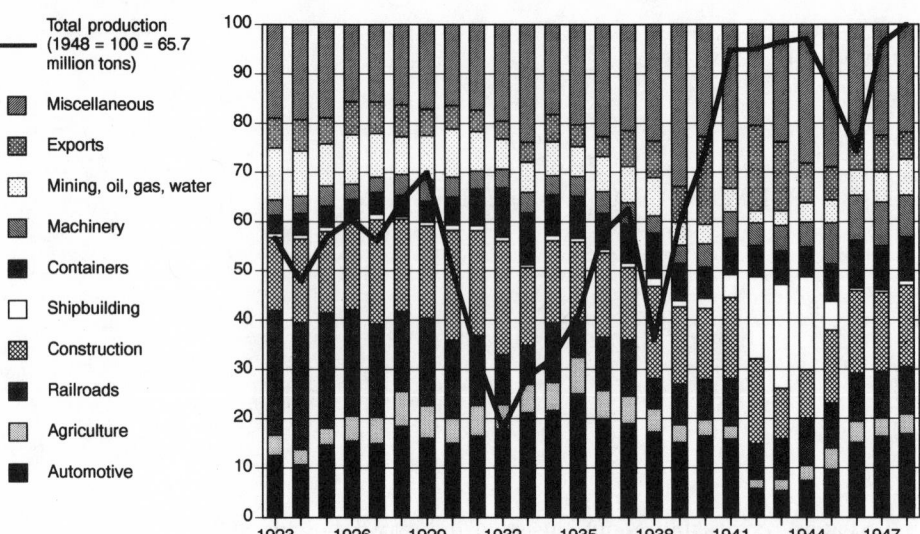

FIG. 7.1. Steel consumption by major sectors, 1923–48. Industry subgroups included in major sectors:

Miscellaneous: furniture and stoves; domestic appliances; refrigerators; office equipment; bolt, nut, and rivet makers; forgers; pressed and formed metal manufacturers.

Mining, Oil, Gas, Water: oil, gas, water, mining, lumbering, and quarrying industries.

Machinery: makers of machinery, hand tools, electrical machinery and equipment.

Containers: those made from light steel products, predominantly tin cans.

Shipbuilding: ships, boats, barges.

Construction: fabricators, building contractors, concrete reinforcing companies, building hardware and trim companies, concrete bar jobbers, highways, boiler and tank makers, power developments, containers made from heavy steel products.

Railroads: trackwork, cars and locomotives, parts, railroad buildings and bridges.

Agriculture: implements, equipment, other farm uses, tractors in some years.

Automotive: automobiles, trucks, parts, and so forth (tractors in some years).

Data from U.S. Steel Corporation, T.N.E.C. Papers (New York: U.S. Steel, 1940), 1:328–29, 335–36; U.S. House of Representatives, Hearings before the Select Committee on Small Business, Steel: Acquisitions, Mergers, and Expansion of 12 Major Companies, 1900 to 1950 (Washington, D.C.: Government Printing Office, 1950), 79–80.

producers and several steel companies gained access to the rich Mesabi ores. But the federal government lost its long-running anti-trust case (1912–20) against the steel combine. In its final decision, the Supreme Court reasserted its "rule of reason" stating that size in and of itself was not illegal and that in cases where no illegal acts had been committed mere size could not be punished.[3]

In short, U.S. Steel was big but not illegal. Its price leadership resulted from its complete dominance in the core markets for steel, as figure 7.2 makes clear. Indeed, many steelmakers had grown comfortable with U.S. Steel's overriding policy of price and technical stability, which permitted them to create or develop markets where the combine chose not to compete, and they testified to the court in favor of the combine. The real price of stability, as outlined in earlier chapters, was the stifling of technological innovation. The principal sources of technical innovation in the steel industry were and continued to be companies outside the orbit of U.S. Steel. The persistence of this curious pattern is illustrated here in the story of the rise of two new alloys (silicon and stainless steels) and a discussion of U.S. Steel's experience with industrial research. Vigorous user-producer relations continued to be the essence of technical innovation. I suggest that their flagging presence is a little appreciated but underlying weakness of the postwar steel industry.

R&D and Technical Change

The remarkable electrical properties of steels containing silicon were first investigated in England in the mid-1880s. Acute demand for such a steel came with the expansion of alternating-current electric power in the United States over the next two decades. Thomas Edison's direct-current Pearl Street Station (1882) served a tiny area in central Manhattan, because direct current can economically be sent only a mile or two. Alternating current, however, can be generated many miles from its point of consumption and can be sent over high-voltage transmission lines without unacceptable losses. Nearly every application—generating, stepping voltages up or down, powering motors, lighting of all types—requires passing alternating current through coils of wire wound around iron or steel cores. Silicon steel proved to be a key material in making possible such a system of generation and distribution; once again, it resulted from collaboration between producers and users.

Electrical manufacturers had known since the 1880s that laminated sheets of iron or steel helped minimize the heat generated by "magnetic friction" (hysteresis losses and eddy currents inside the

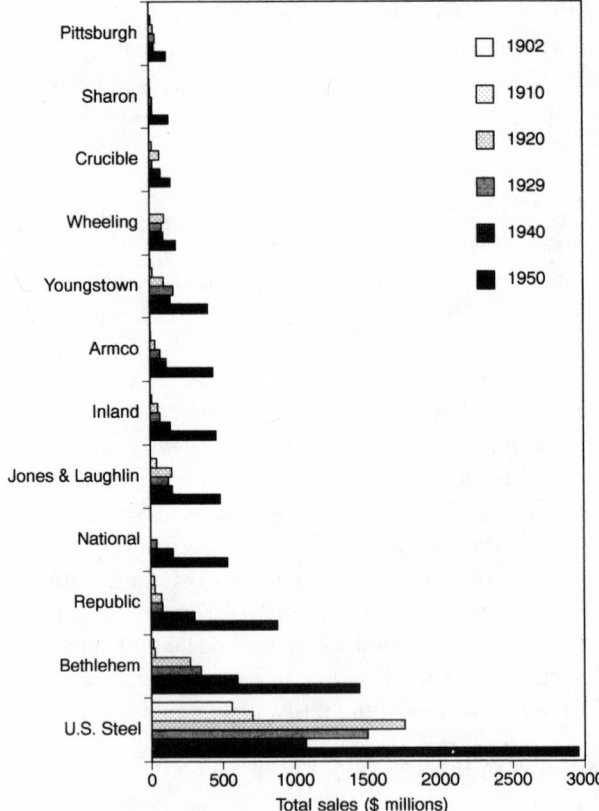

FIG. 7.2. Total sales of leading steel companies, 1902–50. Note that
1902 is the first year that sales figures for U.S. Steel are available. The
companies (with dates of incorporation) are Crucible Steel Company of
America (1900), Sharon Steel Hoop (1900), Pittsburgh Steel (1901), Inland Steel (1893), American Rolling Mill Company (1899), Republic Iron
& Steel (1899), Wheeling Steel (1920), Jones & Laughlin Steel (1902),
Youngstown Sheet & Tube (1900), Bethlehem Steel (1904), and U.S.
Steel (1901). National Steel was organized in 1929. Sales figures were
taken from the first or closest year available for the entries for Bethlehem
1902 (actually 1905), Jones & Laughlin 1910 (1913), Crucible 1929 (1934),
and Sharon 1910 (1909) and 1920 (1921).

Data from Gertrude G. Schroeder, The Growth of Major Steel Companies, 1900–
1950 *(Baltimore: Johns Hopkins University Press, 1953), 216–27.*

core of the winding caused by electromagnetic fields). Solid cast-iron cores actually became hot from magnetic friction, while laminated cores cut such losses to about 30 percent of the energy transmitted. Still, it was not well understood how, or if, the magnetic properties of these sheets could be improved. To develop this field, as it had done with other challenging consumer problems (see chapter 6), the American Rolling Mill Company, or Armco, dispatched a salesman to Westinghouse in 1902 with the offer of a cooperative research effort to develop a new steel with improved magnetic properties. In the year-long program that resulted, a Westinghouse engineer named Wesley J. Beck worked alongside Armco's metallurgist, Robert B. Carnahan, to develop a silicon steel that boosted electrical efficiency by nearly 50 percent. Sheets of steel containing up to 5 percent silicon have extremely desirable magnetic properties, they found, minimizing losses from "magnetic friction." Also in 1903, however, Robert Hadfield, the Sheffielder already famous for manganese steel, filed a U.S. patent for silicon steel; his success in the patent courts eventually compelled Westinghouse, along with General Electric and American Sheet & Tin Plate (a U.S. Steel subsidiary) to become his licensees. Hadfield himself estimated that by 1908 some quarter-million silicon steel transformers had been installed in the United States.[4] In this case, U.S. Steel followed where others—Armco, Westinghouse, and Hadfield—had led.

Such a pattern of innovation also characterized stainless steel, an invention claimed by as many as ten rivals in Europe and the United States. This versatile steel was commercialized about 1914 for steam-turbine blades and for use in the Sheffield steel cutlery trade, and soon began to be used for airplane engine valves. In 1917–18, the two principal rival patent claims were amalgamated in the American Stainless Steel Company, whose shareholders included the Sheffield-based Firth-Brearley Stainless Steel Syndicate, the American automobile inventor Elwood Haynes, and four American steel producers—including Bethlehem, Carpenter, Crucible, and Firth Sterling. U.S. Steel was not among the shareholders or the early licensees. Reportedly, the first really large commercial order for stainless steel came in 1924 from the Du Pont Company, which used the material for making corrosive nitric acid. Stainless steel later became widely used in the dairy and food industries and for many other applications requiring corrosion resistance.[5]

At the periphery of the newest and most promising alloy steels, dismissive of continuous-sheet rolling, actively hostile to new structural shapes, a price leader but not a technical leader: this was U.S. Steel. What was the company doing with technological innovation?

Because the "industrial research" model never had the same significance in the steel industry as in the chemical and electrical industries, it is sometimes suggested that the industry's weakness in industrial research caused its weakness in technical innovation. Given the modes of technical innovation discussed in this book, there is good reason to doubt any necessary equation of industrial research with technical innovation. Even for that paragon of industrial research, Du Pont, there is solid evidence that acquiring competitors, rather than conducting industrial research, remained its most important source of new technology through 1950.[6] A discussion of U.S. Steel's experience with industrial research helps clarify these issues.

If one defines industrial research in terms of *institutional form*, the pioneering laboratory seems to be one set up in 1908 at Carnegie Steel (as a U.S. Steel subsidiary) by John S. Unger. Unger's career path was classic. After earning a doctorate of science from Western University of Pennsylvania, he joined Carnegie Steel in 1887 as assistant chemist at the Edgar Thomson works. Later he served as chief chemist, assistant superintendent, and superintendent of the Homestead armor department (1892–1904), then as assistant general superintendent of the Homestead works (1905). In 1908 he was named manager of Carnegie Steel's new Central Research Laboratory in Duquesne, Pennsylvania, where he remained until 1935. Unfortunately not enough is known at present about this laboratory. His early research centered on characterizing and specifying the conditions for heat-treating steel, a natural extension of his Homestead experience with the Harvey-Krupp process. In 1912 his effort to characterize heat treatment in general terms was hailed as contributing to "possibly the greatest example of the application of scientific methods to the industry."[7]

By the 1920s U.S. Steel, still a sprawling holding company, nominally controlled many research, testing, and engineering laboratories that had been organized by its independent-minded subsidiaries. Sometime before 1920 its subsidiary, American Sheet & Tin Plate, organized a research facility identified as "one of the first adventures into the modern type of industrial research undertaken and sponsored by a steel-producing firm." Other subsidiaries, such as American Steel & Wire and Federal Steel, organized research laboratories that systematically attacked production problems. In 1928 U.S. Steel took steps to centralize these efforts on the industrial-research model pioneered by General Electric, AT&T, Union Carbide, and others. After detailed consultations with Robert A. Milliken, the 1923 Nobel laureate in physics, and

Frank Jewett, president of AT&T's Bell Telephone Laboratories, the steel combine hired John Johnston, chair of the chemistry department at Yale, to oversee its central scientific research facilities. To develop physical metallurgy, Johnston hired Edgar Bain from Union Carbide's research laboratory. In a converted office building of the Federal Shipbuilding & Drydock Company, located in Kearny, New Jersey, Bain and a group of a half-dozen metallurgists initiated, as he put it, "a few precise fundamental investigations." Bain's detailed memoirs suggest that industrial research resulted mostly in promising scientific work and occasionally in technical innovation, but they reveal that industrial research fostered a mode of inquiry that was largely cut off from users.[8]

Unlike Unger, who had risen through the ranks at steel-works laboratories, Bain's experience before 1928 as an industrial researcher was independent of any single company or indeed any single industry. Bain graduated in 1916 with a master's degree from Ohio State University, after attending a special summer session on metallography and pyrometry at Columbia University, where Henry Howe's successors kept up an active research tradition. After briefly serving as an instructor at the University of Wisconsin, as a chemist at B. F. Goodrich, and in the Chemical Warfare Service, he first experienced industrial research at General Electric's Cleveland Wire division. After the war, from 1919 to 1923, Bain worked on the company's famous process for making ductile tungsten wire for lightbulb filaments. Here he also began investigating the metallography, or microstructure, of high-speed steel used to draw out the tungsten into wire. Elucidating a mechanism for high-speed steel's puzzling property of maintaining its hardness at high heat became a central problem for Bain and other physical metallurgists in the 1920s. After leaving General Electric he helped manufacture high-speed steel at the Atlas Steel Corporation for eighteen months, then joined the Union Carbide research laboratory. Here he worked on stainless steel (the company's "dominating interest"), whose corrosion-resisting properties could be traced to chromium. In both the high-speed and stainless investigations, Bain focused on the microstructural mechanisms that caused steel to harden, the ways in which the various alloy elements influenced the microstructures, and the temperature and time intervals needed for such hardening.[9] This scientific problem—mechanisms of hardening—was a continuing theme in his career.

At U.S. Steel beginning in 1928, Bain spent most of his time in the New Jersey laboratory, where he conducted a series of scientific investigations on the mechanisms of hardening that brought him international scientific recognition. His work culminated in time-temperature-transformation diagrams that graphically portrayed how the rate of cooling influenced the formation of different microstructures of steel. Periodically, he visited production subsidiaries. During one visit to the American Steel & Wire Company laboratory and plant in Worcester, Massachusetts, Bain was asked to troubleshoot a continuous heat-treatment production line for heat-treating clock-spring and razor-blade stock. On-site investigation with a magnet suggested to him that the middle of three lead baths was too hot for forming the desired fine-grained austenite; lowering its temperature solved the problem. After moving to the corporation's New York office in 1935, Bain increasingly became involved with securing patent protection for inventions and coordinating the research efforts among the subsidiary laboratories.

Bain's most notable technical innovation came from the New Jersey laboratory: a new microconstituent of steel later named bainite. The discovery was made while he and his four assistants were investigating a scientific problem, the transformation rates of the microconstituent austenite at constant temperature. One day, as he related, at a lower temperature than used before "a wholly new acicular microstructure made its appearance." They found that its mechanical properties were ideally suited for springs that would bend without breaking even under severe loads. After the process was patented in 1933, the American Steel & Wire works at Worcester designed a continuous process (called austempering) that treated each day several tons of springs and other small carbon-steel products. Later an (unnamed) leading manufacturer specially treated millions of such steel shanks for the shoes worn by U.S. ground forces during World War II. As Bain recalled, "The development of these improved properties was a matter of serendipity, since our initial objective had been solely to gain knowledge about the modes and kinetics of the transformation of austenite at constant temperatures." [10]

The "bainite" story suggests that industrial research could result in technical innovation. It challenges the often repeated claim that U.S. Steel was hopelessly incompetent at research. Still, U.S. Steel's experience with industrial research is difficult to analyze given the limited evidence available. It is significant that Bain's work found practical application in the more specialized units of U.S. Steel, especially American Sheet & Tin Plate. Beginning in the 1930s this subsidiary began a line of research that eventually resulted in Corten steel, a high-strength, low-alloy, corrosion-resisting structural material that was arguably U.S. Steel's most important technical contribution. Clearly Bain's scientific work resulted in meaningful theoretical advances (especially his TTT diagrams, as they were called) and a successful business career. He was promoted in 1943 to vice president for research and technology for Carnegie-Illinois, the largest U.S. Steel subsidiary, and served on the technical committees of the wartime Office of Scientific Research and Development. Foremost among steelmakers' war- time achievements were the so-called National Emergency Steels, whose careful conservation of scarce alloying elements was based in part on efforts by Bain and others during the 1930s to devise a general theory of alloy steels. As vice president for research and technology for all of U.S. Steel in 1956, Bain opened the U.S. Steel Research Center in Monroeville, Pennsylvania, which established a true central research facility.[11]

It is not clear that U.S. Steel's research effort was really designed to promote technical innovation, which it did with selective success. Industrial research may have been intended to serve as a foil against antitrust proceedings, a perennial concern, by appearing to promote industrial activity in the public interest. Or research may have been seen as an investment whose currency of return was in company prestige rather than technological innovation. Bethlehem Steel's research center was criticized on the issue of company prestige.[12] This line of criticism deserves close consideration, for during the 1950s the steel industry failed the most important test of its ability to innovate. "U.S. Steel set standards for the American steel industry that foreign competitors have had little trouble beating," concludes one recent analysis.[13] Nevertheless, the corporation's, and indeed the industry's, broad array of problems cannot be traced solely to a "failure" of the industrial-research model.

To situate the industry's decline more fully, it is helpful to look farther afield. The following pages examine two practical areas—economics and labor relations—implicated in the decline of the steel industry. Investigating the economics of technical change suggests why the then-dominant neoclassical economics was an inappropriate framework for assessing technical change and, accordingly, why the decisions about technology made in the 1950s from within this framework were so spectacularly wrong. Then, an investigation of the industry's dismal labor relations, evident ever since the 1892 Homestead strike, helps explain (if not excuse) the poisoned climate of labor-management relations in the 1950s, which erected an additional barrier to constructive change. The Homestead strike preoccupied a generation of labor historians. It was a defining moment for the U.S. labor movement and as such has exerted an independent effect on labor-management relations.

The chapter then proceeds to gather together the strands of an emerging synthesis of theories of technical change. One strand comes from work in institutional and evolutionary economics that has focused attention on the process of technical change. An exposition of its core concepts suggests that they have much in common with recent theorizing about technical change from historical and sociological perspectives. A second strand picks up the insights that come only from a detailed understanding of the shop floor; analysts of labor have several cautionary insights about technical change. The third and final strand is the dynamics of growth and decline.

Economics and Technical Change

The efforts of economists to understand technological change, in the steel industry as elsewhere, have been handicapped by a lack of analytical tools to study dynamic systems. Despite its leading influence in government policy making, neoclassical economics cannot properly conceptualize technical change. This shortcoming is particularly unfortunate given the premise of this study that technical change has played a major role in national development. Furthermore, absence of an adequate and robust conceptualization of technical change fatally weakens all attempts to develop feasible industrial policies. Part of the problem flows from the empirically dubious assumptions made by neoclassical economic theory. Few real markets behave atomistically; very few real actors possess enough knowledge to make omniscient decisions; no real actors possess both knowledge and decision-making strategies to produce optimum results. Further, neoclassical analysis assumes the presence of so-called equilibrium conditions (perfect competition, full capacity utilization, full employment, no economies of scale, and so forth). Such conditions were rarely present in the real world. For instance, from 1898 to 1925 capacity utilization in the steel industry averaged 66.4 percent, with a peak in 1916 (86.2 percent) and a trough in 1921 (33.9 percent).[14] At bottom, neoclassical economics has no adequate conception of technical change itself. Technology is seen as an exogenous input to the economic system, as if technologies could be simply taken off the shelf, ready made. This "leads to disturbances from equilibrium situations and thereby sets in motion an adjustment process leading to a new equilibrium."[15]

Neoclassical economics models a given technology with a "production function," an abstract curve purporting to show the relationship between, say, capital and labor inputs at varying levels of output. For instance, expanding the output of some technologies requires substantial labor inputs (for example, wrought iron) while expansion of others can be done through capital inputs (for example, Bessemer steel). A different technology can be modeled by drawing a curve showing different relationships among the variables. But the *process* of technical change (except for the act of guessing where the new production-function curve is to be drawn) is not and cannot be modeled.[16] While some input-output relationship undoubtedly exists for a given technology, technical change itself alters these relationships in ways that are not quantifiable except after the fact. The inability to conceptualize technical change is the fatal conceptual gap of neoclassical economics. Neoclassical economic studies of technical change are thus meaningless exercises in number crunching.[17] It is not surprising that such narrowly conceived economic studies failed the industry precisely when a new production technology was at issue.

Economists adopting an institutional approach to technical change have avoided many of these conceptual shortcomings. They have forcefully argued that technological innovations are not discrete events whose consequences can be readily quantified, and that economic consequences flow from institutional dynamics that are vastly more complicated than those postulated by neoclassical theory. Technological innovations, writes Bela Gold, "engender readjustments in economic relationships through a variety of intermediary linkages, within a network of interacting pressures, covering an expanding horizon of operations, over extended periods of time, and accompanied by a variety of independent but concurrent developments." Such a necessary reconceptualization of technical change will require, according to Martin Fransman, empirical research capable of leading economists (1) to reconceptualize the company as more than merely a profit-maximizing entity; (2) to analyze the microscopic and macroscopic interactions of the company and its environment; (3) to recognize that technological knowledge is not available without cost to all actors; (4) to appreciate that organizational change is not instantaneous and unproblematic; (5) to analyze the national and international patterns of diffusion and selection of technological knowledge; and (6) to acknowledge the role of the state in shaping technology.[18] This work has made clear that technical change cannot be conceptualized solely in quantitative terms, as the production-function studies have assumed, and that the dynamics of economic institutions themselves are a crucial consideration.

Economists drawing on evolutionary models have at last made steps toward a robust conceptualization of technical change. Rejecting the neoclassical presumptions of static equilibria, they view technical change as a qualitative process, an approach consistent with that of institutional economists. Their long-term perspective centers on dynamics and changing structures. For example, Giovanni Dosi has asked, "Is there any regularity in the functional relationship between the vast number of economic, social, institutional, [and] scientific factors which are likely to influence the innovative process?" These economists conceive technical change as a structured process that is channeled through institutions, including markets, businesses, government agencies, and universities. They appreciate that technical change, especially from a long-run perspective, is not a totally random process but rather one that displays certain regularities and identifiable patterns.[19]

These regularities occur principally because of the nature of search activities (variation) underlying the innovative process. The question is how actors go about finding solutions. While neoclassical economists posit actors that are optimizers or maximizers (making optimum decisions based on complete knowledge), evolutionary economists allow actors to be so-called satisficers, identifying solutions that are "good enough" for the purposes at hand and that are based realistically on a subset of the complete universe of knowledge. Such a model of decision making recognizes that individuals and institutions frequently develop shortcuts in their search strategies, or heuristics, that make some lines of decision making and problem solving much more likely than others. For example, innovation in the U.S. steel industry between 1865 and 1885 centered rather narrowly on improving the Bessemer converter toward a specific set of technical and economic criteria (maximum volume, rapid expansion, quality sufficient for rails) using a subset of available scientific knowledge (chemical rather than physical metallurgy). Similarly, automobile metallurgists in the 1910s focused on trying one new alloying element after the other, invoking the productive heuristic that promising innovations were likely to be found in this manner. It may be suggested that the effort to perfect the open-hearth and electric furnaces effectively blinded steelmakers to alternative lines of problem solving after the 1920s.

The results of such regularities in decision making may be conceptualized as "paradigms" and "trajectories," which offer opportunities for profitable investment and growth of markets over relatively long periods along rather well-defined paths of development and diffusion. A *technological paradigm*, like a scientific one, represents a contextual definition of what the important problems are, what branches of scientific or technological knowledge are relevant, and which materials are to be used. A technological paradigm can be defined "as a 'pattern' for solution of selected techno-economic problems based on highly selected principles derived from the natural sciences. A technological paradigm is both a set of *exemplars*—basic artifacts which are to be developed and improved— and a *set of heuristics*—'Where do we go from here?' 'Where should we search?' 'On what sort of knowledge should we draw?'"

In short, technological paradigms define technological opportunities and suggest how to exploit them. Such paradigms serve to channel innovative efforts in some directions and not others: a *technological trajectory* is the activity of technological change along the economic and technological trade-offs defined by a paradigm. "The crucial hypothesis," writes Dosi, "is that innovative activities are strongly selective, finalized in rather precise directions, often cumulative activities." This view stands in stark contrast to the neoclassical concept of technology as information that is easily applicable, reproducible, and reusable, and permits companies to innovate by drawing on a general pool of technological knowledge. Instead, a particular company's technological future is sharply constrained by its past capabilities. Consequently, technological developments over time tend to be closely related to existing activities, irreversible, and path dependent. Patterns in the selection environment, in which technological alternatives compete, reinforce these tendencies.

These concepts might well lead to a robust theory of technical change. Recent economic, sociological, and historical studies of technical change feature at least three sets of similar concepts and shared insights whose implications merit further development. First, Dosi's concept of a technological paradigm strongly resembles Edward Constant's[20] and also Wiebe Bijker's notion of "technological frame."[21] Further, the working out of paradigms as technological trajectories surely resembles Thomas Hughes's concept of "technological momentum."[22] Second, certain economists have observed that individual technological paradigms may form clusters yielding a kind of meta-paradigm, or "technological regime" whose common sense and rules of thumb dominate engineering and management decisions for decades, affecting the entire economy. Such a concept is complementary to the historian's notion of "technological style" as a broad concept that attempts synthetically to relate a set of specific technical characteristics to their societal context.[23] Third, a common recognition of the interrelatedness of technologies has given rise to several specific concepts, including Thomas Hughes's "technological system" and Latour, Law, and Callon's actor-network model.[24] These concepts are elements in an emergent theory of technical change.[25]

Labor and Technical Change

However promising recent theorizing about technical change may be, such efforts have been hampered by a preference for theorizing at high levels of aggregation (for example, the national economy). Analysts have tended to slight or ignore the details of labor processes and the role of workers in stabilizing or changing technical systems. Consequently, none of these emerging theories seems capable of dealing with industries, such as steel, that are both labor- and capital-intensive. As a corrective, efforts by labor historians to understand technological change "from the bottom up" have yielded some suggestive results. These studies also help explain the industry's difficulties in labor-management relations.

For labor historians, and for many workers, no single event has generated as much passion as the 1892 lockout at Carnegie Steel's Homestead plant. The dramatic events of the five-month struggle are well known and have been discussed elsewhere.[26] They include the tense negotiation between the Carnegie company and the Homestead union over compensation, the company's declaration of a nonunion policy, the flare-up of violence on 6 July when workers and townspeople routed a small army of Pinkerton agents sent by the company to reopen the mill, the arrival of thousands of guardsmen dispatched by the governor, the trial of labor leaders and the collapse of the Amalgamated union. What is of continuing puzzlement and surpassing importance is whether, and to what extent, technological change was either a cause or a consequence of this confrontation. An analysis of these issues has the merit of revealing major currents in the literature on labor and technology generally, including the power of management control, the persistence of worker resistance, and elusive accommodation between labor and management.[27]

An early account emphasizing the power of management control was that of David Brody. Brody focused on the all-embracing "psychology" of mass-production steelmaking that determined the rise of a brutally efficient, nonunion labor system.[28] Accounts following this insight frequently assert that before 1892, skilled workers organized into the Amalgamated Association of Iron and Steel Workers exercised considerable shop-floor control in the steel industry. Although the rise of new steelmaking processes had undermined the union's core of skilled wrought-iron puddlers and boilers, the union remained especially strong among the rolling crews, which negotiated annual contracts in the form of a "sliding scale" that produced handsome salaries while fostering group solidarity. In this context, Katherine Stone has asserted that the "labor system . . . prevented employers from . . . mechanizing their operations" before the 1892 Homestead strike.[29] On this view, Andrew Carnegie and Henry Clay Frick broke the union because it had successfully obstructed efforts to mechanize the productive process. Homestead was indeed the last real stronghold of unionism in the Carnegie empire. The company's Duquesne mill had been nonunion for some time. By 1885, ten years after its opening, technological change at the Edgar Thomson plant had displaced fifty-seven of the sixty-nine men on the heating furnaces and fifty-one of the sixty-three men on the rail mill; later the two union lodges at the mill were disbanded. Carnegie and Frick doubtless hoped for a similar outcome at Homestead.[30]

On this view, as a direct consequence of breaking the union at Homestead, the Carnegie company was able to proceed with its plans to speed up and further mechanize the rolling mills, adopt other labor-saving devices throughout the mills, and arrogate to itself the profits of increased output. Advocates of the management-power thesis dismiss Andrew Carnegie's apologetic public stance as self-serving sophistry. "It was common knowledge that the monumental profits earned by Carnegie Steel in the 1890s grew directly from the defeat of unionism at Homestead," asserts Paul Krause in his book on the Homestead struggle. "Without the encumbrance of the union, Carnegie was able to slash wages, impose twelve-hour workdays, eliminate five hundred jobs, and suitably assuage his republican conscience with the endowment of a library."[31]

Advocates of the persistence of workers' control do not deny corporate power. Rather, they emphasize the many ways in which workers, even in the absence of formal trade unions, continued to exercise a measure of control on the shop floor. David Montgomery's influential *Workers' Control in America* (1979) argued that working-class dynamics emerged from the collaborative nature of industrial work and diverged from capitalistic notions of efficiency. In a later work amplifying these themes, Montgomery devotes his leading chapter to iron- and steelworkers. His interpretation of Homestead explicitly rejects the "golden age" of omnipotent workers posited by Stone, while detailing exactly how and why the Homestead settlement proved so disastrous for trade unionism and workplace relations in the iron and steel industry.[32]

Montgomery emphasizes that on the eve of the Homestead strike the overall membership of the Amalgamated was at an all-time high of 24,000 members, a level of union membership not achieved again until 1934. This was the case even though the union had suffered reverses in the early 1880s. A successful strike three years earlier, in 1889, had given the union official recognition as well as an advantageous settlement on the "sliding scale" that determined overall levels of pay in relation to market prices for iron and steel products. With the 1892 round of negotiations on the sliding scale, however, Carnegie and Frick were determined to break the union. Although only eight hundred of Homestead's most skilled workers were officially members, the Amalgamated's strike decision emptied the mill of the entire thirty-eight-hundred-member work force. For a week following the 6 July battle, the union advisory committee ran the city: patrolling the streets, arranging funerals for the fallen, and even arresting a group of anarchists who came down from Pittsburgh to distribute leaflets. Then eight thousand soldiers of the Pennsylvania National Guard arrived. The following arrests of union leaders on charges of murder, treason, riot, and conspiracy, as well as additional charges against the union advisory committee made by the Pennsylvania supreme court, represented a "crisp and firm declaration that workers' control was illegal."

In detailing the aftermath of the Homestead strike, which included the tripling of steel production within eight years, Montgomery offers a subtle observation on the steel company's new managerial structure, which terminated restrictive union rules and cultivated a hierarchy of fiercely competing individuals, abetted by a wave of emigrants from southern and eastern Europe. He points out that the new management structure "did not represent a radical break with the past . . . It was built on both the technical knowledge of skilled workers, which remained indispensable, and the tradition of promotion within [work] gangs." Nevertheless, Montgomery, like Krause, concludes that the Carnegie company's "prodigious" profits of the 1890s "could not have been earned without Carnegie's triumph over workers' power that had its roots in their accumulated 'skill and knowledge.'"

Advocates of a third position have deemphasized the class-conscious battles, stressing instead the mutual accommodation of workers and managers. The intent is to show the subtle ways in which owners and managers were dependent upon skilled workers, and how both parties participated in suitable and even satisfying accommodation. From this perspective, John Ingham has produced the most sensitive portrait of Pittsburgh iron- and steelworkers during the late nineteenth century. Although he concedes the brutality of the employer-dominated labor system after 1892, he argues that a mutually accommodating labor system existed in Pittsburgh iron and steel mills throughout the 1870s and 1880s, despite instances of strikes. Such an attitude of "honest, if grudging, acceptance of unionism as an expression of the legitimate goals of workers" was distinct from both the militant anti-unionism of Henry Clay Frick and the romantic, paternalist anti-unionism of Daniel J. Morrell, of Cambria Iron in Johnstown. Ingham argues that such an attitude, which included a willingness to negotiate with workers' organizations in a "tough but fair manner," characterized "the majority of upper-class Pittsburgh iron and steel men in the nineteenth century."[33]

Ingham's subtle analysis of the great railroad strike of 1877 and the citywide iron puddlers' strike in 1882 is a helpful corrective to some analysts' determination to see class-conscious revolution in every workers' gesture. Still, his evidence for "examples of goodwill between employer and union" is sometimes fragile. A solid observation is that the "sliding scale" was first suggested by Benjamin F. Jones (of Jones & Laughlin), who in the mid-1860s was seeking a mechanism to automatically adjust workmen's wages to the rise or fall of market prices and thus end the adversarial labor relations plaguing the industry. The sliding scale was later adopted by the Sons of Vulcan and made into the cornerstone of that union's successful strategy in the 1870s and 1880s. Similarly sound is Ingham's reinterpretation of inside contracting, daily and seasonal patterns of work, and other informally arranged aspects of mill and town life, which many labor historians present as evidence of an autonomous working-class culture. Ingham sees them instead as part of "a shared shop culture" of ironmasters and ironworkers. But his "ritualistic" interpretation of confrontations between wrought-iron boilers and management at Jones & Laughlin hangs on brittle evidence.[34]

The dramatic rise of steel production in the 1890s, as earlier chapters have made clear, did not exactly require manufacturers to destroy unions. To his credit, Montgomery grants that the rise of the Bessemer process itself represented a serious threat to wrought-iron puddlers and boilers precisely because manufacturers could expand output by investing capital rather than by hiring additional skilled workers. Similarly, the shift from wrought iron to cast steel at once cut out two preliminary steps (squeezing and roughing) needed to transform the ball of pasty wrought iron into a rollable bar; steel came from the converting house or open-hearth plant as rollable ingots. Wrought-iron workers did not suffer from mechanization of puddling, which failed for excessive fuel consumption.[35] The rapid expansion of the Carnegie company in the 1890s in fact represented its excessive commitment to the Bessemer converter, to say nothing of the Homestead mill's armor profits. The avalanche of raw steel ingots was handled through mechanization, speeding up of rolling, and construction of additional rolling capacity. In the final analysis, then, it is incorrect to assert that because the Carnegie company (and others) rapidly expanded production after breaking the Amalgamated union, such expansion required, and was a direct result of, the company's union busting. Ultimately, the industry's dismal labor policies represented a social choice to retain profits rather than distribute them as wages. "As humanitarians, we might regret" the harmful overwork, wrote the metallurgist Henry Howe. "As managers . . . we would not be justified in diminishing our employers' profits."[36]

The Dynamics of Growth and Decline

The analytical framework of this study offers a middle way between the overaggregated analysis of economics and the underaggregated analysis of labor history. Although we need the generalizations and systematic analysis that naturally emerge from aggregated studies, they must be tempered with the nuance, detail, and contingency that only detailed studies can provide. The question is how to proceed. The approach by consumer sectors, as exemplified in this study, has, I believe, much to recommend it.

What is new about this approach is the focus on the *users* of steel and the attempt to identify patterns based on the interactions of users and producers. This study indicates that such patterns of interaction can be identified, that they change over time, and that the interactions and changes can be meaningfully described, analyzed, and compared within a framework of user sectors. In this view, the "steel industry" does not exist as a monolithic entity; rather, it is composed of various user sectors. As a unit of analysis, user sectors are broad enough to permit meaningful generalization, yet narrow enough to reflect the particulars of time and place. They are a way of grasping both the broad sweep of history and the importance of contingency, individuals, and politics of all sorts.

The Bessemer steel rail sector was so prominent that its characteristics are sometimes taken for those of the industry as a whole. Certainly, the reckless mass production of steel rails was much in evidence for three decades beginning in the 1870s. Among its hallmarks were a supreme attention to quantity production; a sacrifice of quality considerations; the adoption of technologies and processes, including continuous operation, that facilitated these goals; and a distinct shop-floor mentality fostered by the imperatives of rapid throughput. Such a production mode developed in response to the expansive period of railroad building (1870–90) when steel rails were demanded in large volumes and could be of modest quality. A handful of Bessemer mills, not more than a dozen, in the central Pennsylvania, greater Pittsburgh, and Chicago districts together provided enough rails to satisfy the nation's demand. A distinct chemical metallurgy focused on process control also evolved in response to these dynamics. It is no accident that so many of the production-oriented men rapidly promoted by Andrew Carnegie to partnerships—and who initially filled the executive ranks at U.S. Steel—began their careers as chemists at the Edgar Thomson mill. It is also no accident that these same men were purged from U.S. Steel when E. H. Gary finally solidified his control about 1910. Their career trajectories are an index for the rise and fall of a certain mentality of steelmaking.

Such a tyranny of Bessemer technique proved a disaster, however, when users required greater attention to quality. Beginning in the 1880s with the effort to devise a steel suitable for structures, and followed more spectacularly with the rail crisis of the 1900s, the reckless mass production of steel came a cropper. Structural beams made from Bessemer steel were widely discredited by the late 1880s, for the very reason that had made the process so successful for rails. In replacing Bessemer steel for structures, Americans developed a distinctive product: in the United States, if not in Europe, structural shapes were made from basic open-hearth steel, which benefited from a more leisurely pace of production and could be made nearly free from harmful phosphorus. These technical innovations took place in the relatively decentralized structural-steel sector, in which demand was not personalized in the shape of ten or twenty railroad presidents who conducted the yearly acquisition of rails for their road. Instead, orders in the structural sector came from dozens of fabricating firms and scores of architects. An additional clue to the distinctiveness of the structural sector is the prevalence of advertising and its notable form, best expressed by the Carnegie Steel handbooks that defined this sector. The same shortcomings of Bessemer steel were revisited in the 1900s when heavy steel rails began failing for mysterious reasons. The ensuing rail crisis revealed the insane levels to which rail mills had elevated reckless mass production—quickening the flow of metal, raising its temperature, and speeding up the entire rolling process. Again, changing the technical characteristics of the heavy rails required revamping the relationships between rail producers and rail users.

The first sector with a leading consumer willing to pay premium prices for quality was the manufacture of armor plate for battleships, which began in the 1880s and accelerated with the naval rearmament preceding World War I. In this sector the leading consumer, the U.S. Navy, dictated to the steel companies the standards of production and the key technologies. Further, the consumer was personalized in a single individual: the navy secretary himself conducted armor bidding and negotiating. In the total absence of a "market," this sector evolved distinct characteristics, including contract collusion, strategic international activity, and quasi-monopoly price levels. While the evidence for producer cartels in the railroad and structure sectors is persuasive enough, the evidence for pervasive contract collusion in the armor sector is overwhelming. The international technological convergence on Harvey-Krupp armor meant that by about 1900 the armor shops of all the leading companies in the world uncannily resembled one another. Another distinction was the armor sector's quasi-monopoly profits, in support of which barriers to entry were maintained by the active connivance of Pennsylvania's members of Congress, who desired to keep such an employment-rich enterprise from falling to another state. Simply put, there was no sector like the armor sector.

The only other sector demanding quality at virtually any cost was the manufacture of tool steels. Both armor and tool steel were products with whopping price premiums over steel rails. While structural steel commanded a price roughly double that of rails, first-class armor fetched approximately fifteen times the rail price, and the best tool steel cost forty times as much. But tool steels, unlike the blocks of armor that began life as outsized 50-ton ingots, came from crucibles seldom weighing more than 100 pounds. Lavish attention was bestowed on this steel. Crucible steelmakers resisted the insights of scientific metallurgists, preferring instead to rely on the empirical, craft methods that had served them well. Makers of crucible steel in the United States lacked the track record of innovation of their English counterparts, however, and could not compete when electric steelmaking became feasible after 1910. A key consideration was that American crucibles proved unable to meet the huge demand for low-carbon alloy steels from automobile makers.

Indeed, by exerting the imperatives of quality and quantity, the automobile sector was the final consumer to fundamentally change the steel industry. For steelmakers automobiles were a new kind of product: sold directly to individuals and afflicted by unprecedented technical problems. Even more than for structural steel, the consumer orientation of automobiles made advertising imperative. Ford, among others, adopted new alloys for axles and gearboxes and exploited this fact in advertising the Model T. This sector was surprisingly decentralized: besides the so-called big three (Ford, General Motors, and Chrysler), no fewer than thirteen medium-sized automobile manufacturers persisted through the 1930s.[37] This decentralization gave great impetus to standardization. Through standards setting, the smaller companies hoped to achieve some of the advantages of scale possessed by the largest concerns; a key issue was stabilizing the relationship between component suppliers, including steelmakers, and automobile manufacturers. The automobile's challenging technical problems prompted the rise of industrial-research methods in the steel industry's largest consumer. Steel companies attending to these new requirements were forced to be technically innovative.

Another attractive aspect of user sectors is that such an approach naturally places technical and industrial developments into a broader social context. It is no secret that the building of the world's largest steel industry had something to do with the emergence of the United States as a leading economic, military, and political power. The question is how the production and use of steel shaped the contours of national development.

Each chapter of this study provides a piece of the answer to this question. Chapter 1 situated Bessemer steelmaking in the context of western expansion and transcontinental railroad building. Railroads relying solely on domestic wrought-iron rails could never have been built so quickly; railroads relying on imported rails would have threatened the nation's financial independence. It was at least fortuitous that Alexander Holley's father lectured him on the financial evils of importing European rails.[38] Much of the complexion of western expansion can be traced to the rapidity of railroad construction, including the railroad-based agriculture that fed eastern manufacturing cities, the increased military mobility that "solved" the Indian problem, and the perceived shipping injustices that led to the first federal legislation. In this regard, the domestic steel-rail industry was not only the means for building transcontinental railroads but also a necessary condition for rapidly developing the western states and establishing domestic control over financial markets.

The following chapters have shown how steel was bound up with central themes in the nation's development: the structure of skyscraper-based cities, the politics of government-business interaction, the emergence of big business, the reform of metal-working factories, and the rise of automobiles and consumer culture. By accenting developments that were distinctive in American culture, this study suggests how steel helped define a modern nation. For by the 1920s the nation's physical infrastructure was built of steel, its economic infrastructure was built upon the making and using of steel, and its social infrastructure was shaped by the producing and consuming of steel. In short, the United States was a nation of steel.

A focus on user-producer interactions complements and extends the sector-based analysis in this study. For railroads, user-producer interactions were fundamental in selecting and shaping the structure of the early (Bessemer) steel industry as well as its use of a certain type of scientific knowledge. For railroads, as well as for armor, such interactions were of a direct and personal nature. Railroad presidents personally conducted contract negotiations for rails, while the secretary of the navy did the same for armor. In no other sectors did so few individuals have such a dominant role—which is not to say that these individuals always got what they wanted. Even the navy, the sole domestic customer for armor and heavy gun forgings, did not exert a "command" structure over the steel companies. Indeed, much to the frustration of navy secretaries and budget-minded members of Congress, the two leading armor producers (Carnegie and Bethlehem) arranged the bidding for armor to maintain prices and divide contracts throughout the period of naval rearmament. The international technological convergence on Harvey-Krupp armor after 1895 further eroded domestic control over armor procurement. This was true for all the geopolitical rivals (Britain, France, United States, Italy, and Germany). It is fair to say that user-producer interactions in the case of armor began with the navy secretary and extended to steel executives and government officials in half a dozen countries. Achieving central control over such a multicentered system proved impossible.

A quite different mode of user-producer interaction characterized the structural-steel sector. Here the users were dozens of fabricating firms that purchased steel, scores of architects and struc-

tural engineers who designed and specified steel, many construction firms that erected steel buildings, and hundreds of clients who
occasionally expressed a definite preference about steel. Similarly,
the major producers of structural-steel shapes numbered about a
dozen as opposed to the very few armor producers. Moreover, the
crucial influence exerted by city building codes suggests that the
public-policy process, involving many different actors with diverse
agendas, should be considered as an indirect user of structural steel.
Facing this multiplicity of users, steelmakers formed a producers'
cartel for beams that was active during the formative period of this
sector. With price competition out of the question, users therefore
focused on promptness of delivery and quality of product.

The information needs of different sectors also seem to have
varied dramatically in character and mode. In the centralized railroad and armor sectors, information was communicated mostly by
person-to-person interviews, letters, and telegrams, a system that
was feasible because of the small number of decision makers. Significantly, resolving the rail crisis in the reorganized railroads of
the 1900s required new and broader ways to organize and de-
ploy technical knowledge. The decentralized structural-steel sector
needed a different mechanism to mediate between users and producers, especially to exchange technical information. This situation fostered the rise and success of information-rich handbooks.
Compared with trade catalogues, structural-steel handbooks were
packed with detailed technical information and served as early textbooks for the field. No one could have designed a railroad based
on the meager information in a rail catalogue. A similar handbook
for automobiles, published by the Society of Automotive Engineers
in the mid-1920s, reinforces the point that decentralized sectors
had informational needs that could not be met solely by personal
contacts or narrow market signals.

An appreciation of user-producer interactions expands the domain of effective decisionmakers beyond the handful of individuals
in the boardrooms of the largest steel companies. There is no reason
to think that groups isolated from their economic and social context necessarily make good decisions about technologies. In fact,
some of the far-reaching decisions evident in this study resulted
from diversity in decision making. Public policies in the form of
city building codes, as seen in chapter 2, were a determining factor
in the first generation of steel-framed office buildings. The effect
of building codes did not diminish over time.[39] The federal government, quite apart from the Interstate Commerce Act of 1887

and the Sherman Antitrust Act of 1890,[40] was overtly involved with decision making on rails and armor, and indirectly with decisions concerning high-speed steel.

Most influential in expanding companies' effective decisionmaking boundaries were the engineering societies. Unlike trade associations, which typically represent an agglomeration of industry interests, engineering societies have a mixture of public, private, and professional imperatives.[41] The societies were active in all the sectors considered here. The rail crisis at the turn of the century prompted the most dramatic instance of effective cooperation between engineering societies and across companies. Perhaps the best known and broadest of any engineering effort was SAE's successful program to standardize automobile steels. A mix of public and professional imperatives was embodied in Frederick W. Taylor's presidential address before the mechanical engineers in 1906. By revealing his quarter-century of metal-cutting experiments, Taylor stated in effect his primary loyalty to the quasi-public world of engineering as contrasted with the private world of metal-cutting firms. The leading roles in professional engineering played by the people discussed in this book are worth highlighting. Professional engineers, naval officers, and building-code inspectors were among those who contributed to the dynamics of user-producer relations and consequently to the patterning of technical change.

An imbalance in user-producer interactions results in what I have called "forced technology choice." This is a different phenomenon from the open-ended rational assessment of options presumed by neoclassical economics. It is also distinct from the patterning or channeling effects of technological paradigms and trajectories. Forced technology choice can result when a leading user announces a preference for a certain production technology, giving producers the "choice" of adopting that technology or facing failure. A clear instance was the U.S. Navy's pronouncement of preference for certain types of armor and heavy gun forgings. Sometimes, forced technology choice results when options are foreclosed by rivals. An example is the post-1901 rail crisis, when companies cut off from rich Bessemer-grade iron ores were forced to adopt open-hearth furnaces for making rails. Similarly, an imbalance in user-producer relations occupied the Carnegie managers in the late 1890s, when they found that they simply could not get architects and builders to accept structural shapes of Bessemer steel.

These observations suggest a typology of user-producer interactions. *Centralized interactions* exist when a few users (railroads before 1900), or even a single user, interact with a few producers (armor before 1895). Here, so-called market forces are at their weakest while personal interactions, cartel behavior, and trading on such elusive qualities as "reputation" are strongest. *Multicentered interactions* exist when several centralized users (or producers) are meaningfully involved in sector-shaping negotiations (armor after 1895, railroads after 1900). Such arrangements tend to be unstable if some or all of the centers engage in "strategic" behavior, in which their decisions are opportunistic and anticipatory of others' (unknown) actions. Such strategic behavior tends to undermine the bonds of trust and good will, or mutually reinforcing and enhancing opportunism, that can result in the persistence of multicentered interactions over long periods. *Decentralized interactions* exist when no center of power can be identified and many individuals or firms are involved (structures after 1880, factories after 1890). Here, something like a market, as traditionally understood, may have explanatory relevance, especially if markets are viewed as incorporating flows of information. The flow of information often requires intermediary institutions, which may take the form of technical or engineering societies or technical publications. *Direct-consumer interactions* (automobiles after 1910, alloys after 1920) have the special character of anticipating or responding to the perceived desires of the final consumer. Here, direct-consumer advertising is crucial to company strategies, as illustrated by Ford's direct and showroom advertising of vanadium steel and by Armco's direct and collateral advertising of its "99 and $^{87}/_{100}$ percent pure" ingot iron for household appliances. Technological change intimately involves creating, sustaining, or altering user-producer interactions. Accordingly, a typology of user-producer interactions should be more probing than existing typologies if it is to help explain how institutional environments shape and channel technical change.[42]

We now possess a series of powerful insights into the dynamics of technology and social change. Together, these insights offer the realistic promise of being better able, if we choose, to modulate the complex process of technical change. We can now locate the *range of sites* for technical decision making, including private companies, trade organizations, engineering societies, and government agencies. We can suggest a *typology of user-producer interactions*, including centralized, multicentered, decentralized, and direct-consumer interactions, that will enable certain kinds of actions while constraining others. We can even suggest a *range of activities* that are likely to effect technical change, including standards setting, building and zoning codes, and government procurement. Furthermore, we can also suggest a *range of strategies* by which citizens supposedly on the "outside" may be able to influence decisions supposedly made on the "inside" about technical change, including credibility pressure, forced technology choice, and regulatory issues.[43]

Finally, given the centrality of vigorous user-producer interactions in technical innovation, public policies concerning innovation and competitiveness should focus on this crucial dimension. A century after the Sherman Antitrust Act, it is clear that neither antitrust nor regulation, the twin poles of industrial policy in the United States, is a sufficient government instrument to ensure technological innovation and competitiveness. From the perspective of this study, a sensible and feasible public policy would be to promote, monitor, and ensure the vigorous user-producer interactions that are essential for technical change linked to changing user requirements. In retrospect, what was wrong with U.S. Steel was not its size or even its market power but its policy of isolating itself from the new demands from users that might have spurred technical change. The resulting technological torpidity that doomed the industry was not primarily a matter of industrial concentration,[44] outrageous behavior on the part of white- and blue-collar employees,[45] or even dysfunctional relations among management, labor, and government.[46] What went wrong was the industry's relations with its consumers.

The dynamics of decline were already apparent by the 1950s, even though the final crisis for the American steel industry did not come until the 1980s, when a wave of cheap, high-quality imported steel wiped out domestic producers. After 1945 automobiles continued to be the leading sector for steel, consuming vast quantities of rolled sheet steel at high prices. The automobiles' exaggerated tailfins, long bodies, large engines, and heavy chrome-plated bumpers represented a windfall for steel producers. Given the automobile industry's troubles in the 1970s and 1980s, as well as its shift to aluminum and plastics, it is tempting to conclude that the steel industry's problems primarily reflect slack demand from its chief consumer. In fact, however, automobile manufacturers in the 1950s helped undermine the protected nature of the domestic steel market. Strikes during 1947, 1949, 1952, and 1956 convinced many steel consumers, among them automobile manufacturers, that they needed an assured supply of steel, either through placement of advance orders or by imports of steel. Such "hedge" buying of steel reached a peak in 1959, when uncertainties due to pending contract negotiations drove the tonnage of imports (at 7.5 percent of apparent domestic supply) for the first time above the tonnage of exports. Rising imports were also a result of the steel industry's indifference to the demands of smaller consumers, or those with special technical requirements, who were gently told to take their business overseas.[47] Effectively, the industry's rejection of these new demands by users reinforced its technical stagnation.

Such callous treatment of consumers might not have mattered, except for the rise of international competition. For the first half of the twentieth century, imports never exceeded 3.5 percent of domestic supply, while exports fluctuated between 4 percent and 20 percent of total shipments. This favorable balance of trade eroded during the 1950s as German and Japanese steelmakers rebuilt their bombed-out plants with a new production technology, the basic oxygen furnace (BOF), which American steelmakers had dismissed as unproven and unworkable. The BOF (sometimes called the "LD process" after the Austrian firms at Linz and Donawitz that commercialized it in the early 1950s) resembled a large Bessemer converter, except that pure oxygen rather than air was pumped into the liquid metal to refine it into steel. The BOF was faster than the open hearth, requiring just forty minutes for a heat of steel. Using pure oxygen, which became available cheaply and in quantity from the Linde-Frankl process beginning about 1930, had the signal advantage of eliminating the harmful nitrogen compounds that formed whenever air contacted molten steel. Furthermore, installation and operating costs for BOF plants were dramatically less than for open-hearth plants.[48] Nevertheless, the advantages of the BOF seem to have been repeatedly underestimated and its problems overemphasized by U.S. engineers and economists. Accordingly, in the 1950s, when U.S. steelmakers expanded production capacity by 50 percent, nearly all of the expansion was in open-hearth furnaces. With one exception (Jones & Laughlin) the early adopters of BOF technology in the United States were all newcomers (table 7.1).[49]

By the mid-1960s a forced technology choice faced the industry. In some ways the situation resembled Carnegie's predicament with Bessemer structural steel in the 1890s. Put simply, the industry was forced to reverse its commitment to an outmoded technology, and it did so in something of a wild scramble. From 1965 to 1970, no fewer than twenty-four major BOF plants were erected, and now by all the established producers. Yet ever since, American steelmakers have been playing catch-up. By 1970 the BOF accounted for more than 40 percent of total steel production in the United States, surpassing the open hearth, while the comparable figures were 80 percent in Japan and more than 50 percent in West Ger-

TABLE 7.1. Basic oxygen furnaces in United States, 1954–64

Startup Date	Company	Plant Location	No. of Furnaces × Capacity (tons)
1954	McLouth Steel	Trenton, Mich.	3 × 35
1957	Jones & Laughlin	Aliquippa, Pa.	2 × 80
1958	Kaiser Steel	Fontana, Calif.	3 × 110
1958	McLouth Steel	Trenton, Mich.	2 × 110
1959	Interlake, Inc.	Chicago, Ill.	2 × 75
1960	McLouth Steel	Trenton, Mich.	1 × 110
1961	CF&I Steel	Pueblo, Colo.	2 × 120
1961	Jones & Laughlin	Cleveland, Ohio	2 × 225
1962	National Steel	Ecorse, Mich.	2 × 300
1963	Armco Steel	Ashland, Ky.	2 × 160
1963	U.S. Steel	Duquesne, Pa.	2 × 215
1964	Ford Motor Company	Dearborn, Mich.	2 × 250
1964	Wheeling-Pittsburgh Steel	Monessen, Pa.	2 × 200
1964	Wisconsin Steel	South Chicago, Ill.	2 × 140
1964/66	Bethlehem Steel	Lackawanna, N.Y.	3 × 290

Source: William T. Hogan, *Economic History of the Iron and Steel Industry in the United States*, 5 vols. (Lexington, Mass.: Lexington Books, 1971), 4:1546–47.

many, Italy, and Belgium. Beginning in the 1960s a rerun of the BOF affair occurred with continuous casting, the second significant technological advance in postwar steelmaking. U.S. companies have begun to install Swiss and German casting technology. "The Big Steel companies tend to resist new technologies as long as they can," said the president of one upstart mini-mill. "They only *accept* a new technology when they need it to *survive*."[50]

The executives of established U.S. steel companies did more than reject BOF technology—when narrowly conceived economic analysis indicated uncertain profits—they also assisted its transfer to Japan and Germany. They were among the leading backers of reindustrializing rather than pastoralizing postwar Germany, of the Marshall Plan that implemented this goal, and of federal subsidies to Japan's new oxygen steel plants. In total, the U.S. government invested $1.4 billion to reconstruct the steel industry in Europe and Japan. In return American industry secured an assured supply of steel scrap to feed its outmoded open-hearth furnaces.[51]

The only problem was that there were not enough consumers for the standard grades of open-hearth steel and not enough production of the grades of steel that users wanted. Consequently, imported steel figured to a surprisingly large extent in the great postwar economic expansion. The boom in high-rise buildings and highway bridges produced an acute demand for concrete-reinforcing bars, and by 1959 imports of these bars were nearly 40 percent of domestic production. Imports of bars and tool steel shot up from negligible levels in 1945 to more than 1 million tons in 1959. Even plates and sheets—by far the leading U.S. *export* product in the interwar years—accounted for one-tenth of all imports by 1960. From 1957 to 1969, imports of barbed wire, that staple of the western United States, averaged 47 percent of total consumption, while imports of pipe and tubing, used in the oil industry, grew from 2 percent to 16 percent of total consumption. The Benelux countries, which together accounted for 49 percent of total imports in 1958, were progressively displaced by Japan. By 1969 Japan alone accounted for 44 percent of total imports, with steel also arriving from Benelux (15 percent), West Germany (14 percent), France (7.3 percent), the United Kingdom (6.4 percent), and Canada (5.8 percent).[52]

In the "blame apportionment" explanation of the steel industry's decline, enough blame exists to condemn government and management and have some left over for organized labor. Indeed, the aggressive stance of organized labor, itself a reaction to bitter memories of management hostility (choose an evocative date: 1892, 1901, 1914, 1919, 1937), was not conducive to long-term health. The Homestead battle of 1892 left especially deep scars. The very night when the National Industrial Recovery Act was put into effect in June 1933, more than a thousand workers and union supporters attended an organizing rally at Homestead and formed a "Spirit of 1892" Lodge of the long-dormant Amalgamated Association. As late as the 1980s Homestead's local union president philosophized "that the USW should tear a page from the 1892 Homestead strike and treat mill bosses like Pinkerton goons."[53]

The United Steelworkers, like unions in most mass-production industries, gave up their early vision of industrial democracy, in which workers would take a share of responsibility for shop-floor decision making. Instead, a bargain was struck that gave unions handsome contract settlements in exchange for management gaining absolute control over the shop floor. In effect, a fat paycheck was the price of labor-management peace. The piecemeal adjustments of wartime found expression in the Taft-Hartley Act of 1947, which severely circumscribed all labor "activities which obstructed the further development, expansion and productivity of the corporate political economy," even as it entrenched the basic collective-bargaining machinery of the twelve-year-old Wagner Act. The 116-day strike in 1959 was perhaps the last to center explicitly on labor-saving technology (union pickets read "Man versus Machine") as much as wage and security issues. In settlement, the union permanently gained rigid work rules, first written into the contract three years earlier, that appeared to save jobs but in the long run erected yet another barrier to improving work routines and adopting new technologies.[54] In this rigid climate, where meeting the numbers of a production forecast became the only goal that mattered, attention to consumers vanished. According to one production superintendent at U.S. Steel's Homestead plant, the watchword was, "As long as it is hard and grey, ship it."[55]

Ultimately, it is the continuity between 1925 and the present that —frighteningly enough—accounts for the steel industry's demise. Such continuity shows the remarkable tenacity of a technological regime, if you will, that operated systematically to govern technical and social change during the decades following the 1920s. It would be foolish to assert that the industry's decline began in the 1920s, for in that decade the industry began a generation of impressive quantitative growth. Yet the economic success of the industry through the 1960s led to complacency at all levels of the largest companies. One tragic consequence was the flagrant disregard of users' needs and the stagnation in technical capabilities that led directly to industrial decline.[56] Debate will continue over whether the American steel industry's rise to world dominance between 1865 and 1925 furthered the public interest. But it must be clear that its slide to second rank, with the consequent loss of jobs, tax revenues, and community stability, in no way served the public interest. The need still exists for steel buildings, bridges, railways, subways, and washing machines. Steelmaking does not necessarily have to poison the environment. And, as I have suggested above, we no longer lack insight into the process of technical and social change that would allow us to take action. Given the importance of technology in public affairs, we must develop instruments of public policy capable of dealing constructively with the perils and promise of technology.

List of Abbreviations

Journals and Reference Works

AIME Trans.	American Institute of Mining Engineers Transactions
AREMWA Proc.	American Railway Engineering and Maintenance of Way Association Proceedings
ASCE Trans.	American Society of Civil Engineers Transactions
ASME Trans.	American Society of Mechanical Engineers Transactions
ASST Trans.	American Society for Steel Treating Transactions
ASTM Proc.	American Society for Testing Materials Proceedings
DAB	Dictionary of American Biography
INA Trans.	Institution of Naval Architects Transactions
JFI	Journal of the Franklin Institute
JISI	Journal of the Iron and Steel Institute
NCAB	National Cyclopaedia of American Biography
SAE Trans.	Society of Automobile Engineers Transactions
SAEJ	Society of Automotive Engineers Journal
SNAME Trans.	Society of Naval Architects and Marine Engineers Transactions
USNI Proc.	U.S. Naval Institute Proceedings

Archival Material

Note: Archival material is cited in the notes by date, box, folder, and collection.

AC-CMS	Andrew Carnegie—Charles M. Schwab Collection, Historical Collections and Labor Archives, Pennsylvania State University, University Park.
ACLC	Andrew Carnegie Papers, Manuscript Division, Library of Congress, Washington, D.C.
AHH	Alexander H. Holley Papers, Connecticut Historical Society, Hartford.

AICB	Burnham Microfilm Project, Art Institute of Chicago.
AISI	American Iron and Steel Institute Papers (Acc. 1631), Hagley Museum and Library, Greenville, Del.
AJ	Archibald Johnston Papers (Acc. 1770), Hagley Museum and Library, Greenville, Del.
ALH	Alexander L. Holley Papers, Connecticut Historical Society, Hartford.
ALH-SI	Alexander L. Holley Drawings, National Museum of American History, Smithsonian Institution, Washington, D.C.
BS	Bethlehem Steel Papers (Acc. 1699), Hagley Museum and Library, Greenville, Del.
DA	Dankmar Adler Papers (Acc. MiS), Newberry Library, Chicago.
DI	Dowlais Iron Company Collection, Glamorgan Record Office, Cardiff.
DPLB	Burton Historical Collection, Detroit Public Library.
EDMC	Edward De Mille Campbell Papers, Bentley Historical Library, University of Michigan, Ann Arbor.
FA	Ford Archives, Edison Institute, Dearborn, Mich.
FAS	Frank A. Scott Collection (Acc. 3284), Western Reserve Historical Society, Cleveland.
FWT	Frederick W. Taylor Papers, Stevens Institute of Technology, Hoboken, N.J.
FWW	Maryland Steel Company Papers of Frederick W. Wood (Acc. 884), Hagley Museum and Library, Greenville, Del.
HCW	Henry C. Walton Diary at Carnegie Steel Edgar Thomson Works (Acc. 1883), Hagley Museum and Library, Greenville, Del.
HMH	Henry M. Howe Papers, Butler Library, Columbia University, New York City.
HS	Harvey Steel Papers (Acc. 1239), Hagley Museum and Library, Greenville, Del.
JAC	Julius A. Clauss Papers, Bentley Historical Library, University of Michigan, Ann Arbor.
JAG	John A. Griswold Papers, New York State Library, Albany.
J&L	Jones & Laughlin Steel Corporation Papers (Acc. 73:7), Archives of Industrial Society, Hillman Library, University of Pittsburgh.
J&M	William L. B. Jenney and William B. Mundie Papers (Roll 10, AICB).
JWL	John W. Langley Papers, Bentley Historical Library, University of Michigan, Ann Arbor.
KRR	Keith Reeves Rodney Papers (Acc. 1884), Hagley Museum and Library, Greenville, Del.

LIS   Lukens Iron and Steel Co. Papers (Acc. 50), Hagley Museum and Library, Greenville, Del.

MB   Marsh Brothers Papers, Sheffield City Archives.

MEB   Misc. Mss. EB, Historical Society of Western Pennsylvania, Pittsburgh.

MITHNC Misc. Shipbuilding Papers, Hart Nautical Collection, Massachusetts Institute of Technology, Cambridge.

NWL   Nathaniel Wright Lord Papers, Ohio State University Archives, Columbus.

OIS   Oliver Iron and Steel Co. Records (Acc. 64:6), Archives of Industrial Society, Hillman Library, University of Pittsburgh.

PI   Phoenix Iron Papers (Acc. 909), Hagley Museum and Library, Greenville, Del.

PS   Phoenix Steel Papers (Acc. 916), Hagley Museum and Library, Greenville, Del.

PWS   Porter W. Shimer Letters (Acc. 1246), Hagley Museum and Library, Greenville, Del.

RHT   Robert H. Thurston Papers (Acc. 16/2/40), Cornell University, Ithaca, N.Y.

STW   Samuel T. Wellman Papers, Archives of Science and Technology, Case Western Reserve University, Cleveland.

WBD   William B. Dickson Papers, Historical Collections and Labor Archives, Pennsylvania State University, University Park.

WGA   Records of Sir W. G. Armstrong and Company Ltd. (and related companies), 1847–1949 (Acc. 130/1280), Tyne and Wear County Archives, Newcastle-upon-Tyne.

Notes

Preface

1. The analytical insights as well as suggestions for policy instruments are presented in *Managing Technology in Society*, edited by Arie Rip, Thomas J. Misa, and Johan Schot (London: Pinter, 1995).

2. For the distinctive cognitive modes of technologists, see Eugene Ferguson, *Engineering and the Mind's Eye* (Cambridge: MIT Press, 1992); Brooke Hindle, *Emulation and Invention* (New York: New York University Press, 1982); Walter Vincenti, *What Engineers Know and How They Know It* (Baltimore: Johns Hopkins University Press, 1990); and Knut Sørensen and Nora Levold, "Tacit Networks, Heterogeneous Engineers and Embodied Knowledge," *Science, Technology, and Human Values* 17 (Winter 1992): 13–35.

3. See Merritt Roe Smith, ed., *Military Enterprise and Technological Change* (Cambridge: MIT Press, 1985); William Becker, "Managerial Capitalism and Public Policy," *Business and Economic History* 21 (1992): 247–56.

4. For a review of the literature on interactions between science and technology, see John M. Staudenmaier, *Technology's Storytellers* (Cambridge: MIT Press, 1985), 83–120. A penetrating analysis of industrial science is that of David C. Mowery and Nathan Rosenberg, *Technology and the Pursuit of Economic Growth* (Cambridge: Cambridge University Press, 1989). On the "utilities" of science, see Jack Morrell and Arnold Thackray, *Gentlemen of Science: Early Years of the British Association for the Advancement of Science* (Oxford: Clarendon Press, 1981), chap. 7.

5. John Strohmeyer, *Crisis in Bethlehem: Big Steel's Struggle to Survive* (Bethesda, Md.: Adler and Adler, 1986; reprint, New York: Penguin, 1987), 59.

6. On causality see Peter Gay, *Art and Act: On Causes in History—Manet, Gropius, Mondrian* (New York: Harper and Row, 1976), 1–32 (quotation on 1).

7. Contrast David Brody's all-embracing "psychology of steelmaking," discussed in chapter 7.

8. See Alfred D. Chandler, Jr., *Scale and Scope: The Dynamics of Industrial Capitalism* (Cambridge: Belknap Press of Harvard University Press, 1990), 129. It may be appropriate to consider his observation as a hypothesis to be tested, since the two companies he discusses had very different commitments to sales-intensive structural shapes. Carnegie Steel was far more committed than Illinois Steel, which was a latecomer to structures. In 1890 Illinois Steel had a mere 0.35 percent of its output in structural shapes.

9. This basic framework was introduced in my dissertation: Thomas J. Misa, "Science, Technology, and Industrial Structure: Steelmaking in America, 1870–1925" (diss., University of Pennsylvania, 1987).

10. John A. Mathews, "The Electric Furnace in Steel Manufacture," *AISI Yearbook* (1916): 79.

11. On user-producer interactions, see Bengt-Åke Lundvall, "Innovation as an Interactive Process: From User-Producer Interaction to the National System of Innovation," in *Technical Change and Economic Theory*, ed. Giovanni Dosi et al. (London: Pinter, 1988), 349–69. On the important role of users in innovation processes, see Eric von Hippel, *The Sources of Innovation* (New York: Oxford University Press, 1988), 11–27, 106–15; Stuart S. Blume, *Insight and Industry: On the Dynamics of Technological Change in Medicine* (Cambridge: MIT Press, 1992), 261; Paul Israel, *From Machine Shop to Industrial Laboratory: Telegraphy and the Changing Context of American Invention, 1830–1920* (Baltimore: Johns Hopkins University Press, 1992); and Janet T. Knoedler, "Early Examples of User-Based Industrial Research," *Business and Economic History* 22, no. 1 (1993): 285–94.

12. For the recent literature on cartels, see Thomas S. Ulen, "Railroad Cartels before 1887: The Effectiveness of Private Enforcement of Collusion," *Research in Economic History* 8 (1983): 125–44; C. Knick Harley, "Oligopoly Agreement and the Timing of American Railroad Construction," *Journal of Economic History* 42 (December 1982): 797–823; Daniel Barbezat, "A Price for Every Product, Every Place: The International Steel Export Cartel, 1933–39," *Business History* 33 (October 1991): 68–86; Richard A. Lauderbaugh, "Business, Labor, and Foreign Policy: U.S. Steel, the International Steel Cartel, and Recognition of the Steel Workers Organizing Committee," *Politics and Society* 6 (1976): 433–57; Steven B. Webb, "Tariffs, Cartels, Technology and Growth in the German Steel Industry, 1879–1914," *Journal of Economic History* 40, no. 2 (1980): 309–29; Lon L. Peters, "Are Cartels Unstable? The German Steel Works Association before World War I," *Research in Economic History*, suppl. 3 (1984): 61–85.

13. For details on the labor process and systematic analysis of union activity, see James Holt, "Trade Unionism in the British and U.S. Steel Industries, 1880–1914: A Comparative Study," *Labor History* 18 (1977): 5–

35; John A. Fitch, *The Steel Workers* (1910; reprint, Pittsburgh: University of Pittsburgh Press, 1989); Charles Rumford Walker, *Steel: The Diary of a Furnace Worker* (Boston: Atlantic Monthly Press, 1922); Christopher L. Tomlins, *The State and the Unions: Labor Relations, Law, and the Organized Labor Movement in America, 1880–1960* (Cambridge: Cambridge University Press, 1985).

14. For a bottom-up narrative of the industry as a whole, see William T. Hogan, *Economic History of the Iron and Steel Industry in the United States*, 5 vols. (Lexington, Mass.: Lexington Books, 1971); for a top-down economic analysis, see Peter Temin, *Iron and Steel in Nineteenth-Century America* (Cambridge: MIT Press, 1964). A compact survey is W. David Lewis, *Iron and Steel in America* (Greenville, Del.: Hagley Museum and Library, 1976).

15. Carl W. Condit, book review in *Railroad History* 142 (spring 1980): 111–14.

16. Col. W. A. Starrett, *Skyscrapers and the Men Who Build Them* (New York: Charles Scribner's Sons, 1928), 161–62.

17. Defects in British ammunition are stressed by N. J. M. Campbell, *Jutland: An Analysis of the Fighting* (Annapolis: Naval Institute Press, 1986).

18. See Naomi R. Lamoreaux, "Bank Mergers in Late Nineteenth-Century New England: The Contingent Nature of Structural Change," *Journal of Economic History* 51 (1991): 537–57. Chapter 4 also addresses a dispute between Lamoreaux and Chandler about the effect of U.S. Steel's control of iron ore. Compare Chandler, *Scale and Scope*, 135–38, 751 n.128; and Naomi R. Lamoreaux, *The Great Merger Movement in American Business, 1895–1904* (Cambridge: Cambridge University Press, 1985), 144–55.

19. Quoted by David A. Walker, *Iron Frontier: The Discovery and Early Development of Minnesota's Three Ranges* (St. Paul: Minnesota Historical Society Press, 1979), 227.

20. See Vincenti, *What Engineers Know*, chap. 5; and L. Kops and S. Ramalingam, eds., *On the Art of Cutting Metals: 75 Years Later* (New York: American Society of Mechanical Engineers, 1982).

21. Peter J. Ling, *America and the Automobile: Technology, Reform and Social Change* (Manchester: Manchester University Press, 1990), 13.

1. The Dominance of Rails, 1865–1885

1. John Hoyt Williams, *A Great and Shining Road: The Epic Story of the Transcontinental Railroad* (New York: Times Books, 1988), 234, 260–63; Maury Klein, *Union Pacific: Birth of a Railroad, 1862–93* (Garden City: Doubleday, 1987), 218–19.

2. Wesley S. Griswold, *A Work of Giants: Building the First Transcontinental Railroad* (New York: McGraw-Hill, 1962), 307–12 (quotation on 308).

3. Griswold, *A Work of Giants*, 309.

4. James McCague, *Moguls and Iron Men: The Story of the First Trans-*

continental Railroad (New York: Harper and Row, 1964), 304–8 (quotation on 306); Williams, *A Great and Shining Road*, 262–63; Griswold, *A Work of Giants*, 311.

5. Edwin L. Sabin, *Building the Pacific Railway* (Philadelphia: J. B. Lippincott, 1919), 200–204 (quotation on 202).

6. Klein, *Union Pacific: Birth of a Railroad*, 220–28 (quotation on 221).

7. On Fritz, see Lance E. Metz and Donald Sayenga, "The Role of John Fritz in the Development of the Three-High Rail Mill 1855–63," paper presented at the 1990 conference of the Society for Industrial Archeology, held in Quebec. American Iron and Steel Association, *Annual Statistical Report* (Philadelphia, 1879), 17, 32, gives total rail production (net tons of 2,000 lbs.) and total railroad mileage constructed; for later statistics, see figure 1.6.

Year	Rails	Miles
1855	138,674	1,654
1856	180,018	3,642
1857	161,918	2,487
1858	163,712	2,465
1859	195,454	1,821
1860	205,038	1,846
1861	189,818	651
1862	213,912	834
1863	275,768	1,050
1864	335,369	738
1865	356,292	1,177
1866	430,778	1,716

8. Klein, *Union Pacific: Birth of a Railroad*, 137. For the labor-intensive manufacture of wrought iron, see chapter 2 below as well as John N. Ingham, *Making Iron and Steel: Independent Mills in Pittsburgh, 1820–1920* (Columbus: Ohio State University Press, 1991), 34–36; and David Montgomery, *The Fall of the House of Labor: The Workplace, the State, and American Labor Activism, 1865–1925* (Cambridge: Cambridge University Press, 1987), 14–17.

9. Only part of the imports can be traced to a revision in the tariffs on steel rails. In 1870 tariffs stood at a punishing 45 percent ad valorem, around $48 per ton; in 1871 the rate was eased to a flat fee of $28 per ton. Imports vanished after 1874 on sharply falling domestic demand (fig. 1.6).

10. See "'Bessemerizing' and the Nitrate Steel Process," *Engineer* 25 (7 February 1868): 100.

11. *Dictionary of National Biography*, s.v. "Bessemer, Sir Henry"; Geoffrey Tweedale, "Bessemer, Sir Henry," *Dictionary of Business Biography* (London: Butterworths, 1984), 1:309–12.

12. In the near absence of reliable documents, apart from the Dowlais Iron papers, historians must interpret Bessemer's always boastful and

sometimes unreliable *Autobiography*. Bessemer's personal papers were apparently destroyed; see Alan Birch, *The Economic History of the British Iron and Steel Industry, 1784–1879* (New York: Kelley, 1968), 320 n.19; Jeanne McHugh, *Alexander Holley and the Makers of Steel* (Baltimore: Johns Hopkins University Press, 1980), 76 n.1. Notable for comparing Bessemer's account with primary documents is Alan Birch, "Henry Bessemer and the Steel Revolution," *Nachrichten aus der Eisen-Bibliothek* (Schaffhausen) 28 (1963): 129–36, and 30 (1964): 153–59; and Edgar Jones, "The Transition from Wrought Iron to Steel Technology at the Dowlais Iron Company, 1850–90," in *The Challenge of New Technology: Innovation in British Business since 1850*, ed. Jonathan Liebenau (Aldershot: Gower, 1988), 43–57. On Bessemer as a mechanical inventor, see Bessemer, *Autobiography*, 4–137, 329–32; Ernst F. Lange, "Bessemer, Göransson and Mushet: A Contribution to Technical History," *Memoirs of the Manchester Literary and Philosophical Society* 57, no. 17 (1913): 2–5, 33–38. Bessemer gained a total of 117 patents.

13. Bessemer, *Autobiography*, 134–42 (quotations on 130, 136–37); Lange, "Bessemer," 4–5.

14. Blister steel resulted from baking bars of iron that contained essentially no carbon in furnaces filled with charcoal; the process, known as cementation, required as long as a week before the desired amount of carbon in the charcoal diffused into the iron bars, producing blisters on their surface. Blister steel melted in clay crucibles became the "crucible steel" that made Sheffield famous. See Geoffrey Tweedale, *Sheffield Steel and America: A Century of Commercial and Technological Interdependence, 1830–1930* (Cambridge: Cambridge University Press, 1987), chap. 2.

15. Bessemer, *Autobiography*, 142; and Tweedale, "Bessemer," 310.

16. Bessemer, *Autobiography*, 142–44 (quotation on 144); Henry Bessemer, "The Origin of the Bessemer Process," *Iron Age* 58 (3 December 1896): 1065–71; Lange, "Bessemer," 4–5. In modern terms, the oxygen in the air and the silicon and carbon in the metal combined in a violent combustion; such a chemical understanding was yet to emerge.

17. Bessemer, *Autobiography*, 144–54; Henry Bessemer, "On the Manufacture of Cast Steel, Its Progress, and Employment as a Substitute for Wrought Iron," *Engineer* (London) 20 (15 September 1865): 173–74.

18. W. T. Jeans, *The Creators of the Age of Steel* (New York: Charles Scribner's Sons, 1884), 44 (quotation); Bessemer, *Autobiography*, 154–68, fig. 93 (Nasmyth letter). For the first licenses, see H. Bessemer and R. Longsdon, agreement with Dowlais Iron Co., 27 August 1856; Bessemer and Longsdon to G. T. Clark, 28 August 1856, 5 December 1856; Bessemer and Longsdon to Dowlais Iron Co., 3 September 1856; all in D/DG, Section C, Box 2, DI. Madeleine Elsas, ed., *Iron in the Making: Dowlais Iron Company Letters, 1782–1860* (Cardiff: Glamorgan County Council and Guest Keen Iron and Steel, 1960), 193–94, 205–6. Jones, "Wrought Iron to Steel at Dowlais Iron," 51, gives the Dowlais license at £20,000.

19. Lowthian Bell, "Invention of the Bessemer Process," MS 26, August

1903, and Bell to H. M. Howe, 23 March 1899, 3: Patent Applications of Robert Hadfield for the Manufacture of Manganese Steels 1896, HMH.

20. Bessemer, *Autobiography*, 121, 170–74, 182–84; Lange, "Bessemer," 7–12. The precise role of Henry and Percy remains unclear. Lowthian Bell ("Invention") doubted that even Blaenavon pig iron could be made into Bessemer steel.

21. W. M. Lord, "The Development of the Bessemer Process in Lancashire, 1856–1900," *Newcomen Society Transactions* 25 (1945–47): 165–67; W. H. Chaloner, "John Galloway (1804–94), Engineer of Manchester, and his 'Reminiscences,'" in *Industry and Innovation*, ed. D. A. Farnie and W. O. Henderson (London: Frank Cass, 1990), 99–121, especially 116–17; Bessemer, *Autobiography*, 171–77, 213–15, 334; Lange, "Bessemer," 13–17; Bessemer, "On the Manufacture of Cast Steel," 173. Total capital for the Sheffield works was £12,000, of which Bessemer and Longsdon subscribed about £6,000; W. and J. Galloway, £5,000; and William D. Allen, £500. See Tweedale, "Bessemer," 310–11. Bessemer (*Autobiography*, 334) listed the profit (or loss) for the Sheffield works:

Year	Profit (Loss)
1858	(£729)
1859	(1,093)
1860	923
1861	1,475
1862	3,685
1863	10,968
1864	11,827
1865	3,949
1866	18,076
1867	28,622

22. Lord, "Development," 169–74; Bessemer, *Autobiography*, 148–51, 178–81, 256–90; Lange, "Bessemer," 18–20.

23. According to the Queen's Hotel anecdote, Bessemer's share of the partnership was 80 percent and Longsdon's, 20 percent. Bessemer, *Autobiography*, 177, 234–37, 334; Lange, "Bessemer," 28–30. Ironically, Bessemer tried without success to interest the War Department and the Admiralty in his process; Bessemer, *Autobiography*, 184–255.

24. For others' experiments treating molten iron with blasts of air or steam similar to Bessemer's, see Theodore A. Wertime, *The Coming of the Age of Steel* (Leiden: E. J. Brill, 1961), 284–87; J. A. Cantrell, *James Nasmyth and the Bridgewater Foundry: A Study of Entrepreneurship in the Early Engineering Industry* (Manchester: Manchester University Press, 1984), 123–26. Bessemer, *Autobiography*, 168–70, 291–303 (quotation on 303); Lange, "Bessemer," 21–24.

25. Although Pennsylvania Steel may have experimented with basic Bessemer steel in 1884, only one American firm, Pottstown Iron in the

early 1890s, is known to have manufactured significant amounts of basic Bessemer or Thomas steel. See K. C. Barraclough, *Steelmaking: 1850–1900* (London: Institute of Metals, 1990), 234–35, 239–41.

26. On the importance of railroads for the European makers of Bessemer steel, see Ulrich Wengenroth, *Enterprise and Technology: The German and British Steel Industries, 1865–95* (Cambridge: Cambridge University Press, 1994), 31–58. Yet the statistics Wengenroth gives (on pp. 49, 113, 139, 141, 155) reveal the essential point. The year 1882 was *the peak year* between 1871 and 1886 for steel rails in both Germany (379,000 tons consumed, 564,000 tons produced) and Great Britain (502,000 tons consumed, 1,200,000 tons produced). In that year exports accounted for 32.9 percent of German production and 59.4 percent of British production. By comparison, in the United States 1882 was *a* peak year for steel rails, with 1,460,000 tons produced and 1,909,000 tons consumed. (The U.S. long ton is 1.016 European metric tons.)

27. See Harley, "Oligopoly Agreement," 797–823.

28. Alexander L. Holley, *A Treatise on Ordnance and Armor* (New York: D. Van Nostrand, 1865), 104. Holley's early career can be traced in his letters (1851–61) in ALH, and in McHugh, *Alexander Holley and the Makers of Steel*.

29. James Dredge, "Sir Henry Bessemer," *ASME Trans.* 19 (1898): 881–964 (quotation on 939). For Holley's copartnership with Winslow and Griswold, see A. L. Holley to John A. Griswold, 25 April 1866, 2:4, JAG.

30. Leland Stanford and Charlie Crocker were born in Troy, while Theodore D. Judah was raised in Troy, attended the Troy School of Technology, and learned railroad engineering on the Troy and Schenectady road; see Williams, *A Great and Shining Road*, 29, 38, 51. Around 1864 Troy was primarily a center for rerolling rails, along with Worcester, Elmira, Buffalo, Cleveland, Indianapolis, and Chicago; Kenneth Warren, *The American Steel Industry, 1850–1970: A Geographic Interpretation* (Oxford: Clarendon Press, 1973), 26, 89–91.

31. Winslow and Griswold to A. H. Holley, 29 April 1867, and A. H. Holley to Mrs. A. H. Holley, 9 May 1867, AHH.

32. R. W. Raymond, *Memorial of Alexander Lyman Holley, C.E. LL.D.* (New York: AIME, 1884), 71; Robert W. Hunt, "The Original Bessemer Steel Plant at Troy," *ASME Trans.* 6 (1885): 61–70, especially 69; Hunt, "History," 204, 214.

33. Metz and Sayenga, "Role of John Fritz," 15.

34. Testimony of the Kelly brothers' workmen, collected in 1857 for the patent-interference case against Bessemer, confirms that Kelly experimented in 1847 and again in 1851 with blowing air into molten iron and that he claimed to have invented a new process, but the testimony denies that he achieved a workable process. See State of Kentucky, Lyon County, *Interference, William Kelly v. Henry Bessemer*, 13 April 1857, 1:28, *Patent Suit Kelly v. Bessemer 1857*, AISI; and Thomas J. Misa, "Kelly, William," *American National Biography* (Oxford University Press/American Council of Learned Societies, forthcoming). For discussion of the Kelly-Bessemer

controversy, see Thomas J. Misa, "Controversy and Closure in Technological Change: Constructing 'Steel,'" in *Shaping Technology/Building
Society: Studies in Sociotechnology*, ed. Wiebe E. Bijker and John Law
(Cambridge: MIT Press, 1992), 109–39. For an analysis of Kelly's experiments, see Misa, "Science, Technology and Industrial Structure," 4–11,
27–29; and Robert B. Gordon, "The 'Kelly' Converter," *Technology and
Culture* 33 (1992): 769–79.

35. Holley as quoted in "The Invention of the Bessemer Process," *Engineering* 6 (27 March 1896): 414.

36. Continuing legal proceedings beyond the 1866 settlement that set
up the Bessemer Association are suggested in A. L. Holley to John A.
Griswold, 7 July 1869, 22 December 1869, 2:4, JAG.

37. For the third and fourth quarters of 1873, the Pneumatic Steel Association issued dividends totaling $36,500 and $20,000; see Pneumatic
Steel Association to Executors of John A. Griswold, and Z. S. Durfee
to Executors of the late John A. Griswold, 27 January 1873, 5:67, JAG.
In 1888 the cartel tabulated the pounds of raw materials used, the "tons
on which apportionment is based," and total tons treated for the eleven
member companies (Cambria Iron, Cleveland Rolling Mill, Troy Iron and
Steel, North Chicago Rolling Mill, Union Steel, Pennsylvania Steel, St.
Louis Ore and Steel, Carnegie Bros., Bethlehem Iron, Lackawanna Iron
and Coal, Joliet Steel); see Bessemer Steel Company, tables listing tonnage
of metal converted by Bessemer process, October 1886, April/May 1887,
April 1888, 1:5, Correspondence 1886–87, FWW. In 1897 the Bessemer
Steel Association required a deposit of $5,000 and credited its members
at 10¢ per ton shipped; see Bessemer Steel Association to Maryland Steel,
29 January 1897, 17:5, Maryland Steel General Correspondence 1897,
FWW. For the rail producers' cartel active during 1901–15, see chapter 4.

38. James A. Ward, *J. Edgar Thomson: Master of the Pennsylvania*
(Westport, Conn.: Greenwood Press, 1980), 177 (quotation); McHugh,
Alexander Holley and the Makers of Steel, 211–32; Joseph F. Wall, *Andrew
Carnegie* (New York: Oxford University Press, 1970), 309–11, 229. Six hun-
dred thousand dollars was about one-third of the total capitalization; see
Steven W. Usselman, "Running the Machine: The Management of Tech-
nological Innovation on American Railroads, 1860–1910" (diss., University
of Delaware, 1985), 89–97.

39. If the Pennsylvania Railroad officers only knew of the competing
mills' lower cost for rails, wrote Andrew Carnegie, "surely the further
development of Penna Steel Co could be checked. It is absurd for that
company to spend more money to make more rails in that territory. Bethlehem and Scranton can each beat her in cost and [Cambria] can send
rails to seaboard past her at a profit" (Andrew Carnegie to William Shinn,
29 March 1879, vol. 4, ACLC). Still, the link between railroad and steel
mill endured. "As you know . . . the Pennsylvania [Railroad] people as
a rule buy rails from the concerns located on their lines, and which they
control, including the Pennsylvania Steel Co., and the Maryland Steel Co."
(A. Johnston to C. M. Schwab, 20 May 1907, 2:33, AJ).

40. Andrew Carnegie to R. H. Thurston, 26 October 1888, Folder 1888, RHT. On the Pennsylvania's managerial reforms, see Charles F. O'Connell, Jr., "The Corps of Engineers and the Rise of Modern Management, 1827–56," in *Military Enterprise and Technological Change*, ed. Merritt Roe Smith (Cambridge: MIT Press, 1985), 88–116, especially 111–14. On Carnegie's career with the Pennsylvania, see Pennsylvania Railroad General Order 10, 21 November 1859, vol. 1, ACLC; letters to Enoch Lewis in vols. 1–3, ACLC; and the "farewell address," Andrew Carnegie to Pittsburgh Division PRR (circular letter), 28 March 1865, vol. 3, ACLC.

41. Andrew Carnegie, *Autobiography* (Boston: Houghton Mifflin, 1920), 129–30; Wall, *Andrew Carnegie*, 325–30, 342, 356–58; Joseph D. Weeks, "Biographical Notice of William Powell Shinn," *AIME Trans.* 21 (1893): 394–400. "Every manager in the mills was naturally against the new system. Years were required before an accurate system was obtained," admitted Carnegie in his *Autobiography* (129–30).

42. Andrew Carnegie to William Shinn, 30 November 1875, vol. 4, ACLC; Ward, *J. Edgar Thomson*, 189–216.

43. For startup production data, see entries in HCW. For the organization of the Edgar Thomson works, see Carnegie, McCandless and Co., articles of copartnership, 5 November 1872, and Edgar Thomson Steel Co., Limited, articles of limited partnership, 7 November 1874, vol. 4, ACLC. For background on the works, see Wall, *Andrew Carnegie*, 309–11; Arthur Pound, "Two Centuries of Industry: The Rise of Manufactures in Western Pennsylvania," chap. 21, pp. 8–12, MS 1936, MEB; P. Barnes, "Note upon the Cost of Construction of the Converting Works of the Edgar Thomson Steel Company," *AIME Trans.* 6 (1877): 195–96; Barnes, "Memorandum Relating to the Construction Account of the Rail Mill of the Edgar Thomson Steel Company," *AIME Trans.* 7 (1878): 77–78.

44. C. E. Dutton, in Raymond, *Memorial of Alexander Lyman Holley*, 29. Data on the Edgar Thomson works converting vessel is from HCW, November 1875–January 1876.

45. Raymond, *Memorial of Alexander Lyman Holley*, 135 (quotation); A. L. Holley and Lenox Smith, "The Works of the Edgar Thomson Steel Company (Limited)," *Engineering* 25 (19 April 1878): 295–97. For Holley's spatial contortions to reconstruct the Bessemer mill at Troy, see Holley to John A. Griswold, 10 December 1868, 18 and 22 January and 4 March 1869, 2:4, JAG.

46. Holley and Smith, "Works of the Edgar Thomson Steel Company," 295–97.

47. "We find in Harrisburg that boilers over steel furnaces don't make steam enough to pay" (A. L. Holley to John A. Griswold, 22 January 1869, 2:4, JAG).

48. A. L. Holley, "Cooper's Hot Blast Stove for Cupolas, Utilizing Waste Heat from Converters, Adapted to the Edgar Thomson Steel Works," 29 September 1879, Edgar Thomson drawer, ALH-SI.

49. James Gayley, testimony, 20 May 1896, in extracts from patent suits *Carnegie Steel v. Cambria Iron* and *Carnegie Steel v. Maryland Steel*, 1–9,

and 6 June 1895, 30–42 (quotation on 34), 10:1, Maryland Steel Lawsuits 1893–1910, FWW (hereafter cited as Gayley Testimony). The responsibilities of the chemists at the Edgar Thomson works are detailed in the 1875–80 diary, HCW.

50. Gayley Testimony, 6 June 1895, 36.

51. Gayley Testimony, 20 May 1896, 8–9.

52. See certain of Holley's drawings for the Edgar Thomson works that have a second set of dimensions (inked in red) for the Vulcan Iron Works, Folder "Edgar Thomson," ALH-SI; Robert Forsyth, "The Bessemer Plant of the North Chicago Rolling Mill Company at South Chicago," *AIME Trans.* 12 (1883): 254–74; Andrew Carnegie to William Shinn, 4 April 1879, vol. 4, ACLC.

53. See Misa, "Science, Technology and Industrial Structure," 30–31.

54. Alexander L. Holley, "Tests of Steel," *AIME Trans.* 2 (1873): 116–22 (quotation on 117–19); Holley, *Bessemer Machinery* (Philadelphia: Merrihew, 1873), 1 (quotation); Holley, *The Bessemer Process and Works in the United States* (New York: D. Van Nostrand, 1868), 19 (quotation); Holley to John A. Griswold, 29 January 1869, 2:4, JAG. See also Lance E. Metz, "The Arsenal of America: A History of Forging Operations of Bethlehem Steel," *Canal History and Technology Proceedings* 11 (1992): 233–94, especially 284 n.36.

55. On Philadelphia's analytical chemists, see W. Ross Yates, *Joseph Wharton: Quaker Industrial Pioneer* (Bethlehem, Pa.: Lehigh University Press, 1987), 48–49, 94; Thomas J. Misa, "The Changing Market for Chemical Knowledge: Applied Chemistry and Chemical Engineering in the Delaware Valley, 1851–1929," *History and Technology* 2 (1985): 245–68; the advertisement for Britton's lab in Box 3, NWL; and J. Blodget Britton, "The Determination of Combined Carbon in Steel by the Colorimetric Method," *AIME Trans.* 1 (1871–73): 240–42.

56. By 1880 chemical specification of rails was standard in America, according to a leading European authority: "The specification for rails, since the introduction of the use of steel, is becoming almost overdone—in America chemically, with the stipulation of only one certain chemical composition in the rails" (C. P. Sandberg, "Rail Specifications and Rail Inspection in Europe," *AIME Trans.* 9 [1880]: 205). "In America the control and inspection of the quality of steel is overdone in another direction, viz., by chemical analysis, so that steel, even for rails, is now nearly always chemically analyzed and the composition stipulated in contracts" (C. P. Sandberg et al., "Iron and Steel Considered as Structural Materials—A Discussion," *AIME Trans.* 10 [1881]: 406). Dudley's work is analyzed in Usselman, "Running the Machine," 295–300, 317.

57. For contemporaneous comment, see W. Mattieu Williams, *The Chemistry of Iron and Steel Making* (London: Chatto and Windus, 1890), chap. 9.

58. Phoenix Works Diary, 18 and 20 September 1873, 67: Phoenix Works Diary (1870–79), PS.

59. Peter Temin, *Iron and Steel in Nineteenth-Century America* (Cambridge: MIT Press, 1964), 270, 274–75, 278, 284–85.

60. Alexander L. Holley, "Bessemer Machinery," *JFI* 94 (1872): 252–65 (quotation on 252–54, emphasis in original), 391–99; Holley, *Bessemer Machinery*, 2–3; Henry M. Howe, "The Nomenclature of Iron," *AIME Trans.* 5 (1876): 515–37, especially 515. The earliest announcement of the fusion classification I have found was by the inventor of another high-temperature steelmaking process (open hearth); see C. W. Siemens, "The Regenerative Gas Furnace as Applied to the Manufacture of Cast Steel," *Journal of the Chemical Society* (London), n.s. 6 (1868): 279–308, especially 284.

61. Julia Ward Howe wrote this verse of her son, known as a youthful prankster: "God gave my son a palace, / And a kingdom to control; / The palace of his body, / The kingdom of his soul." On Henry Howe's family, see Laura E. Richards and Maud Howe Elliott, *Julia Ward Howe, 1819–1910* (Boston: Houghton Mifflin, 1925), 156 (quotation); and Deborah Pickman Clifford, *Mine Eyes Have Seen the Glory: A Biography of Julia Ward Howe* (Boston: Atlantic Monthly Press/Little, Brown, 1979), 164–65.

62. "For fifty years a struggle waged between scientific metallurgists who wanted to call low-carbon iron 'wrought iron,' and manufacturers who wanted to call it 'steel' if it had low carbon and no slag [that is, had been fused]" (Bradley Stoughton, *The Metallurgy of Iron and Steel*, 4th ed. [New York: McGraw-Hill, 1934], 44). Stoughton was Howe's assistant at Columbia.

63. For Howe, steel was "a compound or alloy of iron whose modulus of resilience can be rendered, by proper mechanical treatment, as great as that of a compound of 99.7 per cent. iron with 0.3 per cent. carbon can be by tempering" (Henry M. Howe, "What Is Steel?" *Engineering and Mining Journal* 20 [11 September 1875]: 258).

64. Alexander L. Holley, "What Is Steel?" *AIME Trans.* 4 (1875): 138–49 (quotation on 138–40).

65. See Howe's table of classification results, reproduced in Misa, "Controversy and Closure," 131; Howe, "Nomenclature of Iron," 516 (quotation); Frederick Prime, Jr., "What Steel Is," *AIME Trans.* 4 (1875): 328–39 (quotation on 332).

66. William Metcalf, "Can the Commercial Nomenclature of Iron Be Reconciled to the Scientific Definitions of the Terms Used to Distinguish the Various Classes?" *AIME Trans.* 5 (1876): 355–65, especially 357; Holley, "What Is Steel?" 147). Metcalf later published several papers on metallurgy including a "classic" in *ASCE Trans.* 16 (1887): 283–389.

67. See Alex W. Bealer, *The Art of Blacksmithing* (New York: Funk and Wagnalls, 1969), 44–45, 146.

68. "Discussion on Steel Rails," *AIME Trans.* 9 (1880): 529–608, especially 551; *Engineering and Mining Journal* 25 (8 June 1878): 396. "Everybody connected with the steel trade knows how irregular are the duty rates, and desires more clearness and simplicity," noted A. Greiner ("Nomenclature of Steel," *Engineering and Mining Journal* 23 [3 March 1877]: 138).

69. John B. Pearse, "The Manufacture of Iron and Steel Rails," *AIME Trans.* 1 (1872): 162–69 (quotation on 163); John Swann, *An Investor's Notes on American Railroads* (New York: Putnam, 1887), 35–37. Holley ("What Is Steel?" 142) observed that the Pennsylvania Railroad "specifies 0.35 carbon steel for its rails, meaning by 'steel,' that it shall be homogeneous or cast." Howe admitted as much: "A Bessemer rail . . . having no welds to yield to the incessant pounding, usually lasts till it is actually worn out by abrasion. Hence, railway managers do not care very much about the degree of carburization of rails said to be steel, provided they are absolutely weldless, and a steel rail has come to mean with them a weldless rail instead of a hard rail. They are, in general, willing to receive all the products of the Bessemer converter as steel, provided they are not too brittle. Were their pleasure alone to be consulted, freedom from welds might be the most convenient ground for the classification of iron" (Howe, "What Is Steel?" 259).

70. Other Carnegie executives had chemistry backgrounds. Charles L. Taylor, an 1876 graduate of Lehigh University, was chemist for Cambria and Homestead. Homer D. Williams, chemist for Cambria Iron, Joliet Steel, and other concerns, was superintendent of Homestead's Bessemer department and president of Carnegie Steel (1915–25). E. Fred Wood, a chemist at Homestead, later became first vice president of International Nickel Co. (1902–19). Data from entries in *DAB* and *NCAB*, and from William B. Dickson, comp., *History of Carnegie Veteran Association* (Montclair, N.J.: Mountain Press, 1938).

71. For this definition, see the lecture notebook of Scott Turner, September 1900, EDMC. The problematic boundary was only between *wrought* iron and steel; the boundary between *cast* iron (more than 2 percent carbon and not forgeable at red heat) and steel was not under dispute. For the continuing importance of analytical chemists, see the voluminous correspondence with iron companies in PWS (Porter W. Shimer was an analytical chemist in Easton, Pa.) and NWL (Nathaniel Wright Lord was a chemistry professor at Ohio State University). For chemical metallurgy in the manufacture of boiler plates, see correspondence in 259 and 260, and laboratory reports in 310, LIS.

72. This is clear in a sample of twenty-seven English-language metallurgical textbooks drawn from the Eisen-Bibliothek, Schaffhausen, Switzerland, and from the University of Chicago's Crerar Library. Before 1880 the carbon classification dominated (100%); during 1880–99 the sample was split evenly between carbon and fusion classifications (several writers gave both); by 1900–1909 the fusion classification (60%) triumphed over the carbon classification (27%) and the emerging microstructural classification of metallography (13%).

73. Henry M. Howe, *The Metallurgy of Steel*, 2nd rev. ed. (New York: Scientific Publishing, 1891), 1.

74. Creed Haymond, *The Central Pacific Railroad Co.: Its Relations to the Government* (San Francisco: H. S. Crocker, 1888), 43 (quotation),

155; Richard C. Overton, *Burlington West: A Colonization History of the Burlington Railroad* (Cambridge: Harvard University Press, 1941), 283, 287, 291, 303 (quotation), 347, 366 (quotation), 381. Later the Burlington issued circulars in French, but aimed them primarily at families of Alsace-Lorraine (in Iowa and Nebraska, the railroad promised, they could "escape from the crushing heels of the German Empire's militarism"). For the activities of railroad land offices, see also William S. Greever, *Arid Domain: The Santa Fe Railway and Its Western Land Grant* (Stanford, Calif.: Stanford University Press, 1954), 43–52, 103–12; L. L. Waters, *Steel Trails to Santa Fe* (Lawrence: University of Kansas Press, 1950), 218–60; Duane P. Swanson, "The Northern Pacific Railroad and the Sisseton-Wahpeton Sioux: A Case Study in Land Acquisition" (M.A. thesis, University of Delaware, 1972).

75. A. L. Holley to John A. Griswold, 24 May 1866, 2:4, JAG. Charles B. Dew (*Ironmaker to the Confederacy: Joseph R. Anderson and the Tredegar Iron Works* [New Haven: Yale University Press, 1966], 318–19) notes that "the depression of 1873, not the Civil War, stunted the growth of the South's largest industrial plant." W. David Lewis, "Joseph Bryan and the Virginia Connection in the Industrial Development of Northern Alabama," *Virginia Magazine of History and Biography* 98 (1990): 613–40; Mark W. Summers, *Railroads, Reconstruction, and the Gospel of Prosperity: Aid under the Radical Republicans, 1865–77* (Princeton: Princeton University Press, 1984), 213–36; Maury Klein, *History of the Louisville and Nashville Railroad* (New York: Macmillan, 1972), 264–77.

76. Summers, *Railroads, Reconstruction, and the Gospel of Prosperity*, 97 (quotation), 123, 213–36, 278–79; John F. Stover, *The Railroads of the South, 1865–1900: A Study in Finance and Control* (Chapel Hill: University of North Carolina Press, 1955), 11–12, 99–121, 155–85, 282.

77. E. G. Campbell, *The Reorganization of the American Railroad System, 1893–1900* (New York: Columbia University Press, 1938), 10–13.

78. See William Cronon, *Nature's Metropolis: Chicago and the Great West* (New York: W. W. Norton, 1991), 55–259, especially 92.

2. The Structure of Cities, 1880–1900

1. Beginning in 1923, the Auditorium Building's owners fought a legal battle with the titleholders to the land beneath it over whether the building could legally be destroyed. Ultimately, what saved the Auditorium was the U.S. Supreme Court decision in 1932 affirming that the original ninety-nine-year lease forbade the building's owners to remove the structure from the land. For this and the Auditorium's later history, see James W. Wright, "The Chicago Auditorium Theater, 1886–1966" (M.A. thesis, Michigan State University, 1966), 108–16; Wilbur Thurman Denson, "A History of the Chicago Auditorium" (diss., University of Wisconsin, 1974), 191–265; Carol Baldridge and Alan Willis, "The Business of Culture," *Chicago History* (spring/summer 1990): 33–51.

2. Transcript of Auditorium Trial testimony, 2 February 1925, Adler &

Sullivan Papers, Roll 6 (frame 16), AICB (hereafter cited as Auditorium Trial); Theodore Turak, *William Le Baron Jenney: A Pioneer of Modern Architecture* (Ann Arbor: UMI Research Press, 1986), 280–83.

3. See Claude Bragdon's unpaged foreword to the original edition of Louis H. Sullivan, *The Autobiography of an Idea* (New York: American Institute of Architects, 1924).

4. See Charles E. Gregersen, *Dankmar Adler: His Theatres and Auditoriums* (Athens, Ohio: Swallow Press, 1990); Carl W. Condit, *The Chicago School of Architecture: A History of Commercial and Public Building in the Chicago Area, 1875–1925* (Chicago: University of Chicago Press, 1964), 69–78; Robert Twombly, *Louis Sullivan: His Life and Work* (New York: Viking, 1986), 178–92.

5. In 1889 structural shapes comprised 154,000 tons of steel and 123,000 tons of iron; see Peter Temin, "The Composition of Iron and Steel Products, 1869–1909," *Journal of Economic History* 23 (1963): 447–76, especially 451.

6. Auditorium Trial, frame 17.

7. Ibid., frame 72; Dankmar Adler, autobiography MS (1893–94), Folder 5, DA; Denson, "History of the Chicago Auditorium," 61.

8. Auditorium Trial, frame 73; Denson, "History of the Chicago Auditorium," 62–63, 99; Twombly, *Louis Sullivan*, 173, 183–84. My account is manifestly at odds with the statement of Condit (*Chicago School of Architecture*, 76) that "there were no consulting engineers on the Auditorium. Except for minor details, Adler did the whole job."

9. Strobel's career is discussed more fully later in this chapter. Marburg joined Keystone in 1885 as a bridge designer immediately upon his graduation from Rensselaer Polytechnic Institute. After leaving Keystone, he designed steel structures at Phoenix Bridge during 1887–89, at Edgemoor Iron from 1889, and in Chicago for Carnegie Steel during 1891–92. He became professor of civil engineering at the University of Pennsylvania in 1893. *NCAB*, s.v. "Marburg, Edgar," 42:142.

10. Auditorium Trial, frames 73–75; William B. Mundie, "Skeleton Construction: Its Origin and Development," MS 1932, William Mundie Papers, Roll 23, AICB.

11. See Carnegie Brothers & Co., *Pocket Companion* (Pittsburgh: Carnegie Brothers, 1881). The Carnegie company began issuing structural handbooks in 1872, with a new edition (and new title) every few years. They are cited here simply as "[company name] handbook" with the year in parentheses. Auditorium Trial, frames 73–75; Joseph F. Wall, *Andrew Carnegie* (New York: Oxford University Press, 1970), 488.

12. Auditorium Trial, frames 76–87; Sullivan, *Autobiography of an Idea*, 311–12 (quotation). In "The Tall Office Building Artistically Considered," first published in 1896 and collected in *Kindergarten Chats* ([New York: George Wittenborn, 1947], 208), Sullivan illustrated this line of reasoning—and gave the world his most memorable phrase: "It is the pervading law of all things organic, and inorganic, of all things physical and metaphysical, of all things human and all things superhuman, of all true mani-

festations of the head, of the heart, of the soul, that the life is recognizable in its expression, that form ever follows function."

13. Sullivan, *Autobiography of an Idea*, 312 (quotation); Auditorium Trial, frame 88.

14. Frank A. Randall, *History of the Development of Building Construction in Chicago* (Urbana: University of Illinois Press, 1949), 127; Joseph Kendall Freitag, *Architectural Engineering with Special Reference to High Building Construction* (New York: Wiley, 1895), 113. The question of cast iron versus wrought iron or steel was unsettled in 1890, according to the twelve-page discussion of the issue in Phoenix Iron Co., "Phoenix Column Construction as Adapted for Buildings," 253 Engineering Dept.—Phoenix Column Construction 1890, PS. The question was still not settled in 1895, as Freitag observed (*Architectural Engineering*, 113): "A discussion as to the relative values of cast *versus* wrought [that is, rolled iron or steel] should hardly seem necessary at the present time, but the repeated use of the cast-iron column in ten- to sixteen-storied buildings . . . shows that the questionable economy of cast columns does still, in the opinion of some architects, compensate for the dangers incident to their use."

15. Engineering authorities gave steel an advantage in strength over wrought iron ranging from a very small amount up to 35 percent, depending upon application (for example, tension versus compression) and dimension (for example, short versus long members). William H. Birkmire (*Skeleton Construction in Buildings* [New York: Wiley, 1894], 35) noted for steel columns: "Experiments thus far upon steel struts indicate that for lengths up to 90 radii of gyration their ultimate strength is about 20 per cent. higher than for wrought-iron. Beyond this point the excess of strength diminishes until it becomes zero at 200 radii. After passing this limit the compression resistance of steel and wrought-iron seems to become practically equal."

16. Wendall Bollman, a former B&O Railroad engineer and then independent bridge builder, apparently originated the specially rolled, segmented, wrought-iron shapes that were riveted together to form columns, and showed the design to S. J. Reeves. Patents were later taken out in Reeves's name. See David Plowden, *Bridges: The Spans of North America* (New York: Viking, 1974), 65. The Phoenix column was the subject of a series of patent suits and countersuits during 1863–69; see Jacob Hays Linville File—Litigation with Phoenix Iron Co. 1863–69, PS.

17. Thomas C. Clarke, a partner of the firm, estimated profit on the Girard bridge to be $270,000, a tidy sum considering that the bonded liability for the project was $300,000; in securing the contract, some $70,000 in "commissions" was obligated with Clarke requesting that "authority be given to me to expend this $70,000 as my own vouchers without giving details" (Thomas C. Clarke to S. J. Reeves, 7 December 1872; John Griffin to S. J. Reeves, 9 December 1872, S. J. Reeves correspondence 1872–78, PI). Paul F. Paskoff, s.v. "Reeves, Samuel J.," in *Iron and Steel in the Nineteenth Century*, ed. Paul F. Paskoff (New York: Facts on File, 1989), 288–92.

18. Phoenix Bridge Co., Index of Orders, p. 127, 701 Phoenix Bridge—Index of Orders (1877–1904), PS; Joseph Harriss, *The Tallest Tower: Eiffel and the Belle Epoque* (Boston: Houghton Mifflin, 1975), 10.

19. Paskoff, "Reeves, Samuel J.," 291. The company's responses to the depression are detailed in General Superintendent's reports (1872–1922), PI.

20. For a detailed account as well as wonderful illustrations, see William Fullerton Reeves, *The First Elevated Railroads in Manhattan and the Bronx of the City of New York* (New York: New York Historical Society, 1936), 6–9, 55–129. The Phoenix columns on the Greenwich line are described in Chamber of Commerce of the State of New York, *Rapid Transit in New York City* (New York, 1905), 46.

21. John Griffin to S. J. Reeves, 31 October 1877, and John Griffin to David Reeves, 31 October 1878, General Superintendent's reports (1872–1922), PI; Phoenix Bridge Co., Index of Orders, 75–77, 701 Phoenix Bridge—Index of Orders (1877–1904), PS; Reeves, *First Elevated Railroads in Manhattan*, 30–32.

22. John Griffin to S. J. Reeves, 14 February and 4 March 1878, S. J. Reeves correspondence 1872–78, PI; John Griffin to David Reeves, 1 November 1880, General Superintendent's reports (1872–1922), PI.

23. Isaac Reeves, 20 June 1875, 68 Notebook of Isaac Reeves (1873–75), PS.

24. The iron puddlers' intolerance of heat is suggested in Phoenix Works Diary, 9 June 1874, 67 Phoenix Works Diary (1870–79), PS, which records temperatures of 93°F at 1 P.M. and 90° at 6 P.M. The same document notes that "puddlers quit work this evening on account of the heat." On "aristocrats," see J. A. Clauss, "The Romance of Iron and Steel," 3, MS 1934, 1: Background—Romance of Iron and Steel, JAC. The 1873 strike can be followed in Phoenix Works Diary, 1 May 1873, 67 Phoenix Works Diary (1870–79), PS; John Griffin to S. J. Reeves, 1 November 1873, General Superintendent's reports (1872–1922), PI.

25. John Griffin to David Reeves, 1 November 1882, General Superintendent's reports (1872–1922), PI; Phoenix Iron Co., "Notice to the Employees; by order of David Reeves," 24 February 1882, broadsides, Hagley Museum and Library, Greenville, Del.

26. The Safe Harbor Iron Works (of Lancaster County, Pa.) was founded and controlled by a slightly different partnership than that which controlled Phoenix Iron (of neighboring Chester County). Safe Harbor was owned and operated by the partnership of Reeves, Abbott & Co., organized in 1848, which included David and Samuel Reeves as well as Joseph Pancoast and Charles and George Abbott; the partners incorporated in 1855 as the Safe Harbor Iron Works. Safe Harbor's principal mill, installed by John Fritz before his Cambria days, produced the new T-rail used by the Pennsylvania Railroad; it also produced Griffin guns during the Civil War. In 1865 a flood destroyed the canal that had linked the mill to markets and ore supplies. The mill stood idle until 1877, when the puddle furnaces were restarted, and in 1880 the Columbia & Port Deposit Railroad laid a spur to

the puddle rail mill. The Reeves family operated Safe Harbor until 1894. Frank Whelan, "Phoenix Iron Works" and "Reeves, Abbott & Company," in *Iron and Steel in the Nineteenth Century*, ed. Paskoff, 277–78, 293–94.

27. John Griffin to S. J. Reeves, 1 November 1872, and John Griffin to David Reeves, 1 November 1882, General Superintendent's reports (1872–1922), PI; Phoenix Iron Co., 217 Accounts—Ledgers (1881–1902), 120–21, PS.

28. See the reports of William H. Reeves to David Reeves, 27 October 1883, 1 November 1884, 31 October 1885, and 30 October 1886, General Superintendent's reports (1872–1922), PI.

29. Phoenix Iron Co., 217 Accounts—Ledgers (1881–1902), 119, 124, PS. Costs of the steel plant are detailed in Fred Heron, memo, 5 April 1889, vol. 267, PS. The company's original $1.15 million bond issue was at 7 percent interest, while dividends on capital stock were typically 6 percent, and was paid to the company's stockholders, principally the Reeves family; the consolidated mortgage was also at 6 percent. Finances can be followed in Phoenix Iron, Board of Directors minutes, 52 Administrative—Minute Book—Board of Directors (1856–96), PS; see entries for 24 December 1883, 26 December 1884, 28 February 1885, 30 August 1890.

30. N. W. Shed, "Notes on the Manufacture of Open Hearth Bridge Steel," *AIME Trans.* 18 (1889): 88–90.

31. A general "minimum" specification for structural steel was given in 1895 by Freitag (*Architectural Engineering*, 204): 0.10 percent of phosphorus; ultimate tensile strength of 60,000–68,000 pounds per square inch; elastic limit of no less than one-half the ultimate strength; elongation of no less than 20 percent in 8 inches; reduction in area not less than 40 percent at point of fracture. His "desired" specification, however, set ultimate strength between 65,000 and 68,000 pounds per square inch (see table).

Shed's steel fared slightly better against an 1893 general specification for bridge steel compiled by the civil engineer F. H. Lewis: for all steels, phosphorus should be less than 0.10 percent, and less than 0.08 percent for all Bessemer or basic steel; for "soft steel" (such as rivets), ultimate strength of 54,000–62,000 pounds per square inch, elastic limit of at least 32,000 pounds per square inch, elongation of at least 25 percent in 8 inches, and reduction in area of at least 45 percent; for "medium steel" (such as eyebars and main pins), ultimate strength of 60,000–70,000 pounds per square inch, elastic limit of at least 35,000 pounds per square inch, elongation of at least 20 percent in 8 inches, and reduction in area of at least 40 percent. See J. B. Johnson, C. W. Bryan, and F. E. Turneaure, *The Theory and Practice of Modern Framed Structures* (New York: Wiley, 1894), 497–98.

In 1890 a general specification for bridge steel did not prescribe tensile strengths, leaving these to the contract, and simply gave limits within which the steel should fall: test bars must have tensile strength within 4,000 pounds of that specified in the contract, the elastic limit must be not less than one-half the tensile strength, the percentage of elongation must be not less than 1,200,000 ÷ tensile strength (in pounds per square inch), and percentage of reduction must be not less than 2,400,000 ÷ tensile strength (in pounds per square inch). See Carnegie handbook (1890), 213.

Phoenix Steel: Physical Tests of Open Hearth Steel (italic entries pass Freitag's "desired" specification)

Carbon (Percent)	Tensile Strength (lbs. per sq. in.)	Reduction of Area (Percent)	Elongation (Percent)
0.22	69,260	45.4	22.50
0.16	*67,500*	*46.3*	*24.25*
0.19	*66,700*	*49.3*	*23.75*
0.21	*66,700*	*53.8*	*24.50*
0.18	64,100	62.6	26.75
0.18	61,200	61.0	31.25
0.13	60,800	58.7	24.25
0.17	60,320	54.1	30.00
0.15	60,000	55.1	26.25
0.15	57,100	65.3	25.00

Note: It is impossible to predict from carbon content which samples have the "desired" tensile strength of 65,000–68,000 lbs. per sq. in.

32. Fred Heron to F. A. Tencate, 5 August 1889, and Fred Heron to Charles Horn, 7 August 1889, vol. 267, PS.

33. See the June reports of William H. Reeves to David Reeves, 1890–97, in General Superintendent's reports (1872–1922), PI.

34. See Phoenix Iron Co., "Phoenix Column Construction as Adapted for Buildings," 253 Engineering Dept.—Phoenix Column Construction 1890, PS.

35. Cooper-Hewitt's New Jersey Steel and Iron (in 1889) and Phoenix (in 1890) were able to roll a 20-inch iron I-beam and a 15-inch steel I-beam. Conversely, Carnegie Steel (in 1890) was able to roll a 20-inch steel I-beam and a 15-inch iron I-beam.

36. Analyses of the Home Insurance Building are in Condit, *Chicago School of Architecture*, 83–87 (quotation on 83); Randall, *Building Construction in Chicago*, 105–7; Gerald R. Larson and Roula Mouroudellis Geraniotis, "Toward a Better Understanding of the Evolution of the Iron Skeleton Frame in Chicago," *Journal of the Society of Architectural Historians* 46, no. 1 (1987): 39–48; Larson, "The Iron Skeleton Frame: Interactions between Europe and the United States," in *Chicago Architecture, 1872–1922: Birth of a Metropolis*, ed. John Zukowsky (Chicago: Prestel/Art Institute of Chicago, 1987), 39–56; Thomas E. Tallmadge et al., "Was the Home Insurance Building in Chicago the First Skyscraper of Skeleton Construction?" *Architectural Record* 76 (August 1934): 113–18, and ad section, 32; Turak, *William Le Baron Jenney*, 237–63; Theodore Turak, "Remembrances of the Home Insurance Building," *Journal of the Society of Architectural Historians* 44, no. 1 (1985): 60–65.

37. Jenney's Home Insurance Building was not solely supported by its metal structure; 18 percent of the column weight was carried by the stone

piers that surrounded (and fireproofed) the iron columns. This made it possible to add two more stories in 1890; in addition, a brick party wall (not part of the original design) was mandated by city building inspectors.

38. W. L. B. Jenney, "Iron and Steel," 1–2, MS 1896, William L. B. Jenney and William B. Mundie Papers, AICB (hereafter cited as J&M Papers).

39. Turpin C. Bannister, "Bogardus Revisited—Part I: The Iron Fronts; Part II: The Iron Towers," *Journal of the Society of Architectural Historians* 15 (1956): 12–22; 16 (1957): 11–19. Henry Ericsson, *Sixty Years a Builder* (Chicago: Kroch, 1942), 216–19.

40. W. L. B. Jenney to *Engineering Record*, 29 May 1895, J&M Papers; W. L. B. Jenney, "Iron and Steel," 2–3, MS 1896, J&M Papers. Jenney gave many accounts of the Home Insurance Building, especially after 1892 when Leroy Buffington threatened to sue Jenney (and virtually all other architects, constructors, and owners of skeleton buildings) for infringing his 1888 patent; see L. S. Buffington to W. L. B. Jenney, 14 May 1892, J&M Papers. Buffington's priority claims are refuted in Turak, *William Le Baron Jenney*, 262–63, 355 n.60; and Condit, *Chicago School of Architecture*, 82 n.6.

41. W. L. B. Jenney to *Engineering Record*, 29 May 1895, J&M Papers. Jenney's design process can be followed in W. L. B. Jenney, "Home Insurance Instructions," notebook, 3 April–27 May 1884, Roll 9 (frames 413–58), J&M Papers.

42. G. W. Cope, *Iron and Steel Interests of Chicago* (Chicago, 1890), 26–27; Andrew Carnegie, *Triumphant Democracy* (1886; reprint, New York: Doubleday, 1933), 42–43 (quotation).

43. W. L. B. Jenney to *Times Herald* (New York), 27 August 1895, J&M Papers.

44. Condit, *Chicago School of Architecture*, 71.

45. See, for example, "Elevation of Truss 'I'," Chicago Stock Exchange, Adler & Sullivan Papers, Roll 6 (frame 87), AICB.

46. W. L. B. Jenney, "Iron and Steel," 2–3, MS 1896, J&M Papers; Donald Hoffmann, "John Root's Monadnock Building," *Journal of the Society of Architectural Historians* 26, no. 4 (1967): 269–77, especially 272.

47. Dankmar Adler, "Foundations of the Auditorium Building, Chicago," *Inland Architect* (March 1888): 31–32; William J. Fryer, "A Review of the Development of Structural Iron," in *A History of Real Estate, Building and Architecture in New York City* (New York: Real Estate Record Association, 1898), 455–87, especially 479; *DAB*, s.v. "Smith, William Sooy."

48. Ericsson, *Sixty Years a Builder*, 217; Turak, *William Le Baron Jenney*, 72–74.

49. Sullivan, *Autobiography of an Idea*, 316; William B. Mundie, addendum to W. L. B. Jenney, "Iron and Steel," MS 1896, J&M Papers; Auditorium Trial, frames 16 and 91; A. F. Reichmann et al., "Memoir of Charles Louis Strobel," *ASCE Trans.* 102 (1937): 1491–94; *DAB*, s.v. "Strobel, Charles Louis"; D. Adler to his brother, 15 December 1897, DA (Saltzstein collection); *DAB*, s.v. "Corthell, Elmer Lawrence."

50. W.L.B. Jenney, "Iron and Steel," MS 1896, J&M Papers.

51. Auditorium Trial, frame 72.

52. Sullivan, *Autobiography of an Idea*, 252–53, 298.

53. W.L.B. Jenney to H. D. Bennett, 5 August and 7 August 1879, and W.L.B. Jenney to Jas. Shearer, 5 August 1879, William L. B. Jenney Papers, Roll 11-A, AICB.

54. On Adler's investigation of European theaters, see D. Adler to Dila Kohn Adler (his wife) [18 August 1888] (on *S.S. Servia*), 20 August (London), 24 August (Glasgow), 26 August (Edinburgh), 29 August (London), 1 September (Cologne), 23 September (Vienna), and 24 September (Paris), Folders 1, 2, and 4, DA. "Capt. Jones and I had had much to say about riding third class," Adler reported on 20 August. "We had fully made up our minds to be democratic—but we had spent a good bit [of] time looking through and examining cars, locomotives, track, etc. [at the Liverpool station] . . . The result of our examination of cars and the way they were used was however that each of us, unknown to the other, had given up his resolve to be democratic, for each bought a first class ticket."

55. Mundie, "Skeleton Construction," 87.

56. W.L.B. Jenney, "The Wind Pressure in Tall Buildings of Skeleton Construction," MS 1895, J&M Papers; W.L.B. Jenney to *Times Herald* (New York), 27 August 1895, J&M Papers.

57. John P. Comer, *New York City Building Control, 1800–1941* (New York: Columbia University Press, 1942); William J. Fryer, "The New York Building Law," in *A History of Real Estate, Building and Architecture in New York City*, 287–98.

58. In 1893–94, formulas for column strengths were not yet standardized, and "every authority has his own treatment of them" (Corydon T. Purdy, "Iron and Steel Tall Building Construction," in *The Theory and Practice of Modern Framed Structures*, 450–51). According to W.L.B. Jenney (lecture to Art Institute of Chicago, 18 May 1896, J&M Papers), Gordon's formula was "the one most commonly used" to compute the strength of columns:

$$W = \frac{f}{1 + \dfrac{L^2}{r^2 a}}$$

$$\text{where } r^2 = \frac{\text{moment of inertia}}{\text{area of cross-section}}$$

where W = breaking load in pounds per square inch of cross-section of columns; f = ultimate crushing strength of short blocks of the given material; a = factor deduced by experiment and depending on the nature of the *ends* of the column (for square ends, 40,000–42,000); L = length of the columns between supports (in inches); r = least radius of gyration of the cross-section of the pillar where r^2 = moment of inertia / area of cross section.

59. Freitag, *Architectural Engineering*, 218–20.

60. Fryer, "Review of the Development of Structural Iron," 466; William H. Birkmire, *The Planning and Construction of High Office Buildings*, 2nd ed. (New York: Wiley, 1900), 98–111 (quotation on 111).

61. Fryer, "Review of the Development of Structural Iron," 464, 467–73 (quotation on 470); Fryer, "The New York Building Law," 289; Turak, *William Le Baron Jenney*, 301–2.

62. Turak, *William Le Baron Jenney*, 237, 286–95; Joseph Siry, *Carson Pirie Scott: Louis Sullivan and the Chicago Department Store* (Chicago: University of Chicago Press, 1988), 28–33, 46–51. According to Condit (*Chicago School of Architecture*, 90–91), Jenney's Fair Store had an area in plan of 55,000 square feet, which dwarfed the 2,268 square feet of the Tower Building or the 1,776 square feet of the Lancashire Insurance Building. Siry reproduces an 1897 advertisement indicating that the nine-story Fair Store had a total area of 677,500 square feet, or 15½ acres, and a volume of just over 10 million cubic feet—larger than the four-story Bon Marché of Paris.

63. Fryer, "Review of the Development of Structural Iron," 464–66.

64. Louis J. Horowitz, *The Towers of New York: Memoirs of a Master Builder* (New York: Simon and Schuster, 1937), 83; Ericsson, *Sixty Years a Builder*, 352.

65. Burnham designed and George A. Fuller raised the Flatiron Building (1901) in Manhattan. Thompson-Starrett of New York and Daniel Burnham of Chicago teamed up to do Union Station in Washington, D.C. (1904). In the 1910s Thompson-Starrett raised the Field Museum, the Conway Building for the Marshall Field Estate, and the Continental and Commercial Bank Building in Chicago; all three were designed by D. H. Burnham & Co. with Ernest Graham as the principal. See Horowitz, *Towers of New York*, 123–32.

66. Ibid., 156.

67. Phoenix Iron Co., 189 Accounts—Ledgers—Philadelphia Office (1884–89), 152, and Phoenix Iron Co., 190 Accounts—Ledgers—Philadelphia Office (1889–95), 182, PS; Millard Hunsiker to Andrew Carnegie, 30 August 1898, vol. 54, ACLC.

68. Andrew Carnegie to W. L. Abbott, 25 July 1888, vol. 10, ACLC.

69. Horowitz, *Towers of New York*, 112, 116 (quotation), 226–32.

70. Andrew Carnegie to W. L. Abbott, 28 October and 29 December 1888, W. L. Abbott to Andrew Carnegie, 16 January 1889, and Andrew Carnegie to W. L. Abbott, June 1889, vol. 10, ACLC.

71. Assuming uniformly distributed loads, taking distances between supports of 20 feet, and correcting for the weight of the beam itself, 15-inch beams weighing 41 pounds per foot could safely support 14.67 tons of net load; 12-inch beams of 41 pounds, 12.26 tons; standard 12-inch beams of 32 pounds, 9.56 tons; the same 12-inch beam section rolled to

36 pounds, 10.16 tons. These examples are from the Carnegie handbook (1890), 57–61. Generally, a factor of safety of eight was used in computing safe loads.

72. Charles S. Clark to Mason D. Pratt, 9 December 1897, 1:10 Correspondence 1897–98, FWW.

73. Compare the first four paragraphs of the Carnegie handbook (1881) and that of Passaic (1886). The 1874 Cooper-Hewitt handbook is the first I have found that not only illustrated its line of structural sections but also provided explanatory notes. The Hagley Museum and Library has an extensive collection of steel company handbooks, filed under the LC call number TA 685.P739.

74. A. F. Reichmann et al., "Memoir of Charles Louis Strobel," *ASCE Trans.* 102 (1937): 1491–94; *DAB*, s.v. "Strobel, Charles Louis."

75. Reichmann et al., "Memoir of Charles Louis Strobel," 1492.

76. See the following volumes, all published by Wiley: William H. Birkmire, *Architectural Iron and Steel* (1891); Birkmire, *Skeleton Construction in Buildings* (1893); J. B. Johnson, C. W. Bryan, and F. E. Turneaure, *The Theory and Practice of Modern Framed Structures* (1894); Joseph Kendall Freitag, *Architectural Engineering* (1895); William H. Birkmire, *The Planning and Construction of American Theatres* (1896); Birkmire, *The Planning and Construction of High Office Buildings* (1898).

77. W.L.B. Jenney to *Engineering Record*, MS 1896, J&M Papers (frame 323).

78. The seventeen buildings can be linked to the Carnegie firm through contractual and/or textual sources.

79. *Engineering News* 46 (25 July 1901): 54 (quotation); Misa, "Science, Technology and Industrial Structure," 218–22; David McCullough, *The Great Bridge: The Epic Story of the Building of the Brooklyn Bridge* (New York: Simon and Schuster, 1972), 372, 382–96, 409–10, 434–51.

80. Charles L. Strobel, "Discussion on Kentucky and Indiana Bridge," *ASCE Trans.* 17 (1887): 177 (quotation); Alfred E. Hunt, "Some Recent Improvements in Open-Hearth Steel Practice," *AIME Trans.* 16 (1887): 693–728 (quotation on 693); George H. Clapp and Alfred E. Hunt, "The Inspection of Materials of Construction in the United States," *AIME Trans.* 19 (1890): 911–32 (quotation on 929).

81. W.L.B. Jenney, "Quality of Steel in Our Important Buildings," MS c. 1892–95, J&M Papers; Clapp and Hunt, "The Inspection of Materials of Construction," 929 (quotation), 930–32.

82. Hunt, "Some Recent Improvements," 694–95 (quotation on 694).

83. Misa, "Science, Technology and Industrial Structure," 222–23; Alexander L. Holley, "Some Pressing Needs of Our Iron and Steel Manufactures" (presidential address to AIME, October 1875), *AIME Trans.* 4 (1875): 77–99; reprinted in *Memorial of Alexander Lyman Holley, C.E. LL.D.*, ed. R. W. Raymond (New York: AIME, 1884), 201–2 (quotation).

84. C. W. Siemens, "The Regenerative Gas Furnace as Applied to the Manufacture of Cast Steel," *Journal of the Chemical Society* (London),

n.s. 6 (1886): 279–308 (quotation on 288); Misa, "Science, Technology and Industrial Structure," 209–11.

85. Hunt, "Some Recent Improvements," 699; H. H. Campbell, "The Open-Hearth Process," *AIME Trans.* 22 (1893): 365, 387–95. The Martins' British patent described a "white heat" (K. C. Barraclough, *Steelmaking before Bessemer*, 2 vols. [London: Metals Society, 1984], 2:108). For the mapping of color-heats to temperatures, see chapter 5.

86. See Misa, "Science, Technology and Industrial Structure," 213–17; and Barraclough, *Steelmaking: 1850–1900*, 194–202.

87. Hunt, "Some Recent Improvements," 693.

88. For these technical developments, see Misa, "Science, Technology and Industrial Structure," 225–28.

89. Ibid., 228–35.

90. For a detailed account of the Thomas-Gilchrist process, see Barraclough, *Steelmaking: 1850–1900*, 203–57. For references, see Misa, "Science, Technology and Industrial Structure," 228 n.2.

91. See Thomas J. Misa, "Reese, Jacob," *American National Biography* (New York: Oxford University Press/American Council of Learned Societies, forthcoming); and Wall, *Andrew Carnegie*, 503–4.

92. Hunt, "Some Recent Improvements," 707–8, 718–24 (quotation on 719).

93. Samuel T. Wellman, "The Early History of Open-Hearth Steel Manufacture in the United States," *Engineering News* 46 (12 December 1901): 448–50, especially 450; Wellman, "The Early History of Open-Hearth Steel Manufacture in the United States," *ASME Trans.* 23 (1902): 78–98, especially 96–97; A. L. Holley to S. M. Felton, 20 July 1879, 6:3, AJ. For Wellman's early work on open-hearth furnaces, see S. T. Wellman to F. W. Ballard, 12 June 1917, STW; Frank Scott, "The Beginnings and Development of Cleveland's Metal Working Industries," *Cleveland Engineering* 19, no. 37 (10 February 1927): 3–11, 16, 18, 20 (copy in 4: Speech 1922, FAS); David B. Sicilia, "Wellman, Samuel Thomas," in *Iron and Steel in the Nineteenth Century*, ed. Paskoff, 359–63. For the original development of basic open-hearth steelmaking in Europe, see Barraclough, *Steelmaking: 1850–1900*, 247–57.

94. H. H. Campbell to F. W. Wood, 22 February 1889, 1:7 Correspondence 1889, FWW. Campbell was named superintendent in 1893 and became general manager after 1901; *Iron and Steel Magazine* 8 (1904): 82–83.

95. H. H. Campbell to F. W. Wood, 22 February 1889 (quotation, emphasis in original), and E. C. Felton to F. W. Wood, 22 February 1889, 1:7 Correspondence 1889, FWW. Campbell could have seen the articles by Charles Walrand, "Om fosfors bortskaffande i magnesia flam-ugn," *Jern-Kontorets Annaler* (Stockholm) (1887): 23–40; and Kurt Sorge, "Om magnesit och dess användning som basiskt, eldfast material," *Jern-Kontorets Annaler* (1888): 118–34.

96. Misa, "Science, Technology and Industrial Structure," 232–33; E. C.

Felton to F. W. Wood, 1 and 5 March 1889, 1:7 Correspondence 1889, FWW.

97. Misa, "Science, Technology and Industrial Structure," 233–35; H. H. Campbell, "The Homogeneity of Open-Hearth Steel," *AIME Trans.* 14 (1885): 358–62; Campbell, "The Physical and Chemical Equations of the Open-Hearth Process," *AIME Trans.* 19 (1890): 128–87 (quotation on 128); Campbell, "The Open-Hearth Process," *AIME Trans.* 22 (1893): 345–511, 679–96.

98. Campbell elaborated: "We may suppose that the violent molecular changes going on in the open-hearth bath during the combustion of carbon effect a continual rearrangement of the elements, and that the particular proportion of the components at any *locus* of the bath will determine the momentary chemical reactions at that point. Everything will be in a transition-state. Phosphate of iron [a theoretical compound in the acid process] must be regarded as composed of phosphoric anhydride and iron oxide; if the latter be set free for an instant in the molecular transmutations just mentioned, it is liable to be reduced by carbon, thus leaving the phosphoric oxide exposed to a reducing action, to which it immediately yields. But in the case of phosphate of lime [the form of phosphorus in the basic process], which is a combination of phosphoric anhydride and oxide of calcium, the latter will be unaffected by any ordinary reducing action, and therefore the phosphoric oxide will not be left, even for a single moment, without a partner" (Campbell, "Open-Hearth," 465).

99. Ibid., 419–46, 462–66, 679–96 (discussion; quotations on 679, 687).

100. On Carnegie's early enthusiasm for basic open-hearth steel, see Andrew Carnegie to William Shinn, 23 March 1879, vol. 4, ACLC; Andrew Carnegie to W. L. Abbott, 28 October 1888 and 16 January 1889, vol. 10, ACLC; and Wall, *Andrew Carnegie*, 502–5.

101. Andrew Carnegie to J. Stephen Jeans, 6 July 1896, and Carnegie to Carnegie Steel–Pittsburgh, 13 July 1896, vol. 38, ACLC; Charles M. Schwab to Carnegie Steel Board of Managers, 6 July 1897, vol. 43, ACLC; Charles M. Schwab to Carnegie Steel Board of Managers, 31 August 1897, vol. 44, ACLC.

102. The 1900 standards permitted Bessemer or open-hearth steel for buildings; "Proposed Standard Specifications for Structural Steel for Bridges and Ships," *ASTM Proc.* 1 (1900): 81–85; "Proposed Standard Specifications for Structural Steel for Buildings," *ASTM Proc.* 1 (1900): 87–92; "Report of Committee A on Standard Specifications for Steel," *ASTM Proc.* 9 (1909): 35–76.

103. Col. W. A. Starrett, *Skyscrapers and the Men Who Build Them* (New York: Charles Scribner's Sons, 1928), 161–62.

104. Dankmar Adler to his family, 29 August 1888, Folder 4, DA.

105. See Anne Millbrooke, "Technological Systems Compete at Otis: Hydraulic versus Electric Elevators," in *Technological Competitiveness*, ed. William Aspray (New York: IEEE Press, 1993), 243–69.

106. Ericsson, *Sixty Years a Builder*, 238 (quotation); Donald Martin Reynolds, *The Architecture of New York City* (New York: Macmillan,

1984), 147–49; Dankmar Adler, "Light in Tall Office Buildings," *Engineering Magazine* 4 (November 1892): 171–96 (quotation on 171); Mona Domosh, "The Symbolism of the Skyscraper: Case Studies of New York's First Tall Buildings," *Journal of Urban History* 14, no. 3 (1988): 321–45.

107. In 1894 *total* fire losses were $128 million, and pig iron production was $75 million; see Freitag, *Architectural Engineering*, 9–23, 129–35. For comparison of yields on railroad bonds and skyscraper financing, see Homer Hoyt, *One Hundred Years of Land Values in Chicago* (Chicago: University of Chicago Press, 1933).

108. W.L.B. Jenney, speech to National Brick Manufacturers Association (quotation; emphasis in original), 24 January 1894, J&M Papers; Turak, *William Le Baron Jenney*, 289; Siry, *Carson Pirie Scott*, 46–51.

109. For the New York Life Building, see W.L.B. Jenney, "Iron and Steel," MS 1896, J&M Papers. The fourth floor of terra-cotta for Adler & Sullivan's Guaranty Building (1894–95) in Buffalo was placed on the completed steel skeleton before the outer walls on the lower three stories were completed; splendid construction photographs are in Birkmire, *Planning and Construction of High Office Buildings*, 2nd ed. (New York: John Wiley, 1900), 107. See also Peter G. Kreitler, *Flatiron: A Photographic History of the World's First Steel Frame Skyscraper, 1901–90* (Washington, D.C.: American Institute of Architects, 1990), 5–9. Other Chicago buildings erected in the same fashion include Jenney & Mundie's Manhattan Building (1889–91), Clinton J. Warren's Unity Building (1892), and D. H. Burnham's Fisher Building (1895).

110. Decade-on-decade growth rates for new building are as follows (percentages, in constant 1913 dollars; data from fig. 2.6):

	1870s–80s	1880s–90s	1890s–1900s	1900s–1910s
Chicago	47.1	106.2	11.0	31.7
Manhattan	145.0	64.4	−6.7	−37.6

3. The Politics of Armor, 1885–1915

1. For the centrality of system integration, see Karl Lautenschläger, "The Dreadnought Revolution Reconsidered," in *Naval History*, ed. Daniel M. Masterson (Wilmington, Del.: Scholarly Resources, 1987), 121–45.

2. According to Sumida's revisionist account, "The battle cruiser *Invincible*, rather than the *Dreadnought*, was the centerpiece of Fisher's capital ship program, and his major goal from at least 1905 onwards was the replacement of the battleship by the battle cruiser." See Jon Tetsuro Sumida, *In Defence of Naval Supremacy: Finance, Technology and British Naval Policy, 1889–1914* (Boston: Unwin Hyman, 1989), 26–27, 37–70, 100, 316–18, 335 (quotation).

3. These official figures are disputed by Oscar Parkes, *British Battleships: Warrior to Vanguard*, rev. ed. (London: Seeley, 1966), 492–96. Parkes states (495) that the *Invincible's* "real protection was only 6 inch

amidships and 4 inch at the bows with no belt [armor] abaft the after turret. Gun positions had 7 inch above the belt and 2 inch below it, with
a 2 inch–1½ inch waterline deck and a 1 inch–¾ inch main deck. As in
the *Dreadnought* the magazines were given 2½-inch side screens below
water." *Invincible's* armor was most effective at ranges less than 9,000
yards (when shell trajectories were nearly flat and would be resisted by
the heavier side armor) and most vulnerable at ranges over 15,000 yards
(when shell trajectories were strongly curved and so would penetrate the
lighter deck armor).

4. N.J.M. Campbell, *Jutland: An Analysis of the Fighting* (Annapolis:
Naval Institute Press, 1986), 60–69; Holloway H. Frost, *The Battle of Jutland* (Annapolis: Naval Institute Press, 1936), 204–5. In addition, I have
drawn on John Keegan, *The Price of Admiralty: The Evolution of Naval
Warfare* (New York: Viking Penguin, 1989), 110–81.

5. Rudyard Kipling, *Sea Warfare* (Garden City, N.Y.: Doubleday, Page,
1917), 206, 211; Richard Hough, *The Great War at Sea, 1914–18* (Oxford:
Oxford University Press, 1983), 219.

6. Sir Julian S. Corbett, *History of the Great War: Naval Operations*,
5 vols. (1920–31), 3:366, quoted in Hough, *Great War at Sea*, 246; Campbell, *Jutland: An Analysis of the Fighting*, 169–70.

7. Frost, *Battle of Jutland*, 328.

8. Arthur J. Marder, *From the Dreadnought to Scapa Flow: The Royal
Navy in the Fisher Era, 1904–19*, 5 vols. (London: Oxford University
Press, 1966), 3:165 (quotation); Admiral Viscount Jellicoe, *The Grand
Fleet, 1914–16: Its Creation, Development and Work* (New York: Doran,
1919), 416 (quotation); Sumida, *In Defence of Naval Supremacy*, 316.

9. Charles H. Fairbanks, Jr., "Choosing among Technologies in the
Anglo-German Naval Arms Competition, 1898–1915," in *Naval History*,
ed. William B. Cogar (Wilmington, Del.: Scholarly Resources, 1988), 127–
42. Fairbanks (129) writes of "the debility of British armor." "There is at
least no firm proof that German armour was superior to British," states
Marder (*Dreadnought to Scapa Flow*, 1:172). "That the maximum range
at which British 9-inch armour was pierced was about 14,600 yards, while
our 12-inch shell were getting through 11-inch plates at 17,500 yards;
13.5 inch shell through 10-inch at about 19,000 yards and 9-inch at nearly
20,000 yards is some evidence of the superiority of our [that is, British]
armour," asserts Parkes (*British Battleships*, 641). "There is good reason to
believe that British armour was superior to German armour . . . German
armour of 10 and 11 inches thickness was pierced by British heavy shells,
but only once was British armour above 9 inches penetrated by German
shell," states Hough (*Great War at Sea*, 278).

10. See Campbell, *Jutland: An Analysis of the Fighting*, especially 384–
87.

11. German ships featured completely watertight compartments, making them nearly unsinkable even when flooded with thousands of tons
of water. British armor-piercing shells frequently exploded on contact,

making a dreadful but harmless racket, rather than exploding *after* having penetrated the armor, and British antiflash devices that should have isolated the gun turret above from its magazine below were removed to allow rapid reloading.

12. Walter R. Herrick, Jr., *The American Naval Revolution* (Baton Rouge: Louisiana State University Press, 1966), 3; Richard D. Glasow, "Prelude to a Naval Renaissance: Ordnance Innovation in the United States during the 1870s" (diss., University of Delaware, 1978).

13. Glasow, "Prelude to a Naval Renaissance," 7–17; Lance C. Buhl, "Mariners and Machines: Resistance to Technological Change in the American Navy, 1865–1869," *Journal of American History* 61 (1974): 703–27; Elting E. Morison, *Men, Machines and Modern Times* (Cambridge: MIT Press, 1966), 98–122.

14. Glasow, "Prelude to a Naval Renaissance," 58–59, 73–100, 117–18, 174–232 (quotation on 222–23), 239–58.

15. Ibid., 246–48, 269–74; Taylor Peck, *Round-Shot to Rockets: A History of the Washington Navy Yard and U.S. Naval Gun Factory* (Annapolis: Naval Institute Press, 1949), 172–79; Benjamin Franklin Cooling, *Gray Steel and Blue Water Navy: The Formative Years of America's Military Industrial Complex, 1881–1917* (Hamden, Conn.: Archon, 1979), 33–35; Ben Baack and Edward John Ray, "Special Interests and the Nineteenth-Century Roots of the U.S. Military-Industrial Complex," *Research in Economic History* 11 (1988): 153–69, especially 147.

16. Glasow, "Prelude to a Naval Renaissance," 249–54; Cooling, *Gray Steel and Blue Water Navy,* 41–48; Peck, *Round-Shot to Rockets,* 180–83.

17. No satisfactory history of the Bethlehem company exists. This account draws on the sources cited in Misa, "Science, Technology and Industrial Structure," 75 n. 1; and on Lance E. Metz, "The Arsenal of America: A History of Forging Operations of Bethlehem Steel," *Canal History and Technology Proceedings* 11 (1992): 233–94.

18. Frank B. Copley, *Frederick W. Taylor: Father of Scientific Management,* 2 vols. (New York: Harper, 1923), 1:101 (quotation); W. Ross Yates, *Joseph Wharton: Quaker Industrial Pioneer* (Bethlehem, Pa.: Lehigh University Press, 1987), 143–47.

19. Yates, *Joseph Wharton,* 70–176, 219–26, 254–63.

20. John Fritz, *Autobiography of John Fritz* (New York: American Society of Mechanical Engineers, 1912), 184 (quotation); Glasow, "Prelude to a Naval Renaissance," 126–27, 277–80; Cooling, *Gray Steel and Blue Water Navy,* 41–43; "Captain William Henry Jaques," *SNAME Trans.* 24 (1916): 215. For additional detail on Bethlehem's transfer of Schneider and Whitworth technology, see Metz, "Arsenal of America," especially 246–63.

21. "It was known that our Navy Department preferred and expected that the [Schneider] method be followed," stated Bethlehem's armor superintendent Russell W. Davenport ("Production in the United States of Heavy Steel Engine, Gun and Armor Plate Forgings," *SNAME Trans.* 1 [1893]: 73–74).

22. Yates, *Joseph Wharton*, 256–57. Bethlehem spent about $1.5 million for the armor and gun-forging plants together; the navy spent $4 million on the Bethlehem contract. See Office of Naval Intelligence, *Papers on Squadrons of Evolutions and the Recent Development of Naval Matériel, Information from Abroad: General Information Series V* (Washington, D.C.: Government Printing Office, 1886), 323; Davenport, "Production of Heavy Forgings," 71–75; "The Bethlehem Hammer," *Iron Age* 52 (13 July 1893): 60.

23. Yates, *Joseph Wharton*, 259–61; Cooling, *Gray Steel and Blue Water Navy*, 65–75.

24. Davenport, "Production of Heavy Forgings," 74–75; "The Bethlehem Hammer," *Iron Age* 52 (13 July 1893): 60.

25. Davenport, "Production of Heavy Forgings," 72–73; W. H. Jaques, "Description of the Works of the Bethlehem Iron Company," *U.S. Naval Institute Proceedings* 15, no. 4 (1889): 531–40, especially 538–39; "The Bethlehem Hammer," 62; Office of Naval Intelligence, *Papers on Squadrons*, 322–23.

26. Cooling, *Gray Steel and Blue Water Navy*, 72–76; Joseph F. Wall, *Andrew Carnegie* (New York: Oxford University Press, 1970), 417.

27. The final deliveries from Whitworth were in February and March 1889. See A. Johnston, contract with Whitworth & Co., 18 January 1886, 1889, 20:11, AJ; Benjamin Franklin Cooling, *Benjamin Franklin Tracy: Father of the Modern American Fighting Navy* (Hamden, Conn.: Archon, 1973), 91–93.

28. Andrew Carnegie, *Triumphant Democracy* (1886; reprint, New York: Doubleday, 1933), 5, 7, 301–5; Wall, *Andrew Carnegie*, 645–49, 689–713. The best account of technical change at Homestead is Mark M. Brown, "Technology and the Homestead Steel Works, 1879–1945," *Canal History and Technology Proceedings* 11 (1992): 177–232, especially 187–200.

29. *Quid:* at Bethlehem's request, Penrose along with Sen. Eugene Hale and Rep. Alston G. Dayton killed a certain amendment to the 1905 navy bill that threatened to complicate the armor business. See E. M. McIlvain to C. M. Schwab, 27 February 1905, 2:32, AJ. *Quo:* "At a meeting between Mr. Schwab, Senator P-n-o-e on Saturday, the 6th instant, it was determined that the Corporation would contribute toward the coming election expenses, to the amount of $5,000" (A. Johnston, memo, 8 October 1906, 2:28, AJ [Bethlehem-Penrose file]). On Penrose's remarkable career as "the nation's greatest political boss," see Walter Davenport, *Power and Glory: The Life of Boies Penrose* (New York: G. P. Putnam's Sons, 1931), 118 (quotation); Robert Douglas Bowden, *Boies Penrose: Symbol of an Era* (New York: Greenberg, 1937).

30. "We have found the secret" for Harvey armor, crowed Carnegie in 1894 to White, who with Blaine lobbied the Russian government on behalf of the company (A. Carnegie to A. D. White, 16 April 1894, vol. 24, ACLC). See also Lieut. C. A. Stone to Millard Hunsiker, 1894, and Millard Hunsiker to Andrew Carnegie, 1 May 1894, vol. 24, ACLC.

31. Cooling, *Gray Steel and Blue Water Navy*, 54–55, 101, 104, 119, 150; Dean C. Allard, Jr., "The Influence of the United States Navy upon the American Steel Industry, 1880–1900" (M. A. thesis, Georgetown University, 1959), 110; W. H. Emory to Andrew Carnegie, 2 July 1890, vol. 11, ACLC; Wall, *Andrew Carnegie*, 647–48.

32. A. Carnegie to George Lauder, 9 March 1895, vol. 31, ACLC; C. M. Schwab to E. M. McIlvain, 19 March 1906, and A. E. Borie to C. M. Schwab, 16 May 1904, 2:32, AJ.

33. A. Carnegie to R. P. Linderman, 25 February 1895, and R. P. Linderman to A. Carnegie, 26 February 1895, vol. 31, ACLC. The length and breadth of the Carnegie-Bethlehem concord may be sampled in Board of Managers minutes, 20 July 1897, vol. 43, ACLC (joint decision not to bid at U.S. Navy's maximum armor price of $300); J. D. Hagenbuch, memo, 20 May 1907, and J. D. Hagenbuch to A. Johnston, 15 May 1907, 14:30, AJ (Carnegie and Bethlehem to bid usual price allowing Midvale to gain a contract by selective undercutting); J. D. Hagenbuch, memo of telephone conversation, 12 October 1908, 7:36, AJ (jointly boosting armor price if U.S. Navy would increase ballistic test); J. D. Hagenbuch, memo of telephone conversation, 20 November 1908, 7:36, AJ (splitting armor order with Navy Department's informal approval); H. S. Snyder to A. Johnston, 19 June 1908, RG109:2, BS (on anticompetitive bidding techniques); A. Johnston to E. G. Grace, 19 July 1906, 1:15, AJ (division of armor order); A. Johnston to C. M. Schwab, 13 April 1911, 1:16, AJ (finally rejecting a renewed international armor cartel, Schwab replied, "Let the whole matter rest for the present. We have nothing to gain just now").

34. See Paul M. Kennedy, *The Rise of the Anglo-German Antagonism, 1860–1914* (London: Allen and Unwin, 1980), 237, 396, 410–31; Robert K. Massie, *Dreadnought: Britain, Germany, and the Coming of the Great War* (New York: Random House, 1991), xxiv (quotation), 256.

35. U.S. House of Representatives Report no. 1178, "Appropriations for the Naval Service," 51st Cong., 1st Sess. (1 April 1890), 16; Cooling, *Benjamin Franklin Tracy*, 87–88; Lewis Nixon, "The New American Navy," *Cassier's Magazine* 7 (1894–95): 107–11.

36. American warships impressed, if they did not alarm, the English naval establishment. See J. H. Biles, "Some Recent War-Ship Designs for the American Navy," *INA Trans.* 32 (1891): 40–66; Biles, "Ten Years' Naval Construction in the United States," *INA Trans.* 43 (1901): 1–22. The German navy, too, through its attachés in Washington, London, and Paris, closely followed these developments; see Gary E. Weir, *Building the Kaiser's Navy: The Imperial Navy Office and German Industry in the von Tirpitz Era* (Annapolis: Naval Institute Press, 1992), 26.

37. Contrast John Scott Russell, "On the True Nature of the Resistance of Armour to Shot," *INA Trans.* 21 (1880): 69–95; and the discussion following J. Barba, "Recent Improvements in Armour Plates for Ships," *INA Trans.* 32 (1891): 138–63.

38. J. F. Meigs, "Late Developments in Ordnance and Armor," *SNAME Trans.* 9 (1901): 201.

39. Cooling, *Benjamin Franklin Tracy*, 89–90 (quotation). "American trials are characterised by conditions rather more favourable to the plate than to the shot; while in France . . . the reverse has been the case. In England . . . the shot and the plate appear to be equally matched," stated C. E. Ellis ("Recent Experiments in Armour," *INA Trans.* 35 [1894]: 221). For explicit discussion of armor test conditions determining results, see "Official Tests of Armor Plate," *American Machinist* 23 (17 May 1900): 459–60.

40. Philip Hichborn, *Report on European Dock-Yards* (Washington, D.C.: Government Printing Office, 1886), 5–7, 83–85; William H. Jaques, "Modern Armor for National Defense," in U.S. Senate, *Report of the Select Committee on Ordnance and War Ships* (Washington, D.C.: Government Printing Office, 1886), 197–99; and "Armor for Ships of War," *Iron Age* 45 (20 March 1890): 468.

41. Weir, *Building the Kaiser's Navy*, 18, 30–32; Julius von Schütz, *Gruson's Chilled Cast-Iron Armour* (London: Whitehead, Morris and Lowe, 1887), 6–8.

42. Hichborn, *European Dock-Yards*, 81–85 (quotation on 85).

43. Several factual errors mar Cooling's two accounts of the September 1890 Annapolis test: the nickel-steel plate was not "Harveyized"; the test rounds were not "completely blunted" by the nickel-steel plate; this was not the first test to use "larger than 6-inch shells"; and the test did not alone produce "definitive" results. Cooling may have confused the September 1890 trial with preliminary trials of Harvey armor plate in June 1890 and February 1891; a full-scale competitive trial of May 1891 featured the first Harveyized nickel-steel plate. See Cooling, *Gray Steel and Blue Water Navy*, 96; Cooling, *Benjamin Franklin Tracy*, 94–95. For reasons discussed later, these details are critical to understanding the purpose of this armor test. The Kimberly Board Report is L. A. Kimberly, "Report of the Board on the Competitive Trial of Armor Plates," *U.S. Naval Institute Proceedings* 16, no. 4 (1890): 629–45 (hereafter cited as Kimberly Board Report).

44. See, for instance, Thomas P. Hughes, *Networks of Power* (Baltimore: Johns Hopkins University Press, 1983), 30.

45. Cooling, *Benjamin Franklin Tracy*, 90, 95–96; "Armor Plates," *Scientific American* 63 (26 July 1890): 57; "Nickel in Steel," *Iron Age* 45 (20 February 1890): 291. The paper fanning interest in nickel was James Riley, "Alloys of Nickel and Steel," *JISI* 1 (1889): 45–77. For background on Ritchie, see *New York Times* (29 September 1890): 1; "The Nickel Mines of Canada," *JISI*, special volume (1890): 359–60.

46. Carnegie to W. L. Abbott, undated (possibly spring 1889), quoted in Wall, *Andrew Carnegie*, 646.

47. Tracy, quoted in W. L. Abbott to Andrew Carnegie, 23 June 1890, vol. 11, ACLC.

48. "If a mixture of nickel in the steel improves it, of course that can be put in without the slightest trouble" (Andrew Carnegie to Secretary of the Navy B. F. Tracy, 5 July 1890, vol. 11, ACLC). On the nickel license,

see W. L. Abbott to F. Rey (London), 30 May 1890, Carnegie Steel (Pittsburgh) to Andrew Carnegie (London), 6 June, and F. Rey to Carnegie Phipps, 16 June, vol. 11, ACLC. On Hunsiker, see Dickson, *History of Carnegie Veteran Association*, 77.

49. Kimberly Board Report, 631; "Important Tests to Be Made of Heavy Armor Plating," *Scientific American* 63 (13 September 1890): 160. The photographs are clearly reproduced in W.H.H. Southerland, "The Armor Question in 1891," in *The Year's Naval Progress*, Office of Naval Intelligence, Information from Abroad: General Information Series X (Washington, D.C.: Government Printing Office, 1891), 279–337, especially 306–25.

50. "Testing Armor Plates for Our War Vessels," *Scientific American* 63 (27 September 1890): 192 (quotation); "The Armor Plate Tests," *New York Times* (22 September 1890): 4.

51. *New York Times* (23 September 1890): 1. The four 6-inch shells produced bulges in the back of the steel plate of 3.5, 3.0, 2.6, and 2.6 inches, and in the back of the nickel-steel plate of 5, 6, 4, and 4 inches; Kimberly Board Report, 643.

52. Six months after the Annapolis trial, J. F. Hall, a representative of William Jessop & Sons and a member of the nickel syndicate, told his fellow members of the (British) Institution of Naval Architects: "This supply of nickel was a question that troubled me very considerably, and it was not at all bettered by the unfortunate publicity given to the trials of the nickel steel plate at Annapolis. I may say that M. Schneider, and everyone connected with nickel steel patents as far as we know, are now amalgamated with myself, and at the present time we have a very strong Nickel Steel Syndicate in England, the Chairman of which is Sir James Kitson, of Leeds, who is also President of the Iron and Steel Institute. When this trial at Annapolis took place I was on my way to America with the Iron and Steel Institute. Sir James Kitson was also on board, and, as I understand, a telegram was sent to America stating that we were on our way there to purchase all the nickel that could be found in the States and Canada or anywhere else, the result of which was that extraordinary Bill which was passed through Congress in about two days, empowering Mr. Secretary Tracey [*sic*] to purchase one million dollars' worth of nickel before we landed in New York. The result of all this was that for a time our efforts to purchase nickel were frustrated very considerably, because the people who had nickel mines and had nickel to sell naturally wanted a very considerably higher price than they did before they knew what it was going to be used for" (J. Barba, "Recent Improvements in Armour Plates for Ships," *INA Trans.* 32 [1891]: 152–53 [discussion]).

53. *New York Times* (22 September 1890 [dateline London 21 September by cable]): 3; *Dictionary of National Biography*, s.v. "Kitson, James," suppl. 1901–11, 404–5; *Iron Age* 46 (25 September 1890): 487.

54. *New York Times* (23 September 1890 [dateline Annapolis 22 September]): 1. The data in the Kimberly Board Report corroborate this account.

55. *New York Times* (24 September 1890 [dateline Washington 22 September]): 2.

56. *New York Times* (25 September 1890): 4. The legislative history of the resolution (H.R. 228) can be followed in the *Congressional Record*, 51st Cong., 1st Sess. (1890): 10365, 10398, 10451–52, 10457, 10459–63, 10554–55, 10648, 10745–46. President Harrison signed the resolution on 30 September.

57. *Scientific American* 63 (4 October 1890): 208; *Iron Age* 46 (2 October 1890): 533.

58. *New York Times* (29 September 1890): 1 (quotation), 5; *Iron Age* 46 (11 December 1890): 1046; Wall, *Andrew Carnegie*, 646–47; F. W. Taussig, *The Tariff History of the United States*, 6th ed. (New York: G. P. Putnam's Sons, 1914), 251–84.

59. Kimberly Board Report, 640–45.

60. On the theme of technological testing, see Trevor Pinch, Malcolm Ashmore, and Michael Mulkay, "Technology, Testing, Text: Clinical Budgeting in the U.K. National Health Service," in *Shaping Technology/ Building Society: Studies in Sociotechnical Change*, ed. Wiebe E. Bijker and John Law (Cambridge: MIT Press, 1992), 265–89; Donald MacKenzie, "From Kwajalein to Armageddon? Testing and the Social Construction of Missile Accuracy," in *The Uses of Experiment: Studies in the Natural Sciences*, ed. David Gooding, Trevor Pinch, and Simon Schaffer (Cambridge: Cambridge University Press, 1989), 409–35.

61. Barba, "Recent Improvements," 138–63, especially 150–63. The "battle of armor" is made explicit in W. H. Jaques, "Recent Progress in the Manufacture of War Material in the United States," *JISI* 1 (1891): 156–227, especially 167, 221; Office of Naval Intelligence, *Papers on Squadrons*, 239; William H. White, *Modern Warships* (London: Trounce, 1906), 17–18.

62. Ellis, "Recent Experiments," 215–16 (quotation on 215). On British evidence for compound armor, see Misa, "Science, Technology and Industrial Structure," 99–100. Compound armor possessed "great power to throw off oblique blows, a work constantly called for on service, but seldom tested by experiment. The compound plate has probably suffered by being compared with steel under direct attack only" (E. E. Quimby, affidavit, 97, in Circuit Court of the United States, Eastern District of Pennsylvania, Third Circuit, *Harvey Steel Company, Complainant, v. the Bethlehem Iron Company, Defendant, March 1895* [hereafter cited as *Harvey v. Bethlehem*]; copy in Hagley Museum and Library, Greenville, Del.).

63. One American officer suggested that English trials gave compound armor the unrealistic advantage of a rigid backing; Southerland, "Armor Question," 331. American tests seemed to use an oak block 36 inches thick, while British tests used a wood block backed by an iron plate. For an isolated example of testing armor as actually mounted on a ship, see Albert W. Stahl, "An Experimental Test of the Armored Side of *U.S.S. Iowa*," *SNAME Trans.* 3 (1895): 247–57.

64. A. Johnston to R. W. Davenport, 19 January 1899, 13:24, AJ (emphasis in original).

65. William H. White, discussion of L. E. Bertin, "Hardened Plates and Broken Projectiles," *INA Trans.* 39 (1898): 23.

66. Edward W. Very, "The Annapolis Armor Test," *U.S. Naval Institute Proceedings* 16, no. 5 (1890): 622. Bertin, "Hardened Plates," and Ellis, "Recent Experiments," used similar quantitative formulations.

67. See Thomas J. Misa, "Retrieving Sociotechnical Change from Technological Determinism," in *Does Technology Drive History?* ed. Merritt Roe Smith and Leo Marx (Cambridge: MIT Press, 1994), 115–41.

68. G. W. Burleigh to T. W. Harvey, 25 June 1913, 7:19, HS.

69. Weir, *Building the Kaiser's Navy*, 19–56; Baack and Ray, "Special Interests," 158–59; Cooling, *Gray Steel and Blue Water Navy*, 155 (quotations), 166.

70. Weir, *Building the Kaiser's Navy*, 30–31, 196–97; Richard Owen, "Military-Industrial Relations: Krupp and the Imperial Navy Office," in *Society and Politics in Wilhelmine Germany*, ed. Richard J. Evans (London: Croom Helm, 1978), 71–89.

71. See the notebook of seventy-seven patents issued to H. A. Harvey (1854–93) for various inventions, including bolts and nuts, taps and dies, nails, railroad chairs, shavers, and washers, 6:15, HS; Harvey Steel (N.J.) stock subscription, 20 June 1888, 3:5, HS; H. A. Harvey to T. W. Harvey, 18 March 1889, 3:6, HS. For "steelifying," see W. A. Smith, affidavit, *Harvey v. Bethlehem*, 17; Hayward A. Harvey, "Process of Treating Low Steel" (U.S. Patent 376,194, filed 8 December 1886, granted 10 January 1888). Harvey's later American armor patents were "Decrementally-Hardened Armor-Plate" (U.S. Patent 460,262, filed 1 April 1891, granted 29 September 1891) and "Composition for Supercarburizing Steel, &c." (U.S. Patent 498,390, filed 28 January 1893, granted 30 May 1893).

72. J. H. Dickinson, affidavit, *Harvey v. Bethlehem*, 73; H. S. Manning to Harvey Steel (N.J.), 20 May 1889, and W. M. Folger to W. A. Smith, 7 October 1893, 7:18, HS; H. A. Harvey to T. Sturges, 28 May 1889, 3:5, HS.

73. U.S. Navy Ordnance Bureau, *Harvey Armor Plate: Official Trials by the Government of the United States* (1891); copy in Hagley Museum and Library, Greenville, Del.

74. E. E. Quimby to H. A. Harvey, 25 March 1891, 7:20, HS; E. E. Quimby to C. E. Mitchell[?], 21 May 1891, 7:17, HS; W. A. Smith, affidavit, *Harvey v. Bethlehem*, 28–29.

75. For the 6-inch plate tested June 1890, Dickinson gave the following percentages of supercarburization: at a depth of ⅛ inch, 1.08 percent of carbon; at ¼ inch, 0.95; at ⅜ inch, 0.78; at ½ inch, 0.68; at ⅝ inch, 0.60; at ¾ inch, 0.55; at ⅞ inch, 0.50, and at 1 inch, 0.45 (0.45 percent was equivalent to high-carbon for rail steel, while 1.08 percent was beyond any normal commercial steel product, being hard to the point of brittleness). W. A. Smith, affidavit, *Harvey v. Bethlehem*, 49; H. A. Harvey, Jr.,

affidavit, *Harvey v. Bethlehem*, 86–90; J. H. Dickinson, affidavit, *Harvey v. Bethlehem*, 74, 76, 79–81; W. M. Folger to H. A. Harvey, 12 March 1892, 7:17, HS.

76. W. A. Smith, affidavit, *Harvey v. Bethlehem*, 19–27 (quotations on 25, 26; from 1892 report of Ordnance Bureau Chief and 1892 report of Secretary of the Navy). Of the pre-1892 warships, Harveyized nickel steel was used for much of the side armor of the *Maine, Texas, Indiana, Oregon,* and *Puritan;* the turret armor for the *Maine, Puritan,* and *Monadnock;* and the barbette for the *Oregon.*

77. W. A. Smith, affidavit, *Harvey v. Bethlehem*, 30–36 (quotation on 36); H. A. Harvey to E. M. Fox, 12 January 1892, E. M. Fox et al. to Harvey Steel (N.J.), 1 November 1892 (cable), and E. M. Fox to Harvey Steel (N.J.), 2 November 1892, 3:5, HS; W. A. Smith, affidavit, *Harvey v. Bethlehem*, 30; E. E. Quimby, affidavit, *Harvey v. Bethlehem*, 109–10.

78. Joseph Ott to E. M. Fox, 29 April 1893, 3:5, HS; J. H. Dickinson, affidavit, *Harvey v. Bethlehem*, 81; J. H. Dickinson, list of plates treated in Europe (1892–96), 3:5, HS.

79. In 1890 Bethlehem's capitalization had been increased to $5 million; Yates, *Joseph Wharton*, 260. In 1892 Carnegie's capitalization had been increased to $25 million.

80. H. A. Harvey to Harvey Steel (N.J.), 7 April 1893, 7:17, HS; Harvey Steel Co. of Great Britain, Ltd., prospectus, 13 April 1893, 3:6, HS; Harvey Steel Co. of Great Britain, prospectus, 28 April 1893, 7:21, HS.

81. The British Admiralty reported (elliptically) in 1894 that "it has been established that Harveyed plates without nickel in the steel show resistance to modern projectiles as great as any hitherto obtained when nickel was combined with steel in plates also treated by the Harvey process" (quoted in J. H. Dickinson, affidavit, *Harvey v. Bethlehem*, 85).

82. E. M. Fox to Harvey Steel (N.J.), 13 June 1893, 3:6, HS. "The London Harvey Co. . . . owns practically all the stock of the American Harvey Co." (M. Hunsiker to E. M. McIlvain, 27 April 1906, 8:42, AJ).

83. A. Carnegie to J. Wharton, 15 May 1896, Folder 1: A. Carnegie to J. A. Leishman, AC-CMS; G. W. Burleigh to T. W. Harvey, 30 April 1913, 7:19, HS; J. W. Langley, court testimony 1897, Papers Regarding Patent Cases, JWL.

84. See Bethlehem's notes of conversation with the Krupp officers Albert Schmitz (director and chief engineer) and Emil Ehrensberger (engineer in charge of armor plate department), 25 October 1893, 18:1, AJ; and R. P. Linderman to Fried. Krupp AG, 22 September 1897, 18:4, AJ.

85. Andrew Carnegie to J.G.A. Leishman, 27 January 1896, vol. 36, ACLC; H. T. Morris and E.O.C. Acker to R. W. Davenport, 10 April 1901, 19:2, AJ. For the U.S. patents covering the Krupp process, as well as the Harvey-Krupp consolidation, see Misa, "Science, Technology and Industrial Structure," 111.

86. A. Johnston to R. W. Davenport, 24 October 1897, and A. Johnston to O. F. Leibert, 25 October 1897, 18:4, AJ.

87. R. P. Linderman to Fried. Krupp AG, 22 September 1897, A. Johns-

ton to R. P. Linderman, 1897, A. Johnston to R. W. Davenport, 24 October 1897, A. Johnston to O. F. Leibert, 25 October 1897, and A. Johnston to R. W. Davenport, 26 October 1897, 18:4, AJ; A. Johnston to O. F. Leibert, 1 November 1897, 18:1, AJ; A. Johnston and J. E. Tatnall, "Notes on Armor Plate," 20 February 1898, 18: Bound Report, AJ.

88. See Armour Plate Committee minutes, especially 9 June and 24 November 1898, 7, 20, and 23 March 1899, 21 March 1900, WGA.

89. A. L. Colby to R. W. Davenport, 8 September 1900, 13:23, AJ (quotation); E. M. McIlvain to C. M. Schwab, 23 December 1904, 2:32, AJ (quotation); Richard T. Nalle, *Midvale and Its Pioneers* (New York: Newcomen Society, 1948), 17–20.

90. See the tabulation of technical questions in A. Johnston to Emil Ehrensberger, 22 December 1898, 1:12, AJ; Emil Ehrensberger to A. Johnston, 7 February 1899, 1:12, AJ; Bethlehem Steel memo [1907], 13:22, AJ.

91. A. Johnston, memo, 30 October 1907, 7:34, AJ.

92. E. M. McIlvain to Sen. Boies Penrose, 5 March 1906, 2:28, AJ; Misa, "Science, Technology and Industrial Structure," 115–25; Cooling, *Gray Steel and Blue Water Navy,* 174, 223–31.

93. Misa, "Science, Technology and Industrial Structure," 125–32; J. D. Hagenbuch, memo (to A. Johnston), 20 May 1907, 14:30, AJ (quotation); A. Johnston to M. Hunsiker, 4 January 1908, 8:43, AJ (quotation).

94. J. D. Hagenbuch, memo of telephone conversation between W. R. Balsinger (Carnegie's engineer of ordnance) and Archibald Johnston (Bethlehem's vice president), 22 July 1907, 7:34, AJ (quotation); see also A. Johnston and J. D. Hagenbuch, memos, 1907 and 1908, 7:34 and 7:35, AJ. On armor experiments by Bethlehem and Carnegie, see Misa, "Science, Technology and Industrial Structure," 132–33.

95. A long document explicitly and systematically contrasted Bethlehem's new process for "noncemented" armor, as the firm desired it to be patented, with the Krupp process, as outlined in the Ehrensberger and Schmitz patents. Bethlehem also contrasted its new process with the methods it used to manufacture the "modified" Class A Krupp armor; see two Bethlehem Steel memos, "Exhibit No. 1" and "Exhibit No. 3" [c. 1907], 13:22, AJ. Bethlehem enlisted the ordnance chief, Admiral N. E. Mason, to expedite the patent application that spring, but the patent was rejected owing to its similarities to several patents of the English metallurgist Robert Hadfield. H. S. Snyder to A. Johnston, 26 March 1908, RG109:3, BS; H. T. Morris to E. G. Grace, 9 October 1908, 14:26, AJ.

96. A. Johnston, memo, 15 August 1907, 7:34, AJ; C. J. Bonaparte to V. H. Metcalf, 15 August 1907, and W. H. Brownson (Acting Secretary of Navy) to Harvey Steel Company, 22 August 1907, 6:27, AJ.

97. Midvale Steel, tender for 2100 Tonnellate di Piastre da Corazzatura, 2 October 1906, 7:28, AJ; Keith Reeves Rodney, travel diary of 1905 visit to steel mills in Europe, KRR; Franco Bonelli, *Lo sviluppo di una Grande Impresa in Italia: La Terni dal 1884 al 1962* (Turin: Einaudi, 1975), 96–99; Luciano Segreto, "More Trouble Than Profit: Vickers' Investments

in Italy, 1906–39," *Business History* 27 (November 1985): 316–37; Weir, *Building the Kaiser's Navy*, 92, 109–12. Terni's obstructionism against Midvale during 1907–8 is detailed in Misa, "Science, Technology and Industrial Structure," 136–39. The several Anglo-Italian ventures are discussed in Kenneth Warren, *Armstrongs of Elswick* (London: Macmillan, 1989), 69–85, 122–27.

98. For the course of the Harvey and Krupp suits, see Misa, "Science, Technology and Industrial Structure," 126–45. On the Harvey liquidation, see Betts, Sheffield, Bentley & Betts to Harvey United Steel Company, 15 February 1911 (cable), 1:6, AJ; Circuit Court of Appeals, Third District, *Fried. Krupp AG v. Midvale Steel* (four cases nos. 40–43), *Federal Register* 191 (11 October 1911): 588–612, 3:5, HS; U.S. Supreme Court, *United States Government v. Harvey Steel* and *Midvale Steel v. Harvey Steel* (13 February 1913), and Harvey United Steel Co., "Liquidators' Statement of Accounts," 6 November 1913, all in 7:19, HS.

99. Keegan, *Price of Admiralty*, 120 (quotation); A.J.A. Morris, *The Scare Mongers: The Advocacy of War and Rearmament 1896–1914* (London: Routledge and Kegan Paul, 1984), 164–84; Sumida, *In Defence of Naval Supremacy*, 330; Weir, *Building the Kaiser's Navy*, 99–121.

100. On government-industry cooperation on price fixing during the war, see Melvin I. Urofsky, *Big Steel and the Wilson Administration: A Study in Business-Government Relations* (Columbus: Ohio State University Press, 1969), 192–247.

101. See Misa, "Science, Technology and Industrial Structure," 237–93.

102. Bethlehem's merger can be followed in U.S. House of Representatives Select Committee on Small Business, *Steel: Acquisitions, Mergers, and Expansion of 12 Major Companies, 1900 to 1950* (Washington, D.C.: Government Printing Office, 1950), 7–14.

103. Carnegie's armor *revenues* can be calculated. The company's contracts with the U.S. Navy in 1890, 1893, and 1896 were for a total of $7.26 million; see source for figure 3.1. Its armor *profits* are another matter. Although exact unit costs of production were a tightly held secret (references to unit costs of armor in Andrew Carnegie's papers have been neatly clipped out), there is some evidence that in 1898 costs were about $150 per ton (see Wall, *Andrew Carnegie*, 649). If so, an estimate of gross profits based on the contract figures given would be $5.43 million. Although a proper allowance for overhead charges would reduce this figure, the inclusion of foreign armor sales would increase it. In either event, it is of the magnitude discussed by Carnegie as the purchase price for the iron ore lands ($3 million).

104. Pilgrim (pseud.), "The Demands of War Influence the Requirements of Peace," *American Machinist* 29 (19 July 1906): 81–82 (quotation); Cooling, *Gray Steel and Blue Water Navy*, 166 (quotation).

4. The Merger of Steel, 1890–1910

1. On Carnegie's English newspaper venture of 1884–86, see Wall, *Andrew Carnegie*, 414–15, 431–42.

2. Ibid., 272, 283–85, 306, 318–20; Summers, *Railroads, Reconstruction, and the Gospel of Prosperity*, 167–74; Ward, *J. Edgar Thomson*, 179–85.

3. A. Carnegie to H. C. Frick, 7 February 1896, vol. 36, ACLC.

4. Wall, *Andrew Carnegie*, 510–15; Walter F. Walton, *The South Pennsylvania Railroad* (Pittsburgh: American Society of Civil Engineers, 1982); Albro Martin, "Crisis of Rugged Individualism: The West Shore–South Pennsylvania Railroad Affair, 1880–85," *Pennsylvania Magazine of History and Biography* 93 (April 1969): 218–43, especially 229–34. Carnegie personally contributed $2.5 million; Henry Oliver and several other Pittsburgh manufacturers together stumped up an equal amount.

5. Andrew Carnegie, "Pennsylvania's Industries and Railroad Policy," address delivered before the Pennsylvania Legislature, 8 April 1889, reprinted from the *Pittsburgh Commercial Gazette*, 14–15; copy in Hagley Museum and Library, Greenville, Del.

6. Vincent P. Carosso, *The Morgans: Private International Bankers, 1854–1913* (Cambridge: Harvard University Press, 1987), 254–57; Frederick Lewis Allen, *The Great Pierpont Morgan* (New York: Harper and Brothers, 1949), 47–56; Edwin P. Hoyt, Jr., *The House of Morgan* (New York: Dodd, Mead, 1966), 163–72. For Morgan's own account of his effort to impose "harmony" among the rival railroads, see *Commercial and Financial Chronicle* (17 October 1885): 445–46.

7. Carosso, *The Morgans*, 254 (quotation); Martin, "Crisis of Rugged Individualism," 221–29.

8. Carosso, *The Morgans*, 255 (quotation); Walton, *South Pennsylvania Railroad*, 4.

9. Allen, *Great Pierpont Morgan*, 53.

10. Martin, "Crisis of Rugged Individualism," 234–41.

11. Albro Martin, *Enterprise Denied: Origins of the Decline of American Railroads, 1897–1917* (New York: Columbia University Press, 1971), 61; E. G. Campbell, *The Reorganization of the American Railroad System, 1893–1900* (New York: Columbia University Press, 1938), 26–30.

12. Campbell, *Reorganization*, 24–28; Carosso, *The Morgans*, 353, 378–83, 392 (quotation).

13. Campbell, *Reorganization*, 30–144, 207 (quotation); Maury Klein, *Union Pacific: The Rebirth, 1894–1969* (Garden City, N.Y.: Doubleday, 1990), 11–29; Stuart Daggett, *Railroad Reorganization* (Cambridge: Harvard University Press, 1908; reprint, New York: Kelley, 1967), 238–44; Maury Klein, *The Great Richmond Terminal* (Charlottesville: University Press of Virginia, 1970), 235–84; Carosso, *The Morgans*, 383; L. L. Waters, *Steel Trails to Santa Fe* (Lawrence: University of Kansas Press, 1950), 178–96.

14. Carosso, *The Morgans*, 219–73, 352–64; Allen, *Great Pierpont Morgan*, 127 (quotation).

15. Campbell, *Reorganization*, 145–49; Carosso, *The Morgans*, 316–49, 365 (quotation), 366 (quotation).

16. Campbell, *Reorganization*, 149–89; Carosso, *The Morgans*, 363 (quotations), 364, 366–85; Daggett, *Railroad Reorganization*, 357–59.

17. Martin, *Enterprise Denied*, 59; Waters, *Steel Trails to Santa Fe*, 197–217, 338–88.

18. Campbell, *Reorganization*, 190–255; Klein, *Union Pacific: The Rebirth*, 11–68, 30–181; Carosso, *The Morgans*, 383–90.

19. Martin, *Enterprise Denied*, 55–61; Campbell, *Reorganization*, 263–71; Carl W. Condit, *The Port of New York: A History of the Rail and Terminal System from the Beginnings to Pennsylvania Station* (Chicago: University of Chicago Press, 1980).

20. Successful reorganizations of seven of the eight largest defaulted railroads were in place by the end of 1897: Southern (May 1893, revised February 1894), Erie (November 1895), Atchison (December 1895), Norfolk & Western (1896), Northern Pacific (August 1896), Philadelphia & Reading (December 1895–March 1897), Union Pacific (October 1897), Baltimore & Ohio (August 1898–July 1899). See Carosso, *The Morgans*, 370–88; Daggett, *Railroad Reorganization*, 29, 216, 355; Naomi R. Lamoreaux, *The Great Merger Movement in American Business, 1895–1904* (Cambridge: Cambridge University Press, 1985), 2. As the following table shows, the wave of manufacturing mergers *followed* these railroad reorganizations.

Year	Manufacturing Mergers	Railroad Reorganizations
1894	—	1
1895	4	2
1896	3	2
1897	6	2
1898	16	0
1899	93	1
1900	21	—
1901	19	—
1902	17	—
1903	5	—
1904	3	—

21. Albro Martin blames the decline of the American railroad industry on regulation, ignoring intermodal competition. A more plausible explanation is suggested by Carosso: "The executives of reconstructed roads were forced to share responsibility for financial policy with the bankers who sat on their boards and had access to the funds the companies required. Because most financiers were conservatively minded, more interested in preserving a reorganized road's solvency than encouraging its future de-

velopment, their influence tended to inhibit the company's managers from responding effectively to technological and market changes and, in time, the rise of new competitors, specifically trucks and automobiles" (*The Morgans*, 389).

22. The standard weight of main-line rail on the Santa Fe was 75 pounds per yard in 1897, 85 in 1902, and 90 in 1910; Waters, *Steel Trails to Santa Fe*, 377–78.

23. Usselman, "Running the Machine," 287; "Report of Committee on Rails," *AREMWA Proc.* 1 (1900): 112–33, especially 112, 121; Robert Trimble et al., "Specifications for Steel Rails: A Discussion of Papers by W. R. Webster and R. W. Hunt," *AIME Trans.* 31 (1901): 967–84, especially 969; "Report of Committee on Rails," *AREMWA Proc.* 5 (1904): 463–80 (quotation on 477).

24. Usselman, "Running the Machine," 311–17, 321–29; "Proposed Standard Specifications for Steel Rails," *ASTM Proc.* 1 (1901): 101–5, especially 103; Robert W. Hunt, "Specification for Steel Rails of Heavy Sections Manufactured West of the Alleghenies," *AIME Trans.* 25 (1895): 653–60, especially 654–55; "Report of Committee on Rails," *AREMWA Proc.* 2 (1901): 188–218, especially 193; "Report of Committee on Rails" (1904), 467; William R. Webster, "The Present Situation as to the Specifications for Steel Rails," *AIME Trans.* 33 (1902): 164–69, especially 166–68. In 1901, ASCE sections dominated the output of five of the nation's largest rolling mills: National Steel rolled 95.0 percent of its tonnage to ASCE sections; Pennsylvania Steel, 93.2 percent; Illinois Steel, 70.0 percent; Cambria Steel, 73.5 percent; Carnegie Steel, 72.0 percent. Two exceptions were Lackawanna Steel (37.2 percent), which rolled to the New York Central's specifications, and Maryland Steel (52.1 percent), which rolled rails for export and to the special sections of the Pennsylvania Railroad's Eastern Division as well as the Baltimore & Ohio.

25. "Report of Committee on Rails" (1900), 116; Robert W. Hunt, "Finishing Temperatures for Steel Rails," *AIME Trans.* 31 (1901): 458–65 (quotation on 460), 967–84; Albert Sauveur, "Structure and Finishing Temperature of Steel Rails," *ASTM Proc.* 2 (1902): 79–96. Thomas H. Johnson, chief engineer of the Cleveland, Cincinnati, Chicago & St. Louis Railway, had learned the arguments made by advocates of physical metallurgy: a rail's resistance to wear was a function of the fineness of the metal grain; fineness of grain was a result of mechanical treatment at proper temperature, and was wholly independent of chemical composition; rail manufacturers had increased the tonnage output "regardless of other considerations," through cutting the rolling time so far as not to allow the metal to cool gradually to a proper finishing temperature while being worked ("Report of Committee on Rails," *AREMWA Proc.* 6 [1905]: 177–78).

26. "Report of Committee on Rails" (1900), 117, 122 (quotation); Robert W. Hunt, "Notes on Rail Steel," *AIME Trans.* 35 (1904): 207–10 (quotation on 207–8).

27. "Report of Committee on Rails" (1900), 118 (quotations), 123, 124; P. H. Dudley, "The Rolling and Structure of Steel Rails," *Metallographist*

6 (1903): 111–29, especially 112–13; Hunt, "Notes on Rail Steel," 207–9. "The speed and momentum," Hunt observed (209), "are against the exercise of the proper kind of care necessary to produce sound ingots of the highest quality of steel."

28. "Report of Committee on Rails" (1900), 113 (quotation); Usselman, "Running the Machine," 318–20; Hunt, "Finishing Temperatures for Steel Rails," 460–63; "Report of Committee on Rails" (1901), 217.

29. "Report of Committee on Rails" (1900), 118; "Report of Committee on Rails" (1901), 200–201, 216; Hunt, "Finishing Temperatures for Steel Rails," 462–63; William R. Webster, "Specifications for Steel Rails," *AIME Trans.* 31 (1901): 449–58 (quotation on 450); "The Kennedy-Morrison Rail Finishing Process," *Metallographist* 4 (1901): 74–81.

30. "Report of Committee on Rails" (1901), 200–201; Simon S. Martin, "Rail Rolling at Lower Temperatures during 1901," *Metallographist* 5 (1902): 191–96; Martin, "Rail Temperatures," *ASTM Proc.* 2 (1902): 75–78; Sauveur, "Structure and Finishing Temperature" (*ASTM Proc.*), 79–96; Dudley, "Rolling and Structure of Steel Rails," 115–22. At Maryland Steel, so Dudley reported, the ingots were given extra care. For instance, the temperature of the soaking pit, which equalized the ingot temperature before the blooming process, was under control of the mill superintendent. The workmen who tended the heating furnace were "checked frequently by photomicrographs of the steel, to prevent over-heating, causing a course granular structure. This is really heat treatment commencing with the ingot" (ibid., 116). For "direct rolling" at Lackawanna, see Thomas E. Leary and Elizabeth C. Sholes, *From Fire to Rust: Business, Technology and Work at the Lackawanna Steel Plant, 1899–1983* (Buffalo: Buffalo and Erie County Historical Society, 1987), 25–29.

31. Edgar Bain, later the top physical metallurgist at U.S. Steel, described the "revelation" of seeing his first photomicrograph of steel at Ohio State in 1910: the microstructure was for the first time "really graspable . . . *here* was the iron carbide in actual platelets, clearly visible" (*Pioneering in Steel Research: A Personal Record* [Metals Park, Ohio: American Society for Metals, 1975], 1 [emphasis in original]).

32. Hunt, "Finishing Temperatures for Steel Rails," 458; Webster, "Specifications for Steel Rails," 450; Sauveur, "Structure and Finishing Temperature" (*ASTM Proc.*), 79–96; Albert Sauveur and H. C. Boynton, "Note on the Influence of the Rate of Cooling on the Structure of Steel," *Metallographist* 6 (1903): 148–55; H. M. Howe, "The Constitution and Thermal Treatment of Steel," *Metallographist* 6 (1903): 85–95; Robert Job, "Steel Rails: Relations between Structure and Durability," *Metallographist* 5 (1902): 177–91. Hunt and others cited William Metcalf's "classic" paper in *ASCE Trans.* 16 (1887): 283. On the use of microstructural metallurgy in railroad discussions, see the comments of Archibald Johnston (Bethlehem Steel) and Thomas H. Johnson (consulting engineer to the Pennsylvania Railroad) in Bethlehem Steel, "Notes from the Proceedings of the American Railway Assn. Meeting, Chicago, April 23rd, 1907," 3 June 1907, 7:33, AJ.

33. Usselman, "Running the Machine," 318–20; Hunt, "Finishing Temperatures for Steel Rails," 462; "Specifications for Steel Rails: A Discussion," *AIME Trans.* 31 (1901): 971–72; Webster, "The Present Situation as to the Specifications for Steel Rails," 167; "Report of Committee on Rails" (1901), 203, 204 (quotation); "Report of Committee on Rails" (1904), 466, 469, 479; Simon S. Martin, "Rail Temperatures," *ASTM Proc.* 2 (1902): 75–78; Sauveur, "Structure and Finishing Temperature" (*ASTM Proc.*), 92–95; "Proposed Modifications of the Standard Specifications for Steel Rails," *ASTM Proc.* 2 (1902): 23–49, especially 42–49; "Report of Committee on Rails" (1905), 190; "Report of Committee on Rails," *AREMWA Proc.* 7 (1906): 549–77, especially 552–53, 562–63. The 1904 AREMWA standard rail specification, which incorporated features of the ASTM and ASCE specifications, required that 33-foot Bessemer steel rails of 85 pounds shrink by no more than 6 inches and that those of 100 pounds shrink by no more than 6⅛ inches. Alternately, it specified that the mills control the finishing temperatures to be no more than 1,600°F for mills rolling from reheated blooms, and no more than 1,750°F for mills rolling direct from ingots. In 1906 a special joint committee from AREMWA and ASCE urged "the desirability of insisting upon . . . a shrinkage clause to control the finishing temperature" as well as a stringent drop-test requirement to ensure mechanical soundness ("Report of Committee on Rails" [1906], 551).

34. The top of the ingot was typically spongy (due to the gases that had not escaped before the metal solidified), piped (due to the hollow core that formed during solidification and contraction), or otherwise inferior (due to the segregation of chemical elements).

35. "Report of Committee on Rails" (1900), 128; "Specifications for Steel Rails: A Discussion," *AIME Trans.* 31 (1901): especially 976, 979; Webster, "The Present Situation as to the Specifications for Steel Rails," 168; "Report of Committee on Rails," *AREMWA Proc.* 3 (1902): 200–220, especially 207; P. H. Dudley, "The Best Metal for Rails," *Iron and Steel Magazine* 9 (1905): 534–37; "Proposed Modifications of the Standard Specifications for Steel Rails," 23–42; "Report of Committee on Rails" (1904), 465, 471–79; "Report of Committee on Rails" (1905), 182–89, 192–94; "Report of Committee on Rails" (1906), 549, 551, 559–62.

36. Robert W. Hunt, "Brief Note on Rail Specifications," *AIME Trans.* 27 (1897): 139–41 (quotation on 140); "Report of Committee on Rails" (1900), 123; Hunt, "Notes on Rail Steel," 209–10. "The combination among the rail makers is resulting in a constant deterioration in the quality of the material we are receiving," wrote one chief engineer (*Metallographist* 6 [1903]: 160). Beginning in May 1901 monthly prices for steel rails remained fixed at $28 per gross ton, while monthly prices for nails, beams, bars, billets, and pig iron fluctuated with market demand; "Fluctuations in the Prices of Crude and Finished Iron and Steel from January 1, 1898, to January 1, 1907," *Iron Age* suppl. (10 January 1907). The price of rails remained constant at $28 from 1902 through 1915; see American Iron and Steel Institute, *Annual Statistical Report for 1930* (New York: AISI, 1931), 96.

37. Specifically, the railroad engineers "normalized" the rail question: it was no longer a "crisis" but a matter of discussing the layout of blank forms to collect information on broken rails, the specifications of standard drop-testing machines, and other routine topics.

38. Technical details of the Talbot process are analyzed in Bethlehem's visitation reports; see A. Johnston, "Notes on a Visit to Pencoyd Iron Works," 21 November 1899, 19:2, AJ; and William Garrett to R. W. Davenport, 25 January 1900, 6:5, AJ. The development of the process can be traced in B. Talbot, "The Open-hearth Continuous Steel Process," *JISI* 57 (1900): 33–108; B. Talbot, "The Development of the Continuous Open-hearth Process," *JISI* 63 (1903): 57–94; G. A. Wilson, "The Talbot Continuous Steel Process and Its Benefits in Steel Making," *Iron and Steel Magazine* 11 (1906): 316–22; and Robert J. Norrell, *James Bowron: The Autobiography of a New South Industrialist* (Chapel Hill: University of North Carolina Press, 1991), xxvii, 96, 120.

39. F. Conlin to Daniel Willard, 11 July 1907, 10:61, AJ. Bethlehem Steel, "Notes from the Proceedings of the American Railway Assn. Meeting, Chicago, April 23rd, 1907," 2, gives comparative cost figures for Bessemer billets and rails:

Year	Bessemer Pig	Conversion and Profit	Billets	Conversion and Profit	Rails
1901	$16.00	$8.00	$24.00	$12.00	$28.00
1907	$23.50	$7.00	$30.50	$4.50	$28.00

40. On Jones & Laughlin, see John N. Ingham, *Making Iron and Steel: Independent Mills in Pittsburgh, 1820–1920* (Columbus: Ohio State University Press, 1991), 91–94; the (unpaged) booklet *Jones & Laughlin Steel Corporation, 1850–1941*, microfilm, J&L; and House Select Committee on Small Business, *Steel: Acquisitions, Mergers, and Expansion*, 17–18.

41. On Bethlehem's Cuban ore, see Misa, "Science, Technology and Industrial Structure," 255. For Bethlehem's campaign to sell open-hearth rails, see F. Conlin to L. E. Johnson (vice president of the Chicago, Burlington & Quincy Railroad), 7 September 1907 (quotation), 10:61, AJ, and a dozen other letters to railroad executives and overseas sales agents in the same file. On the company's financial difficulties during the 1907 panic, see Misa, "Science, Technology and Industrial Structure," 254–66.

42. See Bethlehem Steel, "Notes from the Proceedings of the American Railway Assn. Meeting, Chicago, April 23rd, 1907," 11 and 12 (quotations).

43. David C. Mowery and Nathan Rosenberg, *Technology and the Pursuit of Economic Growth* (Cambridge: Cambridge University Press, 1989), 58 (quotation); a similar point is made and nicely illustrated by Usselman, "Running the Machine," 288–89, 311–12. "Technical principles of far-reaching importance," stated P. H. Dudley, "must be stated, absorbed and digested by hundreds of minds before they can be reduced to practice in a commercial sense for railway interests" (discussion of Sauveur, "Struc-

ture and Finishing Temperature of Steel Rails," *ASTM Proc.* 2 [1902]: 86–87).

44. See the (unheeded) pleas for physical metallurgy in the following articles published in *AIME Transactions:* James C. Bayles, "Microscopic Analysis of the Structures of Iron and Steel," 11 (1882): 261–74; Bayles, "The Study of Iron and Steel," 13 (1884): 15–26; F. Lynwood Garrison, "The Microscopic Structure of Iron and Steel," 14 (1885): 64–75.

45. Hunt, "Finishing Temperatures for Steel Rails," 458; Albert Sauveur, "Structure and Finishing Temperature of Steel Rails," *Metallographist* 5 (1902): 197–202 (quotation on 198–99). See also *Dictionary of Scientific Biography,* s.v. "Sauveur, Albert," 12:126–27; Reginald A. Daly, "Albert Sauveur, 1863–1939," *National Academy of Science Biographical Memoirs* 22 (1943): 121–33. For other applied projects, see Sauveur, "Structure and Finishing Temperature" (*ASTM Proc.*), 79–96. See also Albert Sauveur, "Metallography Applied to Foundry Work," *Iron and Steel Magazine* 9 (1905): 547–53; 10 (1905): 29–32, 309–13, 413–19; 11 (1906): 119–24. See also Sauveur and Jasper Whiting, "The Detection of the Finishing Temperatures of Steel Rails by the Thermo-Magnetic Selector," *Iron and Steel Magazine* 7 (1904): 503–7; Sauveur, "The Casting of Pipeless Ingots by the Sauveur Overflow Method," *Metallographist* 6 (1903): 195–203; Sauveur, "The Practical Value of the Microscope in the Iron and Steel Industry," *AISI Yearbook* (1913): 208–32.

46. Allan Nevins, *Study in Power: John D. Rockefeller—Industrialist and Philanthropist,* 2 vols. (New York: Scribner's, 1953), 2:262. For analysis of the general problem of such deterministic schemes, see Misa, "Retrieving Sociotechnical Change," 115–41.

47. Andrew Carnegie to William Shinn, 1 May 1877, 29 March 1879, vol. 4, ACLC.

48. James Howard Bridge, *The Inside History of the Carnegie Steel Company* (New York: Aldine, 1903), 117–35 (quotation on 117); Wall, *Andrew Carnegie,* 471–72. See Carnegie Brothers & Co., Limited, articles of Consolidated Association, 1 April 1881, vol. 5, ACLC.

49. Bridge, *Inside History,* 117–35 (quotation on 135); Wall, *Andrew Carnegie,* 471–72.

50. Wall, *Andrew Carnegie,* 472. Bridge concurs: "The advantages of industrial consolidation had not, at this date, received any general recognition" (*Inside History,* 135).

51. Carnegie Phipps & Co., articles of limited copartnership, 4 January 1886, vol. 8, ACLC; Wall, *Andrew Carnegie,* 472, 489.

52. See Alfred D. Chandler, Jr., *Scale and Scope: The Dynamics of Industrial Capitalism* (Cambridge: Belknap Press of Harvard University Press, 1990), 129.

53. Wall, *Andrew Carnegie,* 475–76, 486–89, 535; George Harvey, *Henry Clay Frick—The Man* (New York: Charles Scribner's Sons, 1928), 101–2; Bridge, *Inside History,* 150–69.

54. Bridge, *Inside History,* 168. Bridge ascribed the rationalization of this company to Frick, who from January 1889 was chairman of Carnegie

Brothers & Company; in January 1889 he was also elected a director of Carnegie, Phipps & Company but "took no part in administration" of that company (Harvey, *Henry Clay Frick*, 101).

55. Voting was by number of shares each person held; since Andrew Carnegie held more than half the shares he could not be ousted by any combination of his partners. Also, the repurchase at the "book value," which was substantially *less* than the fair market value, advantaged the partnership over the individual (or his estate).

56. See Carnegie Brothers and Henry C. Frick, articles of agreement, 1 January 1887, vol. 10, ACLC; Harvey, *Henry Clay Frick*, 76–92; Wall, *Andrew Carnegie*, 478–86, 489–93, 497, 587 (quotation). Without documentation Burton J. Hendrick tells this apocryphal story: "On a Sunday afternoon [about 1881] Carnegie, with a few friends, was strolling over the hills near his Cresson home, discoursing, as was his habit, on several subjects. Suddenly pausing, apparently struck with a new idea, he remarked quietly but with determination: 'We must attach this young man Frick to our concern. He has great ability and great energy. Moreover, he has the coke—and we need it'" (*The Life of Andrew Carnegie*, 2 vols. [New York, 1932; reprint, New York: Harper and Row, 1969], 1:287).

57. To my knowledge, there is no satisfactory account of Frick's rationalization of Carnegie's steel holdings during 1889–92 (before the Homestead strike) or for that matter during 1893–99. Frick's biographer skips over this period in four vague paragraphs (Harvey, *Henry Clay Frick*, 93–94, 99). Joseph Wall's magnificent biography of Frick's boss devotes little attention to these changes (Wall, *Andrew Carnegie*, 497, 506). Early biographies of Carnegie say little about Frick (Hendrick, *The Life of Andrew Carnegie*). The best account is Bridge, *Inside History*, 167–83, 295–97; however, Bridge takes several pages to say surprisingly little about what exactly Frick did. John N. Ingham ("Frick, Henry Clay," in *Iron and Steel in the Nineteenth Century*, ed. Paskoff, 123–38, especially 129) emphasizes that Frick was responsible for building the Union Railroad that linked up the scattered Carnegie enterprises. All commentators agree that Frick's most advantageous move for the Carnegie company during 1889–92 was his purchase of the Allegheny Bessemer Steel Company (later Duquesne Steel Company), rather than his managerial rationalization. For discussion of the Homestead strike, see chapter 7.

58. Bridge, *Inside History*, 257 (quotation), 268.

59. David A. Walker, *Iron Frontier: The Discovery and Early Development of Minnesota's Three Ranges* (St. Paul: Minnesota Historical Society Press, 1979), 73–100 (quotations on 74, 100); Wall, *Andrew Carnegie*, 587–94.

60. Bridge, *Inside History*, 257–59 (quotations on 259); Wall, *Andrew Carnegie*, 594–97; Walker, *Iron Frontier*, 101–18; Hendrick, *The Life of Andrew Carnegie*, 2:12–14; Terry S. Reynolds, "Oliver, Henry William," in *Iron and Steel in the Nineteenth Century*, ed. Paskoff, 259–62. Oliver's dealings with Frick and the Carnegie concern can be followed in H. W. Oliver to H. C. Frick, 2 June 1891, and Agreement between Carnegie

Brothers, John T. Terry, Henry W. Oliver, and Baltimore & Ohio Railroad, 29 January 1891, Box 1, OIS.

61. Nevins, *Study in Power*, 2:245–75 (quotation on 255).

62. Bridge, *Inside History*, 258–60, 261 (quotation); Wall, *Andrew Carnegie*, 598–601; Walker, *Iron Frontier*, 127–210; Reynolds, "Oliver, Henry William," 262–63.

63. See the memorandum of agreement between Carnegie Steel and Henry W. Oliver, 12 January 1897, 1:1, OIS.

64. H. W. Oliver to H. C. Frick, 27 July 1897, vol. 43, ACLC. Oliver elaborated: "I propose at a risk of using our credit to the extent of $500,000, or possibly one million dollars, to effect a saving, in which our competitors will not share, of four to six million dollars per annum . . . I desire to impress upon you the fact, that if it had not been for our Rockefeller-Mesaba [i.e., Mesabi] deal of last year, with the consequent demoralization in the trade caused by the publication thereof, it would not have been possible for us to now secure the other Range properties I propose to acquire, either by lease or for any reasonable price. We simply knocked the price of ore from $4.00 down to say $2.50 per ton. Now let us take advantage of our action before a season of good times gives the ore producers strength and opportunity to get together by combination" (ibid.).

65. Carnegie's annotations on H. W. Oliver to H. C. Frick, 29 July 1897, vol. 43, ACLC.

66. See Carnegie's extensive annotations on C. M. Schwab to H. C. Frick, 30 July 1897, vol. 43, ACLC.

67. Andrew Carnegie, memorandum of 18 August read into minutes, Board of Managers, 31 August 1897, vol. 43, ACLC.

68. Board of Managers minutes, 3 August 1897, vol. 43, ACLC. In late July Frick recommended purchasing sufficient additional ore to yield a surplus of up to 500,000 tons by next 1 June (a cushion beyond what was necessary to hold the company over the winter, until the next year's shipping season would be fully in swing).

69. Frick continued, "Tilden Pioneer favored. Have Hultsz [Hulst] examine Tilden especially as to manganese. Get option Pioneer lease. Advise leave [wage] bonus alone. Sail Saturday" (H. C. Frick to Carnegie Steel Pittsburgh, 17 August 1897, vol. 43, ACLC). Wall erroneously attributes this cable to Carnegie. In fact, it was not Carnegie (who had just signed the papers to purchase the castle at Skibo) but Frick who was sailing that Saturday. Frick had been at Cluny Castle, in Scotland, on 18 August conferring with Carnegie, Phipps, and Lauder about armor. Schwab explicitly attributed the cable to Frick (C. M. Schwab to A. Carnegie, 20 August 1897, vol. 43, ACLC).

70. H. C. Frick to H. Phipps, Jr., 31 August 1897, vol. 43, ACLC.

71. A. Carnegie to H. C. Frick, 20 September 1897, vol. 43, ACLC.

72. C. M. Schwab to Board of Managers, 6 July 1897, vol. 43, ACLC.

73. D. M. Clemson to C. M. Schwab, 19 July 1897, vol. 43, ACLC. The Board of Managers minutes (13 July 1897, vol. 43, ACLC) report that Carnegie Steel's option to buy ore from the Norrie mines had expired on

13 July 1897, but that Frick, Schwab, and the board all agreed to let the option expire and not make any further ore purchases.

74. C. M. Schwab to A. Carnegie, 20 August 1897, C. M. Schwab to H. C. Frick, 30 July 1897 (the latter extensively annotated by Carnegie), and C. M. Schwab, H. M. Curry, and D. M. Clemson to H. C. Frick (c/o Millard Hunsiker, London), 30 July 1897, vol. 43, ACLC.

75. C. M. Schwab to A. Carnegie, 20 August 1897, vol. 43, ACLC.

76. A. Carnegie to Carnegie Pittsburgh (cablegram), 21 September 1897, and A. Carnegie to H. M. Curry, 27 September 1897, vol. 43, ACLC; Wall, *Andrew Carnegie*, 603–4 (quotation).

77. Wall, *Andrew Carnegie*, 602–7; Walker, *Iron Frontier*, 211–16; Bridge, *Inside History*, 268.

78. Robert Hessen, *Steel Titan: The Life of Charles M. Schwab* (New York: Oxford University Press, 1975), 29–30, 38, 42–58, 68–110, 123–88, 307–10. Hessen reports that Carnegie's confidence in Schwab was undiminished even by a public scandal in 1893–94 concerning irregularities in testing armor plates for the U.S. Navy at the Homestead works, where Schwab was general superintendent.

79. Wall, *Andrew Carnegie*, 784. Hendrick (*The Life of Andrew Carnegie*, 2:129–31) gives a longer version of the speech, noting that Schwab delivered it without notes and never reduced it to writing and that his account (as well as others) is based on Schwab's summary made for the Stanley Committee in 1911 and as a witness in the suit to dissolve U.S. Steel. Compare yet another version in Hessen, *Steel Titan*, 114–16.

80. For the intricate financial machinations necessary to float the mammoth U.S. Steel issue, see Carosso, *The Morgans*, 470–74.

81. Charles M. Schwab, "What May Be Expected in the Steel and Iron Industry," *North American Review* 172 (May 1901): 655–64.

82. See Lamoreaux, *Great Merger Movement;* my account of the motivations behind the formation of U.S. Steel differs from her structural account (62–86), but I agree broadly with her assessment of barriers to entry (142–56).

83. Wall, *Andrew Carnegie*, 616–21, 774–76; Hendrick, *The Life of Andrew Carnegie*, 2:25–34, 123, 125; Francis Lovejoy to Andrew Carnegie, 7 September 1898, vol. 54, ACLC.

84. Wall, *Andrew Carnegie*, 776–80; Carosso, *The Morgans*, 467; Carnegie, *Autobiography*, 152. On Gould see Maury Klein, *The Life and Legend of Jay Gould* (Baltimore: Johns Hopkins University Press, 1986). George Gould's attempts to construct a transcontinental railroad system ran afoul of E. H. Harriman, and in time George was "utterly routed" (Klein, *Life and Legend of Jay Gould,* 486).

85. Carosso, *The Morgans*, 466; Hessen, *Steel Titan*, 112.

86. For the post-1897 rationalization of the transportation system, see Bridge, *Inside History*, 268–74; Harvey, *Henry Clay Frick*, 197–98; Wall, *Andrew Carnegie*, 612–23, 776. Bridge and Harvey portray Frick as the leading figure, a role that Wall attributes to Carnegie.

87. The Pittsburgh, Bessemer & Lake Erie's president wrote Carnegie

that, despite the low freight rate of 3³⁷/₁₀₀ mills per ton-mile, "the railroad has done pretty well. The average train load is much above what we once thought possible especially when you consider . . . the ore cars go back empty" (J. H. Reed to Andrew Carnegie, 7 September 1898, vol. 54, ACLC). I take the phrase "tight coupling" from Charles Perrow, *Normal Accidents: Living with High-Risk Technologies* (New York: Basic Books, 1984), 93–94.

88. Andrew Carnegie to Arthur G. Yates (president of the Buffalo, Rochester & Pittsburgh Railroad), 28 September 1897, vol. 43, ACLC.

89. On Morgan's activities in the 1890s to consolidate and rationalize the industrial system, see Louis Galambos and Joseph Pratt, *The Rise of the Corporate Commonwealth: United States Business and Public Policy in the 20th Century* (New York: Basic Books, 1988), 1–42; Carosso, *The Morgans*, 352–96, 433–64.

90. Allen, *Great Pierpont Morgan*, 171 (quotation); Carosso, *The Morgans*, 467.

91. Wall, *Andrew Carnegie*, 782–83.

92. Walker, *Iron Frontier*, 226–27. Concerning the formation of U.S. Steel, Walker observes, "Among the large holders of Minnesota ore lands, only James J. Hill had been overlooked." And in 1907 U.S. Steel completed an agreement to control Hill's Great Northern properties, the sole remaining large and independently held ore properties in the Lake Superior district (ibid., 272).

93. See Charles A. Gulick, *Labor Policy of the United States Steel Corporation* (New York: Columbia University Press, 1924; reprint, New York: AMS Press, 1968), especially 93–110; David Brody, *Labor in Crisis: The Steel Strike of 1919* (Philadelphia: J. B. Lippincott, 1965), especially 13–44, 147–61; Isaac James Quillen, *Industrial City: A History of Gary, Indiana to 1929* (diss., Yale University, 1942; New York: Garland, 1986), 310–78, 485–504.

94. "For many years after its formation U.S. Steel continued to be a holding company that administered its many subsidiaries though a very small general office. Except for the Carnegie Company, and, after 1907, the Tennessee Coal and Iron Company, these subsidiaries were single-function companies in mining, transportation, coke, metal production, and fabrication. The general office did little to coordinate, plan, and evaluate for the activities of the subsidiaries. Only in foreign purchases and sales was there any clear central direction" (Alfred D. Chandler, Jr., *The Visible Hand: The Managerial Revolution in American Business* [Cambridge: Belknap Press of Harvard University Press, 1977], 361).

95. From 1903 to 1913, the largest standard I-beam offered by the company remained precisely the same: height of 24 inches, flange width of 7 inches, and web thickness of 0.5 inches. In 1913 the Carnegie handbook did list as a "supplemental beam" one with a slightly narrower web (0.39 inches) but with height and flange width identical to the 1903 beam.

96. For a detailed study of how Bethlehem Steel built a structural steel mill to the design of Henry Grey, the inventor of a "universal" mill that

rolled the wide-flanged H-beams, as well as an analysis of the operation of the Pittsburgh-plus pricing system, see Misa, "Science, Technology, and Industrial Structure," 247–91.

97. Some critics of U.S. Steel miss this point; for instance, see Thomas M. Doerflinger and Jack L. Rivkin, *Risk and Reward: Venture Capital in the Making of America's Great Industries* (New York: Random House, 1987), 83–114.

98. On U.S. Steel's acquisition of Tennessee Coal & Iron, see Ida M. Tarbell, *The Life of Elbert H. Gary* (New York: D. Appleton, 1926), 196–201; Norrell, *James Bowron*, xxix–xxxi; Douglas A. Fisher, *Steel Serves the Nation: The Fifty Year Story of United States Steel* (New York: privately printed, 1951), 40–41.

99. Quillen, *Industrial City*, 196–206 and passim.

100. On Schwab at Bethlehem, see Misa, "Science, Technology, and Industrial Structure," 237–42, 260–66.

101. On Gary's efforts to purge the "Carnegie boys," see folders 6:8 and 7:1, WBD; and Gerald G. Eggert, *Steelmasters and Labor Reform, 1886–1923* (Pittsburgh: University of Pittsburgh Press, 1981), 56–76. Biographies of the fifty-one "Carnegie boys" can be found in Dickson, *History of Carnegie Veteran Association*.

5. The Reform of Factories, 1895–1915

1. E. G. Thomsen, "F. W. Taylor—A Historical Perspective," in *On the Art of Cutting Metals—75 Years Later*, ed. L. Kops and S. Ramalingam (New York: ASME, 1982), 1–12; *Engineering News* (13 December 1906): 623; "United Engineering Building—Andrew Carnegie Lays Cornerstone West 39th St.," *New York Times* (9 May 1906): 8; "United Engineering Society Building Dedicated," *New York Times* (17 April 1907): 18.

2. Bruce Sinclair, *A Centennial History of the American Society of Mechanical Engineers, 1880–1980* (Toronto: University of Toronto Press, 1980), 84–94 (quotation on 86); *Engineering News* (13 December 1906): 620.

3. *Engineering News* (6 December 1906): 585.

4. Discussion of Frederick W. Taylor, "On the Art of Cutting Metals," *ASME Trans.* 28 (1907): 31–350, especially 283, 290–91. On the "shop culture" that nurtured this generation of mechanical engineers, including Taylor, see Monte Calvert, *The Mechanical Engineer in America, 1830–1910: Professional Cultures in Conflict* (Baltimore: Johns Hopkins Press, 1967).

5. Discussion of Taylor, "Art of Cutting Metals," 295, 301–3.

6. While high-speed steel spread quickly across the United States and to Europe, the Taylor system of management was installed complete in just six establishments; various portions of the system were installed in at least forty-six industrial companies and two government manufacturing concerns (Watertown Arsenal and Mare Island Shipyard). See Daniel Nelson, "Scientific Management in Retrospect," in *A Mental Revolution: Scientific Management since Taylor*, ed. Daniel Nelson (Columbus: Ohio

State University Press, 1992), 5–39, especially 17–18; Nelson, *Managers and Workers: Origins of the New Factory System in the United States, 1880–1920* (Madison: University of Wisconsin Press, 1975), 71–72.

7. For the hypothesis that Taylor's invention of high-speed-tool steel impelled him to create an arsenal of management techniques to speed work through machine shops and speed up the machinists, see Hugh G. J. Aitken, *Scientific Management in Action: Taylorism at Watertown Arsenal, 1908–1915* (Cambridge: Harvard University Press, 1960; reprint, Princeton: Princeton University Press, 1985).

8. Nonmetals firms employing 6,000–8,000 wage earners were Armour (meat packing) and Amoskeag Mills (textiles). See Nelson, *Managers and Workers,* 7.

9. Ingham, *Making Iron and Steel,* 39 (quotation). According to Marcus A. Grossmann and Edgar C. Bain, "The preparation of high speed steel is carried out most simply by melting the proper ingredients together in a crucible. This was the earliest method of manufacture, and was for many years the most widely used. It is still not uncommon to find it regarded as the best, or at least the safest for uniformity of 'quality.' In the United States at present [i.e., 1931] a very much larger amount of high speed steel is made in the electric furnace than is made in the crucible, but this has become true only in the past decade" (*High Speed Steel* [New York: Wiley, 1931], 1).

10. For the founding of Sheffield steel, see K. C. Barraclough, *Steelmaking before Bessemer,* 2 vols. (London: Metals Society, 1984), 2:1–30, fig. 2; and the sources cited in Misa, "Science, Technology and Industrial Structure," 151 n.1.

11. Barraclough, *Steelmaking before Bessemer,* 2:59–63 (quotation on 63).

12. Geoffrey Tweedale, "Sheffield Steel and America: Aspects of the Atlantic Migration of Special Steelmaking Technology, 1850–1930," *Business History* 25 (1983): 225–39 (quotation on 226); Tweedale, *Sheffield Steel and America: A Century of Commercial and Technological Interdependence, 1830–1930* (Cambridge: Cambridge University Press, 1987), 3–9.

13. S. T. Wellman to F. W. Ballard, 12 June 1917, Folder Correspondence, STW. Wellman placed the date of Anderson, Cook's crucible-steel furnace at 1867. By the late 1870s Pittsburghers were beginning to modify orthodox Sheffield practices at the Black Diamond Works; see Tweedale, *Sheffield Steel and America,* 16–19; Barraclough, *Steelmaking before Bessemer,* 2:226–27, 353–55.

14. Geoffrey Tweedale, "Metallurgy and Technological Change: A Case Study of Sheffield Specialty Steel and America, 1830–1930," *Technology and Culture* 27 (1986): 200–210; Tweedale, "Transatlantic Specialty Steels: Sheffield High-Grade Steel Firms and the USA, 1860–1940," in *British Multinationals: Origins, Management and Performance,* ed. G. Jones (Aldershot: Gower, 1986): 75–95, especially 86–89.

15. T. Holland Nelson, "Comparison of English and American Meth-

ods of Producing High Grade Crucible Steel," *ASST Trans.* 3 (1922–23): 279–98; Tweedale, *Sheffield Steel and America*, 14–15, 37–41.

16. Tweedale, *Sheffield Steel and America*, 57–68.

17. H. S. Snyder, memo, 6 August 1907, 17:2, AJ. In *Sheffield Steel and America* Tweedale asserts that "Sheffield and not America reaped the commercial benefits of the process [high-speed steel] in its early days" (71). He also notes that "the introduction of high-speed steel revitalised Sheffield's tool steel trade with America" (100). But such evidence as he gives (see his table 8.1) tends to confirm my interpretation: that compared with the manufacture of carbon tool steels (thoroughly dominated by Sheffielders), high-speed-tool steels *reduced* Sheffield's share of the U.S. market for tool steels.

18. *Metallographist* 3 (1900): 176, and 5 (1902): 252, 341. See also James J. Mahon, "The Microscope in Crucible Steel Manufacture," *Metallographist* 6 (1903): 165–66; and James A. Aupperle, "An Etching Method for Determining Whether Steel Has Been Made by the Crucible Process," *Iron and Steel Magazine* 11 (1906): 383–85.

19. E. T. Clarage, "The Manufacture of Tool Steel," *American Machinist* 29 (1 November 1906): 573–76 (quotation on 576).

20. Daniel Nelson, *Frederick W. Taylor and the Rise of Scientific Management* (Madison: University of Wisconsin Press, 1980), 21–28. For a charming photograph of Taylor as "Miss Lillian Gray," see Charles D. Wrege and Ronald G. Greenwood, *Frederick W. Taylor: The Father of Scientific Management—Myth and Reality* (Homewood, Ill.: Business One Irwin, 1991), pls. after 210.

21. On the founding of Midvale Steel, see Tweedale, *Sheffield Steel and America*, 114–16; C. D. Wrege and R. G. Greenwood, "The Early History of Midvale Steel and the Work of Frederick W. Taylor, 1865–90," *Canal History and Technology Proceedings* 11 (1992): 145–76; "The William Butcher Steel Works," *JFI* 84 (1867): 293–94; C. M. Woodward, *A History of the St. Louis Bridge* (St. Louis: G. I. Jones, 1881), 68, 79–91, 97–105.

22. Nelson, *Frederick W. Taylor*, 29–46; Taylor, "Art of Cutting Metals," 31–350.

23. Letter from William A. Fannon in *Frederick W. Taylor: A Memorial Volume* (New York: Taylor Society, 1920), 98–108 (quotation on 99); Wrege and Greenwood, "Early History of Midvale Steel," 145 and 163 (quotations); Nelson, *Frederick W. Taylor*, 48.

24. Nelson, *Frederick W. Taylor*, 47–75; Wrege and Greenwood, *Frederick W. Taylor*, 61. On Taylor's move from Midvale to Manufacturing Investment, see the following items in *Frederick W. Taylor: A Memorial Volume:* Sanford E. Thompson, "Stages between Midvale and Bethlehem," 66–71; statement of Admiral Chauncey F. Goodrich, 81; letter from William A. Fannon, 98–108, especially 102–3.

25. "The special dangers to be avoided in hardening each kind of tool must be learned by experience," one Sheffield steelmaker noted. "Some

tools will warp, or 'skeller,' as we say in Yorkshire, if they are not plunged into the water in a certain way. Tools of one shape must cut the water like a knife; those of another shape must stab it like a dagger. Some tools must be hardened in a saturated solution of salt, the older the better; whilst others are best hardened under a stream of running water" (Henry Seebohm, "On the Manufacture of Crucible Cast Steel," *JISI* 2 [1884]: 391); and T. S. Ashton, *An Eighteenth-Century Industrialist: Peter Stubs of Warrington* (Manchester: Manchester University Press, 1939), 5 and passim.

26. Frank B. Copley, *Frederick W. Taylor: Father of Scientific Management*, 2 vols. (New York: Harper, 1923), 1:429–44 (quotation on 432). Mushet's original "self-hardening" steel contained 5.5 percent tungsten and 0.25–0.5 percent chromium; R. W. Davenport to R. P. Linderman, 25 January 1900, 17:2, AJ.

27. Taylor, "Art of Cutting Metals," 38–39, 50–51.

28. Taylor to Cramp, quoted in Copley, *Taylor,* 1:439–41. Taylor's assistant Leighton Lee noted that although the Cramp machine shop resisted the tool-room system, the tool room in the yard "proved a success and later on its size was doubled and the tools from *all* the outside departments were consolidated in the one room—thus doing away with three tool room keepers and their assistants" (L. Lee to F. W. Taylor, 10 August 1898, 34B, FWT).

29. F. W. Taylor to L. Lee, 12 August 1898, 34B, FWT.

30. Nelson, *Frederick W. Taylor,* 76–81 (quotation on 80–81); Nelson, "Taylorism and the Workers at Bethlehem Steel, 1898–1901," *Pennsylvania Magazine of History and Biography* 101 (1977): 487–505.

31. "Many of the schemes proposed by Fred Taylor had a great deal of merit, however . . . he personally did not seem to have the ability to carry them out in a reasonable time. This was due principally to the antagonistic methods used by him in handling men," wrote Archibald Johnston, then Bethlehem's president and formerly its armor department head, who had used Taylor's methods (A. Johnston, memo, 18 November 1909, 11:46, AJ).

32. Nelson, *Frederick W. Taylor,* 81–83 (quotation on 81), 93. F. W. Taylor to R. P. Linderman, 16 August 1898, Lot 241 Knisely, BS. Taylor's remark, "As far as the engineering side of this works go I have at present no authority whatever," is a puzzle (F. W. Taylor to L. Lee, 12 August 1898, 34B, FWT).

33. E. S. Knisely notebooks (1899–1900), Lot 241 Knisely, BS; Nelson, *Frederick W. Taylor,* 83–85.

34. Taylor, "Art of Cutting Metals," 258.

35. From this point my account of the Taylor-White experiments departs from Nelson's (*Frederick W. Taylor,* 86–91). I have relied on the Taylor papers and the previously unused Bethlehem documents. Evidence from the Taylor papers that one of Taylor and White's assistants, Edmund Lewis, may have accidentally heated up a tool to the high heat that produced the hardened high-speed steel is presented in Charles D. Wrege and Ronald G. Greenwood, "Frederick W. Taylor's Work at Bethlehem

Steel, Phase II: The Discovery of High-Speed Tool Steel, 1898," *Canal History and Technology Proceedings* 13 (1994): 115–63.

36. E. S. Knisely, History of Several MESH Tools (22–31 October 1898), 1–2, 17 October 1899, Lot 241—E. S. Knisely: Tool Steel Reports, BS; R. W. Davenport to R. P. Linderman, 25 January 1900, 6, 17:2, AJ; F. W. Taylor to W. H. Wahl, 26 August 1902, 5M, FWT; [J. Welden and M. White], "Reports on Early History of Inventions by Welden and White," 5 October 1899, 32, FWT; M. White to R. W. Davenport, 5 January 1900, 32E, FWT.

37. Knisely, History of Several MESH Tools (22–31 October 1898), 2; Bethlehem Steel, Experiment on Cutting Steel, 31 October 1898 (experiments 208–11), Lot 241—E. S. Knisely: Tool Steel Reports, BS; Taylor, "Art of Cutting Metals," 51 (quotation); R. W. Davenport, "Scheme of Experiment no. 25," 28 October 1898 (quotation); copy in M. White to R. W. Davenport, 5 January 1900, 32E, FWT; R. W. Davenport to R. P. Linderman, 25 January 1900, 17:2, AJ; Welden and White, "Reports," 3.

38. Knisely, History of Several MESH Tools (22–31 October 1898), 3–7; Bethlehem Steel, Experiment on Cutting Steel, 31 October 1898 (experiments 212–15 and following); M. White to R. W. Davenport, 5 January 1900, 3–4, 32E, FWT; data appended to Welden and White, "Reports."

39. "This treatment [of quenching and annealing to Taylor and White's 1906 specifications] is essentially the one still in use today, and is seldom deviated from, except in certain special cases," wrote Grossmann and Bain in 1931 (*High Speed Steel*, 63).

40. M. White to R. W. Davenport, 5 January 1900, 5, 32E, FWT; Welden and White, "Reports," 12.

41. White responded to Davenport's request to date the "air-hardening tool steel treatment invention" by noting that "this invention was made on or before October 31st, 1898" (M. White to R. W. Davenport, 5 January 1900, 32E, FWT). See the discussion in F. W. Taylor to W. H. Wahl, 26 August 1902, 5M, FWT; M. W. Welsh, memo, 22 November 1898, Lot 241 Knisely, BS.

42. See M. W. Welsh, Report on Heat Treating and Lathe Testing Tool Steel, 22 November 1898 (Midvale), as well as later reports dated 19 December 1898 (Sanderson), 20 December 1898 (Carpenter), 21 December 1898 (Atha), 29 August 1899 (Crescent); all in Lot 241—E. S. Knisely: Tool Steel Reports, BS. For a description of the Midvale trial, see Misa, "Science, Technology and Industrial Structure," 169–73.

43. M. W. Welsh, memo, 22 November 1898, 5–6, Lot 241 Knisely, BS; R. W. Davenport to R. P. Linderman, 25 January 1900, 17:2, AJ. A later account described the lathe as "a 66-inch Bement-Miles lathe . . . belted to an Evans friction-cone countershaft . . . geared to a 40 horsepower Westinghouse motor" (Charles Day et al., "The Taylor-White Process of Treating Tool-Steel," *Metallographist* 6 [1903]: 131).

44. "Apropos of measuring differences of temperature with the eye . . . I have sometimes been impressed with what seemed a very striking dif-

ference in temperature between adjacent bodies, though the difference certainly must have been considerably less than 10°C.," noted the metallurgist Henry Howe (discussion of H. H. Campbell, "The Open-Hearth Process," *AIME Trans.* 22 [1893]: 692). See also Alex W. Bealer, *The Art of Blacksmithing* (New York: Funk and Wagnalls, 1969), 122–24, 149–55.

45. H. M. Howe to Bethlehem Steel, 20 December 1895, and H. Souther to R. W. Davenport, 19 December 1895, 17:2, AJ. For Henry Souther's effort to standardize specifications for automobile steels, see chapter 6.

46. Bethlehem Steel memo, 14 April 1899, 5, Lot 241 Knisely, BS.

47. Concerning individual interpretation, Archibald Johnston, head of Bethlehem's armor department, stated: "We at 5H. never work to colors for the reason that we are unable to find any two persons who will agree as to the color of a heat" (A. Johnston to F. W. Taylor, 17 July 1899, 17:2, AJ). See also F. W. Taylor to A. Johnston, 15 July 1899, 17:2, AJ.

48. M. White and F. W. Taylor, "Colors of Heated Steel Corresponding to Different Degrees of Temperatures," paper presented to ASME, December 1899 (to form part of *ASME Trans.* 30), 17:2, AJ; Maunsel White and F. W. Taylor, "Colors of Heated Steel Corresponding to Different Degrees of Temperatures," *Metallographist* 3 (1900): 41–43; E. M. McIlvain to C. M. Schwab, 15 November 1901, 2:31, AJ.

49. Henry M. Howe, "The Color-Names for High Temperatures," *Metallographist* 3 (1900): 44, 46.

50. C. W. Waidner, "Methods of Pyrometry," *Iron and Steel Magazine* 8 (1904): 539–45 (quotation on 539–40); Daniel Kevles, *The Physicists* (New York: Knopf, 1977), 66–67, 72–74, 81; and Rexford C. Cochrane, *Measures for Progress: A History of the National Bureau of Standards* (Washington, D.C.: National Bureau of Standards, 1966), 61, 65–68, 78, 111–12, 127–28. These were pioneering efforts. After 1910 the automobile industry rapidly adopted pyrometers and devised an entire technology for heat-treating steel; see chapter 6.

51. R. W. Davenport to R. P. Linderman, 25 January 1900, 3, 17:2, AJ.

52. Ibid., 5–6.

53. Ibid., 3–4; compare *American Machinist* 25 (3 July 1902): 959.

54. Circuit Court of the United States (New Jersey), *Bethlehem Steel Company v. Niles-Bement-Pond Company,* 29 January 1909, 29, 198/1 MB (hereafter cited as *Bethlehem v. Niles-Bement-Pond*); *Iron Trade Review* 33 (9 August 1900): 17; "The Taylor-White Process of Treating Steel," *American Machinist* 23 (16 August 1900): 784.

55. R. W. Davenport to R. P. Linderman, 25 January 1900, 8, 17:2, AJ. See F. W. Taylor and M. White, "Metal-Cutting Tool and Method of Making Same," U.S. Patents 668,269 and 668,270 (February 1901).

56. R. W. Davenport to R. P. Linderman, 25 January 1900, 10–11, 17:2, AJ; F. W. Taylor to Capt. C. F. Goodrich, 16 July 1900, 60A1, FWT; F. W. Taylor to W. H. Wahl, 11 March 1901, 5M, FWT.

57. F. W. Taylor to Capt. C. F. Goodrich, 16 July 1900 (quotation),

60A1, FWT; R. W. Davenport to R. P. Linderman, 25 January 1900, 9–10, 17:2, AJ; E. S. Knisely to M. White, 6 September 1900, Lot 241 Knisely, BS.

58. *American Machinist* 23 (31 July 1900): 756; *Iron Trade Review* 33 (9 August 1900): 18; *Iron Age* 66 (2 August 1900): 19.

59. Franz Reuleaux, "Bericht über die Sitzung vom 5. November 1900," copy with translation, 15 January [1901], 3–4, 92D, FWT; D. V. Merrick to Kempton Taylor, 16 August 1915, 12T, FWT.

60. Hessen, *Steel Titan,* 147; Copley, *Taylor,* 2:153, 156, 158–62, 164 (quotation); Trigger (pseud.), "The Riddle of the Sphinx," *American Machinist* 25 (20 March 1902): 416. Confusingly, Schwab soon sold Bethlehem to a Morgan-backed syndicate and then regained the company in 1904.

61. H. S. Snyder, memo, 6 August 1907, 17:2, AJ. The sixteen companies (with exceptions to the standard $3,000 license fee) were: Link Belt Engineering, Allentown Rolling Mills, Deering Harvester ($5,000), Lodge & Shipley Machine Tool, Farrel Foundry & Machine ($4,000), Richmond Locomotive & Machine Works ($4,000), Pennsylvania Railroad, Hewes & Phillips Iron Works, Southwark Foundry & Machine ($4,000), St. Louis & San Francisco Railroad, Lehigh Valley Railroad, Morgan Engineering ($5,300), Cleveland Frog & Crossing, Gueber Engineering, Jeanesville Iron Works, Charles E. Machold ($1,000). Carnegie Steel obtained a license on 2 January 1901 for an undisclosed sum. William Sellers & Company obtained a license directly from Taylor on 31 October 1901 for $6,000.

62. H. S. Snyder, memo, 6 August 1907, 17:2, AJ; A. N. Cleaver to F. W. Taylor, 18 July 1901, 33F, FWT. For the rights to Norway, Sweden, and Denmark, A. E. Johler, through V. H. Meyer, paid $11,250; and for Canadian rights Goldie, McCulloch & Company, Ltd., paid $15,000.

63. See Copy of Description Sent Link Belt Eng. Co. (to treat tools made from air-hardening steel by the Taylor-White process), 1900, 92D, FWT; and Bethlehem Steel, memo, 1900, 17:2, AJ.

64. See Olivier Zunz, *Making America Corporate, 1870–1920* (Chicago: University of Chicago Press, 1990), 125–48; Misa, "Science, Technology and Industrial Structure," 152–53.

65. See *American Machinist* 25 (6 November 1902): ad section 4, 32. See also the following articles in *American Machinist* 26 (21 May 1903): Oberlin Smith, "High Speed Metal Cutting," 730–31, 846–47; H. M. Norris, "The Inefficiency of the Standard Lathe for High Speed Cutting," 838–40; William Lodge, "The Capabilities of High Speed Steel," 841–42; John E. Sweet, "What Are the New Machine Tools to Be?" 1734, 1788.

66. William Lodge, "High-Speed Steels," *American Machinist* 27 (31 March 1904): 428–29; Oscar E. Perrigo, *Modern American Lathe Practice* (New York: Henley, 1907), 231 (quotation); Le Roy Tabor (Eclipse Machine Co., Elmira N.Y.), "Experience with the Latest Steels," *American Machinist* 28 (26 January 1905): 117; A. D. Wilt, Jr., "The Introduction of High-Speed Steel into a Factory," *Engineering Magazine* 27 (1904): 913–20 (quotation on 918–19); Charles Day, "Advanced Practice in Economi-

cal Metal Cutting," *Engineering Magazine* 27 (1904): 549–66 (quotation on 550).

67. J. M. Gledhill, "The Development and Use of High-Speed Steel," *Iron and Steel Magazine* 9 (1905): 19–44, especially 43; James M. Dodge, "Results of the Use of the Taylor System of Shop Management," *American Machinist* 29 (8 March 1906): 326–27; Oberlin Smith, discussion of Taylor, "Art of Cutting Metals," 311.

68. John T. Nicolson and Dempster Smith, *Lathe Design for High- and Low-Speed Steels* (London: Longmans, 1908), v (quotation); Perrigo, *Modern American Lathe Practice*, 223 (quotation).

69. H. S. Snyder, memo, 6 August 1907, 17:2, AJ. Snyder listed the principal importers as the firms of Herman Boker (New York), F. R. Phillips (Philadelphia), E. R. Kent (Chicago), Cann & Saul (Philadelphia), B. M. Jones (New York), and William Jessop (New York).

70. "The Structure of High-Speed Steels," *Engineer* (22 July 1904); reprint, *Iron and Steel Magazine* 8 (1904): 347–49. See H. Le Châtelier, "Aciers rapides à outils," *Revue de métallurgie* 1 (1904): 334; noted by Paul G. Bastien, "The Beginnings of Microscopic Metallography in France and Its Effect on the Physical Chemistry of Alloy Steels, Prior to 1920," in *Sorby Centennial Symposium on the History of Metallurgy*, ed. C. S. Smith (New York: Gordon and Breach, 1965), 181.

71. M. White to F. W. Taylor, 19 February 1906 (quotation), and F. W. Taylor to M. White, 20 March 1907 (quotation), both in 84F, FWT; Michael Sanderson, "The Professor as Industrial Consultant: Oliver Arnold and the British Steel Industry, 1900–1914," *Economic History Review* 31 (1978): 585–600 (quotation on 588); Tweedale, "Metallurgy and Technological Change," 212–13,84F, FWT; H. S. Snyder to A. Johnston, 26 March 1908, RG109:3, BS. The Bethlehem suit against Niles-Bement-Pond can be followed in 9:44–47, AJ.

72. In his final decision Judge Cross excoriated two witnesses, doubtless Taylor and White, who refused to testify until Bethlehem agreed to compensate them. "Conduct of the character above outlined, if not contrary to public policy, is certainly near the border line" (*Bethlehem v. Niles-Bement-Pond*, 30–32).

73. See Nelson, ed., *A Mental Revolution*, 5–39 (quotation on 5). For an analysis of Taylor's narratives, see Martha Banta, *Taylored Lives: Narrative Productions in the Age of Taylor, Veblen, and Ford* (Chicago: University of Chicago Press, 1993), especially 93–99, 113–35.

74. Taylor, "Art of Cutting Metals," 57–58 (quotations). Taylor's address reached the broader engineering community through *Engineering News* (6 December 1906): 580–85. For further "lessons" for young engineers and their educators, see Frederick W. Taylor, "A Comparison of University and Industrial Methods and Discipline" (New York: ASME, 1906).

75. Taylor, "Art of Cutting Metals," 253.

76. Taylor here confounded the two forms of *carbon*, which were central to the hardening theory of "carbonists" like Osmond, with two forms of iron-carbon *crystals*, which both the carbonists and the "allotropists"

like Le Châtelier accepted. Osmond's "hardening carbon" would be best translated as "carbon dissolved in austenite [or martensite]" according to Bain (*Pioneering in Steel Research*, 104).

77. Taylor, "Art of Cutting Metals," 256, 258–59, 260–61, 263–64.

78. For a discussion of American metallurgical investigations of high-speed steel, which took off in the 1920s, see Misa, "Science, Technology and Industrial Structure," 199–201.

79. Discussion of Taylor, "Art of Cutting Metals," 294, 302–3, 315, 342.

80. Taylor, "Art of Cutting Metals," 36.

81. Ibid., 56.

82. The range of Taylor's impact can be sampled in Samuel Haber, *Efficiency and Uplift: Scientific Management in the Progressive Era, 1890–1920* (1964; reprint, Chicago: University of Chicago Press, 1973); Charles S. Maier, "Between Taylorism and Technocracy: European Ideologies and the Vision of Industrial Productivity in the 1920s," *Journal of Contemporary History* 5, no. 2 (1970): 27–61; Judith A. Merkle, *Management and Ideology: The Legacy of the International Scientific Management Movement* (Berkeley: University of California Press, 1980); Edward T. Morman, *Efficiency, Scientific Management, and Hospital Standardization* (New York: Garland, 1988); and Nelson, *A Mental Revolution.*

6. The Imperative of Automobiles, 1905–1925

1. See Floyd Clymer, *Henry's Wonderful Model T, 1908–27* (New York: McGraw-Hill, 1955), 14–16, 22–26. For background on Studebaker–E.M.F., see Donald Finlay Davis, *Conspicuous Production: Automobiles and Elites in Detroit, 1899–1933* (Philadelphia: Temple University Press, 1988), 68–71; Asa E. Hall and Richard M. Langworth, *The Studebaker Century* (Contoocook, N.H.: Dragonwyck Publishing, 1983), 40–47; Stephen Longstreet, *A Century on Wheels: The Story of Studebaker, 1852–1952* (New York: Henry Holt, 1952), 69–83. For background on the automobile stunts, see James Rood Doolittle, ed., *The Romance of the Automobile Industry* (New York: Klebold Press, 1916), 217–18.

2. Clymer, *Henry's Wonderful Model T, 1908–27*, 27.

3. Other points of contact include drop forging, "a tender infant until the advent of the automobile" (R. T. Herdegen, "Forging Practice for Automobile Parts," *ASST Trans.* 3 [1922–23]: 841). For nonferrous metals, see Stuart W. Leslie, *Boss Kettering* (New York: Columbia University Press, 1983), 123–48; T. A. Boyd, *Professional Amateur: The Biography of Charles Franklin Kettering* (New York: E. P. Dutton, 1957), 119–23.

4. Thaddeus F. Baily [president of the Electric Furnace Company of America, Alliance, Ohio], "The Heat Treatment of Steel in Automatic Electric Furnaces," *AISI Yearbook* (1915): 434.

5. See U.S. Congress, House of Representatives, Report of the Subcommittee on Technology to the National Resources Committee, *Technological Trends and National Policy*, House Document No. 360, 75th Cong., 1st Sess. (Washington, D.C.: Government Printing Office, 1937), 350–51, as

quoted in Eldon S. Hendriksen, *Capital Expenditures in the Steel Industry 1900 to 1953* (New York: Arno, 1978), 107.

6. "The war needs stirred some of the doubting Thomases to action and [electric] furnaces were installed in great numbers, particularly in the years 1917 and 1918," stated John A. Mathews ("The Present Status of the Electric Furnace in Refining Iron and Steel," *AISI Yearbook* [1922]: 358). Mathews drew attention to "the splendid results of electric furnace ordnance steels made at Charleston . . . in the largest furnaces in the United States" (363). See also W. L. Ainsworth, "U.S. Naval Ordnance Plant, Charleston, W. Va.," *USNI Proc.* 46, no. 2 (1920): 237–43. For the rise—and fall—of electric steel production in West Virginia compared with other states, see table 6.3.

7. On ALAM (and its rivals), see James J. Flink, *America Adopts the Automobile, 1895–1910* (Cambridge: MIT Press, 1970), 318–31; Davis, *Conspicuous Production*, 5, 129–31; John B. Rae, *American Automobile Manufacturers: The First Forty Years* (Philadelphia: Chilton, 1959), 72–81; Rae, *The American Automobile: A Brief History* (Chicago: University of Chicago Press, 1965), 33–41; George S. May, *R. E. Olds: Auto Industry Pioneer* (Grand Rapids, Mich.: William B. Eerdmans, 1977), 221–23, 298–303.

8. Rae, *Automobile Manufacturers*, 9–12, 79; George V. Thompson, "Intercompany Technical Standardization in the Early American Automobile Industry," *Journal of Economic History* 14 (winter 1954): 1–20, especially 3, 7; David A. Hounshell, *From the American System to Mass Production, 1800–1932: The Development of Manufacturing Technology in the United States* (Baltimore: Johns Hopkins University Press, 1984), 190–208, 214.

9. In 1910 General Motors produced 21 percent of the nation's motor vehicles; Ford, 10 percent; and Studebaker–E.M.F., 8 percent. See Thompson, "Intercompany Technical Standardization," 4–7, 10–11; Rae, *Automobile Manufacturers*, 80; Flink, *America Adopts*, 301–4; Davis, *Conspicuous Production*, 88–91.

10. Henry Souther, "Specifications for Materials," *SAE Trans.* 5 (1910): 168–202 (quotations on 192, 196).

11. Ibid., 196, 197.

12. SAE Iron and Steel Division, "Preliminary Report of the Iron and Steel Division," *SAE Trans.* 6 (1911): 61–102 (quotation on 61).

13. Ibid., 62.

14. Ibid., 94–102.

15. Ibid., 64–92.

16. The nine companies were Crucible Steel Co. of America, Carnegie Steel, Colonial Steel, Midvale Steel, Vulcan Crucible Steel, Halcomb Steel, Bethlehem Steel, Pennsylvania Steel, and Vanadium Sales Co. of America. Representatives of Carpenter Steel, Cambria Steel, and Republic Iron & Steel were expected but did not attend. See SAE Iron and Steel Division, "Second Report of the Iron and Steel Division," *SAE Trans.* 6

(1911): 498–509. This episode has incorrectly been cited as proof of the *general* opposition of steelmakers to standards; see Thompson, "Intercompany Technical Standardization," 15–16.

17. SAE Iron and Steel Division, "Third Report of the Iron and Steel Division," *SAE Trans.* 7, pt. 1 (1912): 32.

18. Ibid., 78.

19. Ibid., 81, 93, 91.

20. SAE Iron and Steel Division, "Fourth Report of Iron and Steel Division," *SAE Trans.* 8, pt. 2 (1913): 35–40, 61–63.

21. The SAE began publishing charts that correlated surface hardness and heat treatments with elastic limit and toughness. The total number of steels was trimmed from seventy-four to forty-two, and the numbers assigned to types were revised. See SAE Iron and Steel Division, "Seventh Report of Iron and Steel Division," *SAE Trans.* 10, pt. 2 (1915): 20; SAE Iron and Steel Division, "Fifth Report of the Iron and Steel Division," *SAE Trans.* 9, pt. 2 (1914): 7–15, 67 (quotation); "Pre-War Limits in Steel Specifications Restored," *SAEJ* 6 (1920): 25.

22. "Physical-Property Charts Much Used," *SAEJ* 15 (1924): 184–85.

23. J. Kent Smith, "Vanadium: Its Services in Automobile Manufacture," 7 March 1907, 10 (quotations), 14, Lot 241 Knisely, BS.

24. John Wandersee, Reminiscences (1952), 20–21, 22 (quotation), Oral History Section, FA; Rae, *Automobile Manufacturers*, 33, 107 (quotation).

25. Wandersee, Reminiscences, 23–25 (quotations); Hounshell, *American System*, 218–22; Allan Nevins, *Ford: The Times, the Man, the Company* (New York: Charles Scribner's Sons, 1954), 462.

26. Henry Ford, "Special Automobile Steel," *Harper's Weekly* 51 (16 March 1907), cited in Flink, *America Adopts*, 287; Clymer, *Henry's Wonderful Model T, 1908–1927*, 21 (quotation), 153, 155.

27. See the booklets in Lot 241 Knisely, BS; and J. Kent Smith, "Steel Differences," *Iron Age* 81 (26 March 1908): 1010–11.

28. Wandersee, Reminiscences, 26.

29. Henry Ford (in collaboration with Samuel Crowther), *My Life and Work* (Garden City, N.Y.: Doubleday, Page, 1922; reprint, New York: Arno, 1973), 18 (quotation), 66. See Hounshell, "Cul-de-sac: The Limits of Fordism and the Coming of 'Flexible Mass Production,'" in *American System*, 263–301; Arthur J. Kuhn, *GM Passes Ford, 1918–38: Designing the General Motors Performance-Control System* (University Park: Pennsylvania State University Press, 1986).

30. Wandersee, Reminiscences, 26, 27.

31. General Motors formalized its research activities by founding the General Motors Research Corporation in 1920. See Leslie, *Boss Kettering*, 97–102.

32. Ralph H. Sherry, "Metallurgy in the Automotive Industry," *SAE Trans.* 12, pt. 2 (1917): 323–50 (quotation on 323); reprinted in *SAEJ* 1 (1917): 246–55.

33. Ibid., 326, 327, 329.

34. Ibid., 350.

35. On the Liberty engines, see Rae, *Automobile Manufacturers*, 123–35; Allan Nevins and Frank E. Hill, *Ford: Expansion and Challenge, 1915–33* (New York: Charles Scribner's Sons, 1957), 65–68; Davis, *Conspicuous Production*, 142.

36. Harold F. Wood, "Application of Liberty Engine Materials to the Automotive Industry," *SAEJ* 5 (1919): 23–30 (quotation on 30), 243–45; Charles M. Manly, "International Aircraft Standards," *SAE Trans.* 13, pt. 2 (1918): 103–11 (quotation on 111); O. E. Hunt, "Probable Effect on Automobile Design of Experience with War Airplanes," *SAE Trans.* 14, pt. 1 (1919): 154–61 (quotation on 155).

37. According to Hunt, "in an airplane engine dependability, light weight, high mean effective [cylinder] pressure, excellent economy, first cost, flexibility, and quietness are controlling factors in about the order listed, while in the case of the car engine the relative order of importance is almost completely reversed and would read dependability, flexibility, quietness, first cost, economy, light weight, and high mean effective pressure" ("Probable Effect," 156).

38. Ibid., 159, 161.

39. Ibid., 160–61.

40. This list is based on the affiliation of authors contributing technical papers on chromium-molybdenum steel from 1920 to 1924; for the list of papers, see Misa, "Science, Technology and Industrial Structure," 331.

41. Nevins and Hill, *Ford*, 416.

42. R. B. Schenk of Buick Motor Company has also been identified as a pioneer in combining measurable mechanical characteristics of automobile alloy steel into a "merit index"; see Bain, *Pioneering in Steel Research*, 259.

43. J. D. Cutter, "A Suggested Method for Determining the Comparative Efficiency of Certain Combinations of Alloys in Steel," *ASST Trans.* 1 (1920–21): 188–90 (quotation on 189–90). Using data from tensile testing, Cutter's merit index used elastic limit and ultimate strength (thousand pounds per square inch), and percentage of elongation in 2 inches and percentage reduction of area:

Merit index = ((elastic limit + ultimate strength) ÷ 2) ×
   percent elongation / (100 − percent reduction)

44. Arthur H. Hunter, "Molybdenum," *AISI Yearbook* (1921): 127–51, especially 135–46; G. W. Sargent, "Molybdenum as an Alloying Element in Structural Steels," *ASTM Proc.* 20 (1920): 5–30, especially 27. For evidence that Hunter expressed a consensus about molybdenum, see Misa, "Science, Technology and Industrial Structure," 333 n.2.

45. Hunter, "Molybdenum," 145.

46. Ibid., 146.

47. George B. Waterhouse, comment on John S. Unger, "Some Metallurgical Developments in the Manufacture of Iron and Steel," *AISI Yearbook* (1912): 83–94 (quotation on 94); Robert R. Abbott [Peerless Motor Car Co., Cleveland], "The Heat Treatment of Automobile Steels," *AISI Yearbook* (1920): 421 (quotation); E. F. Collins [consulting engineer, industrial

heating division, General Electric Co., Schenectady], "Some Observations on Furnaces and Fuels including the Electric Furnace for Heat Treating," *ASST Trans.* 4 (1923): 723 (quotation).

48. A. H. Frauenthal [chief inspector, Chandler Motor Car Co.] and C. S. Morgan [superintendent of heat treating, Automobile Machine Company], "Heat Treatment of Automotive Parts and Description of Equipment Used," *ASST Trans.* 8 (1925): 851–60 (quotation on 857).

49. C. L. Ipsen contrasted the atmospheric conditions surrounding an electric furnace (temperature 79°F, CO_2 4 percent, humidity 60 percent) with a fuel-fired furnace (temperature 121°F, CO_2 7 percent, humidity 23 percent), concluding that "a man's vitality after four hours' work in the atmosphere surrounding the fuel furnace would equal that of the man working in the atmosphere surrounding the electric for eight hours" ("Selection of Electric Furnaces for Steel Treating," *ASST Trans.* 3 [1922–23]: 720–22). See also Ipsen, "The Electric Furnace as It Affects Over-all Cost of Heat Treated Parts," *ASST Trans.* 2 (1921–22): 984–89.

50. J. M. Watson, "The Heat Treatment of Automobile Parts," *ASST Trans.* 6 (1924): 716–27 (quotation on 725–26).

51. Harry E. Hemstreet [Sheldon Spring & Axle Co., Wilkes-Barre, Pa.], "Heat Treatment of Truck Axles," *ASST Trans.* 1 (1920–21): 209–11 (quotation on 211). For conveyor-type heat-treating systems, see also Thaddeus F. Baily, "The Heat Treatment of Steel in Automatic Electric Furnaces," *AISI Yearbook* (1915): 434–45; Ipsen, "Selection of Electric Furnaces," 727–28.

52. Watson, "Heat Treatment of Automobile Parts," 716. For the Hupmobile's history, see Davis, *Conspicuous Production*, 92–93.

53. Watson, "Heat Treatment of Automobile Parts," 720–21.

54. Ibid., 725, 727.

55. American Rolling Mill Company, *The First Twenty Years* (Middletown: American Rolling Mill Company, 1922), 21–67, 155–65.

56. Ibid., 105–26, 181–84.

57. Christy Borth, *True Steel: The Story of George Matthew Verity and His Associates* (Indianapolis: Bobbs-Merrill, 1941), 227–31; American Rolling Mill, *First Twenty Years*, 97–104.

58. George C. Crout and Wilfred D. Vorhis, "John Butler Tytus: Inventor of the Continuous Steel Mill," *Ohio History* 76 (summer 1967): 132–45; Thomas J. Misa, "Tytus, John B.," in *American National Biography*.

59. Crout and Vorhis, "John Butler Tytus," 139.

60. Douglas Alan Fisher, *The Epic of Steel* (New York: Harper and Row, 1963), 144–46.

61. John A. Mathews, "The Electric Furnace in Steel Manufacture," *AISI Yearbook* (1916): 71–90 (quotation on 79); Mathews, "The Present Status of the Electric Furnace in Steel Manufacture," *AISI Yearbook* (1922): 362.

62. Mathews, "Electric Furnace in Steel Manufacture," 84; Joseph Schaeffers, "Electro Steel," *SAE Trans.* 6 (1911): 46–58.

63. William R. Walker, "The Electric Furnace as a Possible Means of Producing an Improved Quality of Steel," *AISI Yearbook* (1912): 67.

64. Mathews, "Electric Furnace in Steel Manufacture," 80.

65. Mathews, "Present Status of the Electric Furnace," 365.

66. See the following articles in *ASST Trans.* 1 (1920–21): E. F. Collins, "Relative Thermal Economy of Electric and Fuel-Fired Furnaces," 217–29; Seth A. Moulton and W. H. Lyman, "Relative Economy of Oil, Gas, Coal and Electric Heated Furnaces," 249–70; T. F. Baily, "The Relative Economy of Electric, Oil, Gas and Coal-Fired Furnaces," 401–2.

7. The Dynamics of Change

1. U.S. Steel reported that in 1937, 941 customers had billings over $100,000 each and accounted for 73 percent of gross sales; in 1939, 663 customers with billings over $100,000 accounted for 68 percent of gross sales. Beyond the integrated steel mills of Ford and International Harvester, several other consumers—including American Car & Foundry, American Locomotive, the Atchison, Topeka & Santa Fe Railroad, Continental Can, Simonds Saw & Steel, and Timken Roller Bearing—had by 1940 built semi-integrated or nonintegrated steel capacity. See U.S. Steel Corporation, *T.N.E.C. Papers*, 3 vols. (New York: U.S. Steel, 1940), 1:381, 386.

2. For the later history of the steel industry, see Richard A. Lauderbaugh, *American Steel Makers and the Coming of the Second World War* (Ann Arbor, Mich.: UMI Research Press, 1980); Paul A. Tiffany, *The Decline of American Steel: How Management, Labor, and Government Went Wrong* (New York: Oxford University Press, 1988); Bela Gold et al., *Technical Progress and Industrial Leadership: The Growth of the U.S. Steel Industry, 1900–1970* (Lexington, Mass.: Lexington Books, 1984); John P. Hoerr, *And the Wolf Finally Came: The Decline of the American Steel Industry* (Pittsburgh: University of Pittsburgh Press, 1988). On the Wagner Act, see Christopher L. Tomlins, *The State and the Unions: Labor Relations, Law, and the Organized Labor Movement in America, 1880–1960* (Cambridge: Cambridge University Press, 1985), 103–243.

3. The course of *United States v. United States Steel Corporation* can be followed in Urofsky, *Big Steel*, 24–26, 78–83, 181–82, 340–43. On the Great Northern iron-ore properties, see Walker, *Iron Frontier*, 220–24; and Gold, *Technical Progress and Industrial Leadership*, 357–58, which cites U.S. Federal Trade Commission, *Control of Iron Ore* (Washington, D.C.: Government Printing Office, 1952), 7–8.

4. American Rolling Mill, *First Twenty Years*, 89–95; Borth, *True Steel*, 124–40; Tweedale, *Sheffield Steel and America*, 62–63, 215. On magnetic eddy currents, see W. Bernard Carlson, *Innovation as a Social Process: Elihu Thomson and the Rise of General Electric, 1870–1900* (Cambridge: Cambridge University Press, 1991), 84–86, 99, 104.

5. A complete account of stainless steel is sorely needed, evaluating the many rival claims and discussing the development and diffusion of the

various stainless steels. Among the sources are Fisher, *Epic of Steel*, 158–62; Tweedale, *Sheffield Steel and America*, 75–83 (on Harry Brearley); Ralph D. Gray, *Alloys and Automobiles: The Life of Elwood Haynes* (Indianapolis: Indiana Historical Society, 1979), 145–66; John Truman, "The Initiation and Growth of High Alloy (Stainless) Steel Production," *Historical Metallurgy* 19, no. 1 (1985): 116–25; and David A. Hounshell and John Kenly Smith, Jr., *Science and Corporate Strategy: Du Pont R&D, 1902–80* (Cambridge: Cambridge University Press, 1988), 276–78.

6. Of twenty-five major product and process innovations at Du Pont between 1920 and 1950, only ten resulted from Du Pont's in-house research, while fifteen resulted from acquiring outside inventions or companies, according to the National Bureau of Economic Research study cited in George Basalla, *The Evolution of Technology* (Cambridge: Cambridge University Press, 1988), 128.

7. John S. Unger, "Some Metallurgical Developments in the Manufacture of Iron and Steel," *AISI Yearbook* (1912): 83–94 (quotation on 94). On Unger's biography, see *Iron Age* (3 October 1935): 37; *Steel* (30 September 1935): 18.

8. Bain, *Pioneering in Steel Research*, 101, 256. The proliferation of laboratories and the scientific naivete of U.S. Steel executives are emphasized by Paul A. Tiffany, "Corporate Culture and Corporate Change: The Origins of Industrial Research at the United States Steel Corporation, 1901–29," paper presented at the annual meeting of the Society for the History of Technology, Pittsburgh, 25 October 1986.

9. Bain, *Pioneering in Steel Research*, 1–97.

10. Ibid., 98–117 (quotation on 116), 260–61.

11. Ibid., 118–261, 226–28 (on Corten steel).

12. John Strohmeyer, *Crisis in Bethlehem: Big Steel's Struggle to Survive* (Bethesda, Md.: Adler and Adler, 1986; reprint, New York: Penguin, 1987), 59.

13. Doerflinger and Rivkin, *Risk and Reward*, 84.

14. Gold et al., *Technological Progress and Industrial Leadership*, 580–83.

15. Nathan Rosenberg's critical comment is quoted in Giovanni Dosi and Luc Soete, "Technological Innovation and International Competitiveness," in *Technology and National Competitiveness*, ed. Jorge Niosi (Montreal: McGill-Queen's University Press, 1991), 91–118 (quotation on 92).

16. "The economist really need not know at all what it feels like to be inside a steel plant—he quantifies technological change by making measurements of output per hour, or output per unit of this, or input per unit of that," stated R. M. Solow, who developed the formal theoretical framework around the production-function concept in the 1950s (quoted in Gold et al., *Technical Progress and Industrial Leadership*, 98). The production-function concept is critiqued in Giovanni Dosi et al., *Technical Change and Economic Theory* (London: Pinter, 1988), 436–38; and Gold et al., *Technical Progress and Industrial Leadership*, 77–95.

17. Some studies nonetheless make for informative reading. See, for in-

stance, the debate over whether it was "rational" for railroads to adopt steel rails, published in *Explorations in Economic History:* Jeremy Atack and Jan K. Brueckner, "Steel Rails and American Railroads, 1867–80," 19 (July 1982): 339–59; C. Knick Harley, "Steel Rails and American Railroads, 1867–80: Cost Minimizing Choice—A Comment on the Analysis of Atack and Brueckner," 20 (1983): 248–57; Ann M. Carlos, "Steel Rails versus Iron Rails: Evidence from Canada," 21 (1984): 169–75. A sorely needed inquiry into railroad companies' actual decisions is Usselman, "Running the Machine," 81–133.

18. Gold et al., *Technical Progress and Industrial Leadership,* 15–101 (quotation on 90–91); Martin Fransman, *Technology and Economic Development* (Boulder: Westview, 1986), 115–30.

19. Giovanni Dosi, "Technological Paradigms and Technological Trajectories: A Suggested Interpretation of the Determinants and Directions of Technical Change," *Research Policy* 11 (1982): 147–62 (quotation on 148). This and the next three paragraphs follow and quote Dosi et al., *Technical Change and Economic Theory,* 9–66, 197–218, 221–38 (quotation on 224–25), 567–79, 590–607. This literature may be sampled in Norman Clark and C. Juma, *Long-Run Economics: An Evolutionary Approach to Economic Growth* (London: Pinter, 1987); and Rod Coombs, P. Saviotti, and V. Walsh, *Economics and Technological Change* (London: Macmillan, 1987).

20. Edward Constant defines a technological paradigm as "an accepted mode of technical operation, the usual means of accomplishing a technical task. It is the conventional system as defined and accepted by a relevant community of technological practitioners. A technological paradigm is not just a device or process, but, like a scientific paradigm, is also rationale, practice, procedure, method, instrumentation, and a particular shared way of perceiving a set of technology. It is a cognition . . . A technological paradigm is passed on as a tradition of practice in the preparation of aspirants to its community membership" (quoted in John M. Staudenmaier, *Technology's Storytellers: Reweaving the Human Fabric* [Cambridge: MIT Press, 1985], 64).

21. Wiebe Bijker uses the concept of "technological frame" to explore how "explicit theory, tacit knowledge, and general engineering practice, cultural values, prescribed testing procedures, devices, material networks and systems" are developed by social groups in relation to a specific artifact. See Wiebe E. Bijker, "The Social Construction of Bakelite: Toward a Theory of Invention," in *The Social Construction of Technological Systems,* ed. Bijker, Thomas P. Hughes, and Trevor J. Pinch (Cambridge: MIT Press, 1987), 159–87, especially 168–74; and Bijker and John Law, eds., *Shaping Technology/Building Society: Studies in Sociotechnical Change* (Cambridge: MIT Press, 1992), 87–94, 179 (quotation).

22. Even after prolonged growth and consolidation, technological systems do not become autonomous, argues Thomas P. Hughes, they acquire momentum. "They have a mass of technical and organizational components; they possess direction, or goals; and they display a rate of growth

suggesting velocity. A high level of momentum often causes observers to assume that a technological system has become autonomous" (Hughes, "The Evolution of Large Technological Systems," in *Social Construction of Technological Systems*, ed. Bijker, Hughes, and Pinch, 76–80 [quotation on 76]). Though a physical metaphor, momentum is a social concept: it points not to the internal logic of technique but to the organizations and people committed by various interests to the system, to manufacturing corporations, research and development laboratories, investment banking houses, educational institutions, technical societies, and regulatory bodies.

23. According to John M. Staudenmaier, "a technological style can be defined as a set of congruent technologies that become 'normal' (accepted as ordinary and at the same time as normative) within a given culture. They are congruent in the sense that all of them embody the same set of overarching values within their various technical domains. For example, it can be argued that the United States, beginning with the U.S. Ordnance Department's 1816 commitment to the philosophical ideal of standardization and interchangeability, gradually adopted a set of normal technologies that incorporate that ideal. From this point of view many distinct technological developments—the machine tool tradition, the growth of standardized and centrally controlled rail systems, the centralization and standardization of corporate research and development, the use of consumer advertising to program individual buying habits, the increasing centralization and complexity of electricity and communications networks, etc.—can be interpreted as participating in a single style, embodying a specific set of values within a specific world view" (*Technology's Storytellers*, 200).

24. See essays by each of these four authors in *Social Construction of Technological Systems*, ed. Bijker et al.; and *Shaping Technology/Building Society*, ed. Bijker and Law. See also Bruno Latour, *Science in Action* (Cambridge: Harvard University Press, 1987).

25. See Thomas J. Misa, "Theories of Technological Change: Parameters and Purposes," *Science, Technology and Human Values* 17 (winter 1992): 3–12.

26. Accounts of the Homestead strike begin with Arthur G. Burgoyne, *Homestead: A Complete History of the Struggle of July, 1892, between the Carnegie Steel Company, Limited, and the Amalgamated Association of Iron and Steel Workers* (Pittsburgh, 1893), on which many writers have relied over the years. See also Paul Krause, *The Battle for Homestead, 1880–92: Politics, Culture, and Steel* (Pittsburgh: University of Pittsburgh Press, 1992); and William Serrin, *Homestead: The Glory and Tragedy of an American Steel Town* (New York: Random House, 1992). Many shorter accounts of the strike exist, besides the ones discussed here; for a new labor-history perspective, see Linda Schneider, "The Citizen Striker: Workers' Ideology in the Homestead Strike of 1892," *Labor History* 23 (1982): 47–66.

27. For a review of the literature suggesting these themes, see Ingham, *Making Iron and Steel*, 96–98.

28. David Brody, *Steelworkers in America: The Nonunion Era* (Cambridge: Harvard University Press, 1960), 1–26.

29. Katherine Stone, "The Origins of Job Structures in the Steel Industry," *Review of Radical Political Economics* 6 (summer 1974): 61–97; reprinted in *Labor Market Segmentation*, ed. R. C. Edwards et al. (Lexington, Mass.: Heath, 1975), 27–84 (quotations on 33–35).

30. Carnegie's activities and motivations are fully discussed in Wall, *Andrew Carnegie*, 537–82.

31. See Krause, *Battle for Homestead*, 359, 361. On the twelve-hour day, David Montgomery (*The Fall of the House of Labor: The Workplace, the State, and American Labor Activism, 1865–1925* [Cambridge: Cambridge University Press, 1987], 41) suggests that twelve-hour shifts were extended to at least one-third of the Homestead workers after 1892. Different occupations within a steel mill varied dramatically in working hours, according to the 1910 statistics presented in Irmgard Steinisch, *Arbeitszeitverkürzung und sozialer Wandel: Der Kampf um die Achtstundenschicht in der deutschen und amerikanischen Eisen- und Stahlindustrie, 1880–1929* (Berlin: Walter de Gruyter, 1986), 173–79. In 1910 the percentages of workers with regular weeks of eighty-four hours (i.e., seven twelve-hour days) were blast furnace (67 percent), Bessemer (17 percent), open hearth (21 percent), puddling (0.14 percent), rolling mill (6 percent), tube mill (0.39 percent). The percentage of all steel- and rolling-mill workers with regular weeks of eighty-four hours was 10.85 percent; with seventy-two hours or more, 43.69 percent; with sixty hours or more, 80.42 percent.

32. For this and the discussion in the next two paragraphs, see David Montgomery, *Workers' Control in America: Studies in the History of Work, Technology, and Labor Struggles* (Cambridge: Cambridge University Press, 1979); and Montgomery, *Fall of the House of Labor*, especially 9–45 (quotations on 39, 45).

33. See Ingham, *Making Iron and Steel*, 98–127. For a tabulation of strikes in greater Pittsburgh during 1867–92, see Krause, *Battle for Homestead*, 404–5.

34. As contrasted with labor historians' "confrontational" interpretation, Ingham gives as evidence for his "ritualistic" interpretation a garbled and incomplete version of a diary entry of B. F. Jones: "Under existing circumstances it is impossible to further resist the demands of the Boilers" (Ingham, *Making Iron and Steel*, 118). But the complete diary entry reveals Jones and the ironmasters to be much less accommodating. In recording a decision to settle an ongoing strike by the boilers since 2 June 1879 by signing the boilers' tonnage scale for another year, Jones writes that a meeting of the manufacturers resolved that "under existing circumstances it is *inexpedient* to further resist the *unjust* demands of Boilers" (B. F. Jones diary, 7 June 1879 entry, J&L; emphasis added).

35. The manufacture of wrought iron was successfully mechanized in the 1930s, according to Michael W. Santos, "Laboring on the Periphery: Managers and Workers at the A. M. Byers Co., 1900–1956," *Business History Review* 61 (spring 1987): 113–33.

36. The sources of stability—the careful cultivation of loyalty in skilled employees, the mobility of unskilled immigrant workers, and the local politics of steeltowns—for this labor system are discussed in Brody, *Steelworkers in America*, 27 (quotation from *Journal of the British Iron and Steel Institute* [1890]: 113), 80–124.

37. See Harold Katz, *The Decline of Competition in the Automobile Industry, 1920–40* (New York: Arno, 1977), 167–241.

38. "New York appears much duller, in business and in every thing, than I have seen it in several years. The [problem] is certain kinds of business *have been* greatly *overdone* in the country—particularly rail-way building, and especially has this been overdone, by going so largely into debt to Europe for iron to build them with. We are indebted over $400,000,000 in Europe and chiefly or largely for rail-road iron, and in R.R. bonds. The iron should have been made at home, and then substantial prosperity would have followed" (A. H. Holley to A. L. Holley, 13 June 1854, Folder 1851–54, ALH; emphasis in original).

39. On the influence of the 1916 New York zoning code on the "set back" skyscraper, see Sally A. Kitt Chappell, "A Reconsideration of the Equitable Building in New York," *Journal of the Society of Architectural Historians* 49 (March 1990): 90–95; and Carol Willis, "Zoning and Zeitgeist: The Skyscraper City in the 1920s," *Journal of the Society of Architectural Historians* 45 (March 1986): 47–59.

40. Alfred Chandler emphasizes the Sherman Act as a direct expression of cultural values: "That legislation and the values it reflected probably marked the most important noneconomic cultural difference between the United States and Germany, Britain, and indeed the rest of the world insofar as it affected the long-term evolution of the modern industrial enterprise" (Alfred D. Chandler, Jr., *Scale and Scope: The Dynamics of Industrial Capitalism* [Cambridge: Belknap Press of Harvard University Press, 1990], 73). Martin Sklar instead stresses the gradual transformation of the act through judicial construction and political debate (Martin J. Sklar, *The Corporate Reconstruction of American Capitalism, 1890–1916: The Market, the Law, and Politics* [Cambridge: Cambridge University Press, 1988], 14 n. 11, 89–154).

41. On the mix of public and private imperatives in professional engineering, see Edwin T. Layton, Jr., *The Revolt of the Engineers* (1971; reprint, Baltimore: Johns Hopkins University Press, 1986).

42. A typology of sectors at a rather high level of aggregation (supplier-dominated, scale-intensive, specialized-suppliers, science-based) is discussed in Dosi et al., *Technical Change and Economic Theory*, 231–32, 302–4, which follows Keith Pavitt, "Sectoral Patterns of Technical Change: Towards a Taxonomy and a Theory," *Research Policy* 13 (1984).

43. These insights are discussed further in *Managing Technology in Society*, ed. Arie Rip, Thomas J. Misa, and Johan Schot (London: Pinter, 1995); as well as in Kurt Fischer and Johan Schot, eds., *Environmental Strategies for Industry: International Perspectives on Research Needs and Policy Implications* (Washington, D.C.: Island Press, 1993).

44. See Walter Adams and Joel B. Dirlam, "Big Steel, Invention, and Innovation," *Quarterly Journal of Economics* 80 (May 1966): 167–89; Alan K. McAdams, "Big Steel, Invention, and Innovation, Reconsidered," *Quarterly Journal of Economics* 81 (August 1967): 457–82; and Janet T. Knoedler, "Market Structure, Industrial Research, and Consumers of Innovation: Forging Backward Linkages to Research in the Turn-of-the-Century U.S. Steel Industry," *Business History Review* 67 (spring 1993): 98–139.

45. See Serrin, *Homestead.*

46. See Tiffany, *Decline of American Steel;* Strohmeyer, *Crisis in Bethlehem.*

47. See Gold, *Technical Progress and Industrial Leadership,* 619, 644–46.

48. In the 1950s the "first cost" per ton of annual capacity was about $15 for the BOF and about $35 for the open hearth, according to J. K. Stone (manager, steel plants development, Kaiser Engineers), as cited in Gold, *Technical Progress and Industrial Leadership,* 556 n.28, 550–52.

49. Total U.S. steelmaking capacity in 1949 was 96.1 million tons, and in 1959, 147.6 million tons; see Gold, *Technical Progress and Industrial Leadership,* 580, 642–45.

50. Richard Preston, *American Steel: Hot Metal Men and the Resurrection of the Rust Belt* (New York: Prentice-Hall, 1991), 86 (emphasis in original). The diffusion of BOF and continuous-casting technology is tabulated in William T. Hogan, *Economic History of the Iron and Steel Industry in the United States,* 5 vols. (Lexington, Mass.: Lexington Books, 1971), 4:1545–47, 1567–68.

51. On the international dimensions of the BOF, see Tiffany, *Decline of American Steel,* 133–34, 169–76.

52. Statistics from Hogan, *Economic History of the Iron and Steel Industry,* 5:1888–89, 2042–44; and Gold, *Technical Progress and Industrial Leadership,* 647–49.

53. In 1901 and 1909 U.S. Steel's strikebreaking dealt the final blows to the Amalgamated union. The infamous Ludlow massacre at the Rockefeller-owned Colorado Fuel and Iron in 1914 left sixty-six people dead, including two women and eleven children. Suppression of the 1919 nationwide steel strike involved beatings by police. Finally, in 1937 came the collapse of the so-called Little Steel strike in which Bethlehem, Republic, Youngstown Sheet & Tube, and Inland "fought [union] organizers with violence, intimidation, and mass firings." See Hoerr, *And the Wolf Finally Came,* 259 (quotation in text) and 263 (quotation in note); Montgomery, *Fall of the House of Labor,* 343–47; H. Lee Scamehorn, *Mill and Mine: The CF&I in the Twentieth Century* (Lincoln: University of Nebraska Press, 1992), 38–55; Serrin, *Homestead,* 193.

54. On the 1959 strike, see Strohmeyer, *Crisis in Bethlehem,* 64–72. On the 1930s vision and 1980s implementation of worker-participation schemes, see Hoerr, *And the Wolf Finally Came,* 155–61, 262–95, 444–46; Serrin, *Homestead,* 250–51. On the Taft-Hartley Act, see Tomlins, *The State and the Unions,* 252–316 (quotation on 316).

55. Serrin, *Homestead*, 303.

56. On the undeveloped theme of decline, see Richard F. Hirsh, *Technology and Transformation in the American Electric Utility Industry* (Cambridge: Cambridge University Press, 1989); Johan Schot, "Technology in Decline: A Search for Useful Concepts; The Case of the Dutch Madder Industry in the Nineteenth Century," *British Journal for the History of Science* 25 (March 1992): 5–26.

Index

Boutelle, Charles A., 114
Bridge, James, 157
bridges: Brooklyn, 75; Eads (St. Louis), 133; Girard Avenue (Philadelphia), 51; Kinzie Street (Chicago), 75; Merchants (St. Louis), 65; Missouri River (Glasgow), 75; and skyscrapers, 64–65
Brinley, Charles A., 180
British Association for the Advancement of Science (BAAS), 9
British Iron and Steel Institute, 113, 317n.52
Britton, J. Blodget, 30
Brody, David, 266
Brown, A. Page, 60
Brown University, 16
Buchman & Deisler (architects), 60
Buffington, Leroy, 63, 305n.40
Buick Motor Co., 345n.42
building codes, 66–69, 352n.39
buildings: Auditorium, 45–50, 64, 66; Carnegie Hall, 47; Chicago Athletic Club, 87; Columbia (New York), 68; Crocker (San Francisco), 60; Exchange (Boston), 70; Fair Store (Chicago), 69, 87; Fisher (Chicago), 311n.109; Flatiron, 44, 88, 307n.65; Fort Dearborn, 66; Government Printing Office, 60; Grand Central Station, 143; Guaranty (Buffalo), 88, 311n.109; Haughwout (New York), 85; Havenmeyer (New York), 69; Home Insurance (Chicago), 60–64, 69, 87, 304n.37; Hoyt (New York), 60; Lancashire Insurance Co. (New York), 68; Library of Congress, 60; Ludington (Chicago), 69; L. Z. Leiter (Chicago), 75; Manhattan (Chicago), 311n.109; Manhattan Life Insurance (New York), 51, 69; Masonic Temple, 64; New York Life (Chicago), 66, 86 (fig. 2.10), 88; New York World, 60, 69; Pearl Street Station, 60; Pennsylvania Station, 143; Pulitzer, 69; Schiller (Chicago), 47; Stock Exchange (Chicago), 64, 85 (fig. 2.9); Tower (New York), 68, 307n.62; Union Trust (New York), 60; United Engineering Societies, 173; Unity (Chicago), 51, 311n.109; Venetian (Chicago), 73

Bullard Machine Tool Co. (Bridgeport, Conn.), 200
Burnham, Daniel H., 44, 45, 65, 88, 307n.65; and steel firms, 66
Burnham & Root (architects), 64–65
Burt, Henry J., 65
Butterley Iron (Derbyshire), 10

Cambria Iron, 4, 20 (fig. 1.8), 22, 26, 51, 97, 102, 269, 294n.37
Cambria Steel, 37, 130, 171, 176, 325n.24, 343n.16
Camden & Amboy Railroad, 16
Campbell, Harry H., 80–82
Campbell, James, 3
Canadian Copper Co., 112, 114
Carbon Steel Co. (Pittsburgh), 232
Carnahan, Robert B., 257
Carnegie, Andrew, xxii, 63, 70, 79, 82, 111, 156–59, 163, 267–68; and Pennsylvania Railroad, 21–23, 133–35; and social Darwinism, 164, 168; and steel merger, 165–69; *Triumphant Democracy* (1886), 102; and United Engineering Societies, 173
Carnegie, Phipps & Co., 80, 102, 156
Carnegie, Thomas, 157
Carnegie Brothers & Co., 48, 156, 157, 158, 294n.37
Carnegie Steel, 130, 149, 158, 164, 325n.24, 340n.61, 343n.16; Central Research Laboratory (Duquesne, Pa.), 258; and Chicago architects, 66, 74; column designs, 73; Conneaut mill, 168–69; Duquesne plant, 267, 330n.57; Edgar Thomson plant, 22, 23–28, 38, 66, 134, 135, 149, 267, 271; Homestead plant, 80, 121, 157, 176; —, armor mill, 102, 103, 332n.78; —, strike at (1892), 266–68; —, structural mill, 49, 60, 82–83; structural steel handbook, 71–74, 170, 333n.95; Union Iron, 21, 48, 156; Union Railroad, 169, 330n.57; vertical integration of, 155–64. *See also* Keystone Bridge Co.
Carpenter, H.C.H., 174, 203, 206
Carpenter Steel, 257, 343n.16
cartel: in armor sector, 106, 118; in iron ore, 162; in rail sector, 19–21, 152, 327n.36; in structural sector, 70; in tool steel, 202

BOOKS IN THE NEW SERIES

The American Railroad Passenger Car, New Series, no. 1, by John H. White, Jr.

Neptune's Gift: A History of Common Salt, New Series, no. 2, by Robert P. Multhauf

Electricity before Nationalisation: A Study of the Development of the Electricity Supply Industry in Britain to 1948, New Series, no. 3, by Leslie Hannah

Alexander Holley and the Makers of Steel, New Series, no. 4, by Jeanne McHugh

The Origins of the Turbojet Revolution, New Series, no. 5, by Edward W. Constant II

Engineers, Managers, and Politicians: The First Fifteen Years of Nationalised Electricity Supply in Britain, New Series, no. 6, by Leslie Hannah

Stronger Than a Hundred Men: A History of the Vertical Water Wheel, New Series, no. 7, by Terry S. Reynolds

Authority, Liberty, and Automatic Machinery in Early Modern Europe, New Series, no. 8, by Otto Mayr

Inventing American Broadcasting, 1899–1922, New Series, no. 9, by Susan J. Douglas

Edison and the Business of Innovation, New Series, no. 10, by Andre Millard

What Engineers Know and How They Know It: Analytical Studies from Aeronautical History, New Series, no. 11, by Walter G. Vincenti

Alexanderson: Pioneer in American Electrical Engineering, New Series, no. 12, by James E. Brittain

Steinmetz: Engineer and Socialist, New Series, no. 13, by Ronald R. Kline

From Machine Shop to Industrial Laboratory: Telegraphy and the Changing Context of American Invention, 1830–1920, New Series, no. 14, by Paul Israel

The Course of Industrial Decline: The Boott Cotton Mills of Lowell, Massachusetts, 1835–1955, New Series, no. 15, by Laurence F. Gross

High Performance: The Culture and Technology of Drag Racing, 1950–1990, New Series, no. 16, by Robert C. Post

A Nation of Steel: The Making of Modern America, 1865–1925, New Series, no. 17, by Thomas J. Misa

Library of Congress Cataloging-in-Publication Data

Misa, Thomas J.
　　A nation of steel : the making of modern America, 1865–1925 / Thomas J. Misa.
　　　　p.　　cm. — (Studies in the history of technology)
　　Includes bibliographical references and index.
　　ISBN 0-8018-4967-5 (alk. paper)
　　1. Steel industry and trade—United States—History.　2. Steel industry and
trade—Social aspects—United States—History.　I. Series: Johns Hopkins studies
in the history of technology.
HD9515.M576　1995
338.4'7669142'0973—dc20　　　　　　　　　　　　　　　　　　　94-38681